WAVE OPTICS

Basic Concepts and Contemporary Trends

WAVE OPTICS

Basic Concepts and Contemporary Trends

Subhasish Dutta Gupta
University of Hyderabad, India

Nirmalya Ghosh
Indian Institute of Science
Education & Research (IISER) - Kolkata, India

Ayan Banerjee
Indian Institute of Science
Education & Research (IISER) - Kolkata, India

CRC Press is an imprint of the
Taylor & Francis Group, an informa business

CRC Press
Taylor & Francis Group
6000 Broken Sound Parkway NW, Suite 300
Boca Raton, FL 33487-2742

First issued in paperback 2019

ISBN-13: 978-1-4822-3773-3 (hbk)
ISBN-13: 978-0-367-37735-9 (pbk)

Library of Congress Cataloging-in-Publication Data

Dutta Gupta, Subhasish.
Wave optics : basic concepts and contemporary trends / Subhasish Dutta Gupta (University of Hyderabad, India), Nirmalya Ghosh (Indian Institute of Science Education & Research, Kolkata, India), Ayan Banerjee (Indian Institute of Science Education & Research, Mohanpur, India).
pages cm
Includes bibliographical references and index.
ISBN 978-1-4822-3773-3 (alk. paper)
1. Physical optics. I. Ghosh, Nirmalya. II. Banerjee, Ayan. III. Title.

QC395.2.W38 2015
535'.2--dc23 2015018963

Visit the Taylor & Francis Web site at
http://www.taylorandfrancis.com

and the CRC Press Web site at
http://www.crcpress.com

To our parents, who have been our inspiration all along

Contents

Preface

Over the past two decades, wave optics has undergone a metamorphosis, becoming a very exciting field. Even within the realms of classical optics, there have been reports of nonstandard and counterintuitive phenomena. These include the prediction and subsequent realization of negative index materials and their use for superresolution. The field of metamaterials or engineered materials has emerged in a big way to take up the challenges of beating the diffraction limit or the Rayleigh criterion. Even the simplest problem of reflection and refraction, albeit with structured beams, offers a novel way of understanding a fundamental notion like spin-orbit coupling. Special attention is given to beams carrying orbital angular momentum. The old problems of Goos-Hänchen and Fedorov-Imbert shifts for spatially finite (in the transverse plane) beams are being looked at from an altogether different angle. Analogous spin-orbit coupling has been observed with tightly focused beams in stratified geometry. Another applied area of research has been the trapping of neutral particles in typical optical tweezer setups. There have been numerous applications in biology and in other related areas. On a different note, a proper understanding of light propagation in layered media—in particular, in periodic structures—has opened up novel means for engineering the dispersion. We can, in fact, control the group velocity leading to fast and slow light via the manipulation of the Wigner delay. The recognition that the stratified media could be the optical prototype of one-dimensional scattering problems in quantum mechanics led to several interesting effects like the Hartman effect and the perfect transmission through reflectionless potentials. There have been experiments to demonstrate all such effects in optics. The issues related to the optical theorem and nonreciprocity also drew a lot of attention. Another major discovery was the counterintuitive report on extraordinary transmission, which resulted from an understanding of a flaw in the scalar Fresnel-Kirchhoff diffraction theory.

We believe that most of the above discoveries, or at least their underlying concepts, are well within the reach of undergraduate or advanced undergraduate students of optics who have some basic knowledge of EM theory and mathematics. The main aim of this book is to show that this is indeed so. After a brief overview of elementary concepts, the readers are exposed to some of the recent trends. Of course, the coverage is definitely not exhaustive or complete. The goal is to pass on the excitement and thrill so that budding students and researchers take interest in what is happening in optics today.

The book has special appeal for Indian students since a great majority of them are not exposed to such ideas in their standard optics courses.

A 'nonstandard' book like this could not have been achieved without help from others. The book resulted from lectures and notes delivered at the University of Hyderabad and IISER Kolkata and especially from a series of Schools and Discussion Meetings on Metamaterials and Plasmonics over several years. A lot of simplified research material is also included. A great many thanks to all our collaborators. Particular thanks are due to some of our students who helped a great deal in preparing the manuscript. Among them, Nireekshan, Jalpa and Shourya merit special mention.

S. Dutta Gupta, Hyderabad

N. Ghosh, Kolkata

A. Banerjee, Kolkata

List of Figures

List of Tables

Chapter 1

Oscillations

Oscillations and vibrations constitute one of the major areas of study in physics. Most systems can oscillate freely. Generally, heavier ones have low oscillation frequency while the lighter ones have large frequencies. The variety of phenomena exhibiting repetitive motions have been discussed nicely by French [1]:

> "Systems can vibrate freely in a large variety of ways. Broadly speaking, the predominant natural vibrations of

small objects are likely to be rapid, and those of large objects are likely to be slow. A mosquito's wings, for example, vibrate hundreds of times per second and produce an audible note. The whole earth, after being jolted by an earthquake, may continue to vibrate at the rate of about one oscillation per hour. The human body itself is a treasure-house of vibratory phenomena."

All the above phenomena have one thing in common, i.e., repetitive motion or periodicity. The same pattern of displacement is repeated over and over again. It can be simple or complicated. Irrespective of the nature of oscillations, the pattern is generally represented by plots where the horizontal axis represents the steady progress of time. Such pictures make it easy to recognize one cycle or one period of oscillation, which keeps on repeating.

1.1 Sinusoidal oscillations

Sinusoidal oscillations take place in a vast majority of mechanical systems. This is due to the fact that in most cases, the restoring force is proportional to the displacement. Such motion is always possible if the displacement is small enough. In general the restoring force F can have the following dependence on the displacement x:

$$F(x) = -(k_1 x + k_2 x^2 + k_3 x^3 + ...). \tag{1.1}$$

For small displacements we can ignore the terms proportional to x^2, x^3 and other higher-order terms. This leads to an equation of motion

$$m\frac{d^2x}{dt^2} = -k_1 x, \tag{1.2}$$

which has a solution of the form

$$x = A\sin(\omega t + \phi_0), \quad \omega = \sqrt{\frac{k_1}{m}}. \tag{1.3}$$

Thus sinusoidal oscillation in simple harmonic motion is a prominent possibility in small oscillations. It could be an approximation (though perhaps a very close one) to the true motion.

The mathematical reasoning for having most of the oscillations as sinusoidal motion is based on the Fourier theorem. According to the theorem, any disturbance that is periodic with period T can be built up from a set of pure sinusoidal oscillations of periods T, $T/2$, $T/3$, etc., with appropriately chosen amplitudes.

A motion like Eq. (1.3) is referred to as simple harmonic motion (SHM). Its basic characteristics are listed below:

1. The motion is confined within $x = \pm A$. A is known as amplitude.

2. The motion has period T, which is the time between successive maxima, or, more generally, between two successive times having the same value of the pair x and $\frac{dx}{dt}$.

Given Eq. (1.3), T must correspond to an increase of 2π in the argument of sine:

$$\omega(t+T) + \phi_0 = (\omega t + \phi_0) + 2\pi \quad \Rightarrow \quad T = \frac{2\pi}{\omega}. \tag{1.4}$$

The known state for displacement x and velocity $v = \frac{dx}{dt}$ at $t = 0$ (or at any other moment) completely specifies later (or earlier) behavior. For $t = 0$,

$$x_0 = A\sin(\phi_0), \tag{1.5}$$

$$v_0 = \omega A\cos(\phi_0), \tag{1.6}$$

which can be solved for the amplitude A and phase ϕ_0:

$$A = \sqrt{x_0^2 + \left(\frac{v_0}{\omega}\right)^2}, \tag{1.7}$$

$$\phi_0 = \arctan\left(\frac{\omega x_0}{v_0}\right). \tag{1.8}$$

Every real oscillation has a beginning and an end. If a SHM starts at t_1 and switches off at t_2, then its mathematical description amounts to three statements:

$$\begin{aligned} -\infty < t < t_1 \quad & x = 0, \\ t_1 \leqslant t \leqslant t_2 \quad & x = A\sin(\omega t + \phi_0), \\ t_2 < t < \infty \quad & x = 0. \end{aligned}$$

Think about the physical and mathematical implications of 'infinite' vs. 'finite.'

1.1.1 Rotating vector representation

Simple harmonic motion can be regarded as the projection of uniform circular motion (see Fig. 1.1). Let a horizontal disk of radius A rotate with uniform angular velocity ω [rad/sec]. Let a peg P be attached to it at the edge and let a parallel beam of light cast the shadow of the peg on the vertical wall. Then, this shadow performs a SHM with period $T = 2\pi/\omega$. The instantaneous

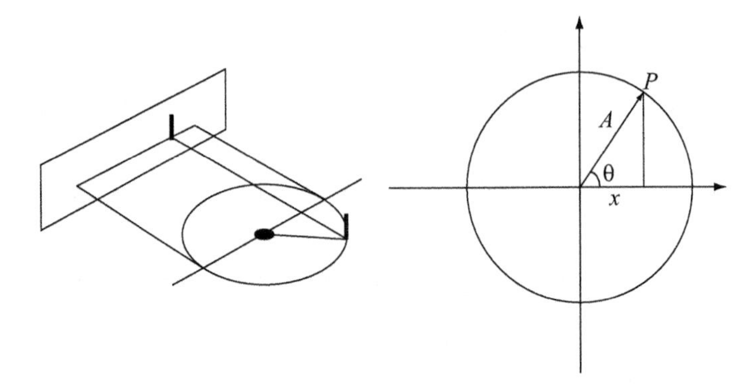

FIGURE 1.1: SHM as a projection of uniform circular motion.

position of point P is determined by radius A and the variable angle θ. As with polar coordinates, we take the counterclockwise direction as positive. Angle θ can be written as

$$\theta = \omega t + \alpha, \tag{1.9}$$

where α is the value of θ at $t = 0$. The displacement x is then given by

$$x = A\cos(\theta) = A\cos(\omega t + \alpha) = A\sin(\omega t + \alpha + \pi/2). \tag{1.10}$$

Thus if SHM is written as in Eq. (1.3), then $\phi_0 = \alpha + \pi/2$.

1.2 Superposition of periodic motions

In many physical situations two or more harmonic oscillations are applied to the same object. A typical situation may correspond to the human eardrum subjected to sound from different sources. Henceforth we will assume the following: **The resultant of two or more harmonic oscillations will be taken to be the sum of individual oscillations**. The associated physical question is indeed very deep. *Is the displacement produced by two disturbances acting together equal to a straightforward superposition of the displacements as they occur separately?* The answer to this can be yes or no depending on whether or not the **displacement is proportional to the force causing it**. If simple addition (superposition) holds, we say that we are dealing with a linear system.

1.2.1 Superposition of two oscillations having the same frequency

The mathematics of this physical situation is discussed in Appendix A dealing with complex notations. Some interesting consequences follow if we look at Eq. (A.29),

$$|Z_0|^2 = A_0^2 = A_1^2 + A_2^2 + 2A_1A_2\cos(\phi_2 - \phi_1), \tag{1.11}$$

which uses the same values for the amplitudes, i.e., $A_1 = A_2 = A$. For the amplitude of the resultant field we then have

$$A_0 = 2A\cos(\delta/2), \quad \delta = \phi_2 - \phi_1. \tag{1.12}$$

It is clear from Eq. (1.12) that the amplitude can vanish for discrete values of the phase difference. It can also achieve the maximum value $2A$ at intermediate points. Thus destructive or constructive interference is a typical signature of superposition of oscillatory phenomena.

1.2.2 Superposition of two oscillations having different frequencies

Let the two oscillations of distinct frequencies be given by

$$x_1 = A_1\cos(\omega_1 t), \tag{1.13}$$

$$x_2 = A_2\cos(\omega_2 t). \tag{1.14}$$

For brevity we dropped the initial phases. For arbitrary ω_1 and ω_2, the resultant displacement $x = x_1 + x_2$ can be complicated—perhaps never repeating itself, for example. The condition for periodicity of the combined motion is that the component motion periods must be commensurable. There should exist two integers n_1 and n_2 such that

$$T = n_1T_1 = n_2T_2. \tag{1.15}$$

The period of the combined motion is given by T obtained for smallest integral values of n_1 and n_2. For example, for $T = 0.02$ sec, $f_1 = \omega_1/(2\pi) = 450$ Hz, $f_2 = \omega_2/(2\pi) = 100$ Hz, and we have $n_1 = 9$ and $n_2 = 2$. See Fig. 1.2 for a visual presentation of this situation.

In the case of commensurable periods, the resultant oscillation can depend on the initial phases. This is depicted in Figs. 1.3 and 1.4 for oscillations having maxima at $t = 0$ (e.g., $A_1\cos(\omega_1 t)$ and $A_2\cos(\omega_2 t)$) and 0 at $t = 0$ (e.g., $A_1\sin(\omega_1 t)$ and $A_2\sin(\omega_2 t)$).

We now analyze the beating effect, assuming the two SHM amplitudes to be the same. In that case we obtain

$$x = A(\cos(\omega_1 t) + \cos(\omega_2 t)) = 2A\cos\left(\frac{\omega_1 - \omega_2}{2}t\right)\cos\left(\frac{\omega_1 + \omega_2}{2}t\right). \tag{1.16}$$

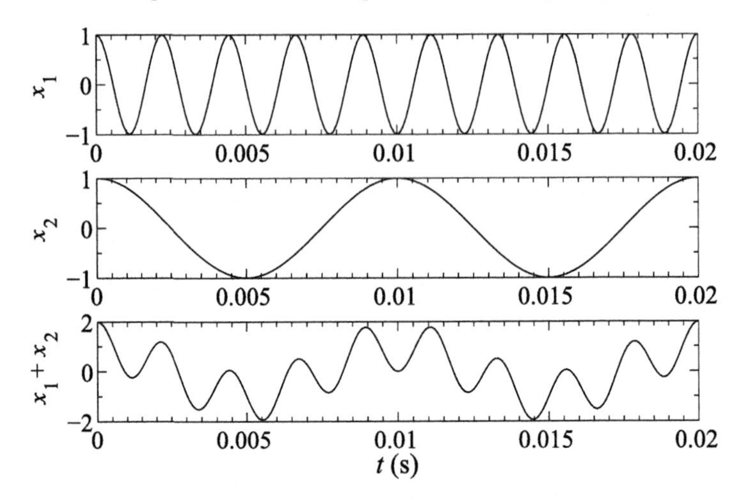

FIGURE 1.2: Superposition of two oscillations with commensurable periods (2:9).

Eq. (1.16) holds for any pairs of frequencies. But beat phenomena are physically meaningful when

$$|\omega_1 - \omega_2| \ll \omega_1 + \omega_2, \tag{1.17}$$

i.e., when the combined oscillation approximates a SHM at the average frequency $(\omega_1 + \omega_2)/2$. The envelope of such oscillations is given by

$$x = \pm 2A \cos\left(\frac{\omega_1 - \omega_2}{2} t\right). \tag{1.18}$$

The beating of two oscillations with two frequencies (600 and 700 Hz) is shown in Fig. 1.5. The time between two successive zeros of the envelope is **one half-period** of the modulating envelope, i.e., $2\pi/(|\omega_1 - \omega_2|)$ because of the $\pm$ sign in front. Thus the beat frequency is simply the difference of the individual frequencies and not half of this frequency.

1.2.3 Combining two oscillations at right angles

Earlier we were discussing superposition of two oscillations in one dimension. We now concentrate on the case when the two oscillations take place along perpendicular directions. Let a point moving in the xy plane experience simultaneously the displacements along x and y as follows:

$$x = A_1 \cos(\omega_1 t + \alpha_1), \tag{1.19}$$

$$y = A_2 \cos(\omega_2 t + \alpha_2). \tag{1.20}$$

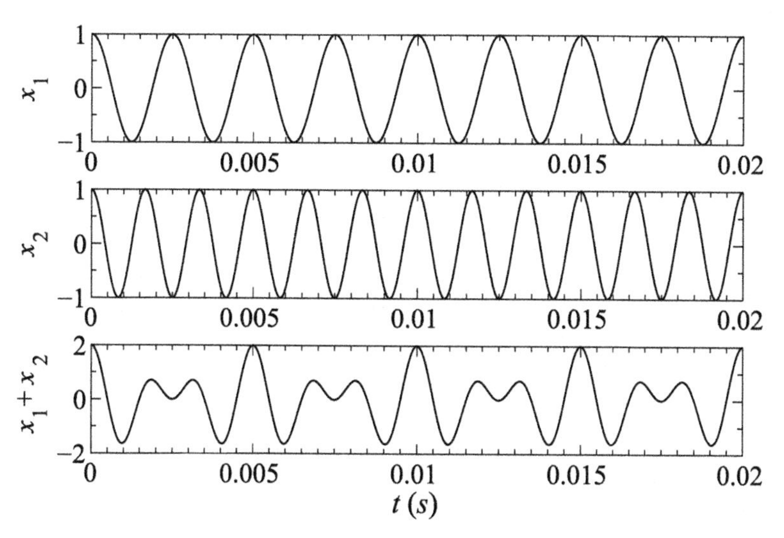

FIGURE 1.3: Superposition of two oscillations with commensurable periods (2:3) with maxima at $t = 0$.

We consider two special cases, namely, (i) when the perpendicular oscillations have the same frequency and (ii) when the frequencies are different.

Perpendicular motions with equal frequencies

With a suitable choice of initial time, Eqs. (1.19) and (1.20) can be written as

$$x = A_1 \cos(\omega t), \tag{1.21}$$

$$y = A_2 \cos(\omega t + \delta). \tag{1.22}$$

Here δ is the initial phase difference, and in this case it is the phase difference at all other times. For different values of δ, we have different relations between x and y:

$$\delta = 0, \quad y = (A_2/A_1)x, \tag{1.23}$$

$$\delta = \pi/2, \quad \frac{x^2}{A_1^2} + \frac{y^2}{A_2^2} = 1, \tag{1.24}$$

$$\delta = \pi, \quad y = -(A_2/A_1)x, \tag{1.25}$$

$$\delta = 3\pi/2, \quad \frac{x^2}{A_1^2} + \frac{y^2}{A_2^2} = 1. \tag{1.26}$$

Though the equations of the ellipse given by Eqs. (1.24) and (1.26) are the same, the first (second) one is drawn clockwise (counterclockwise). For $\delta =$

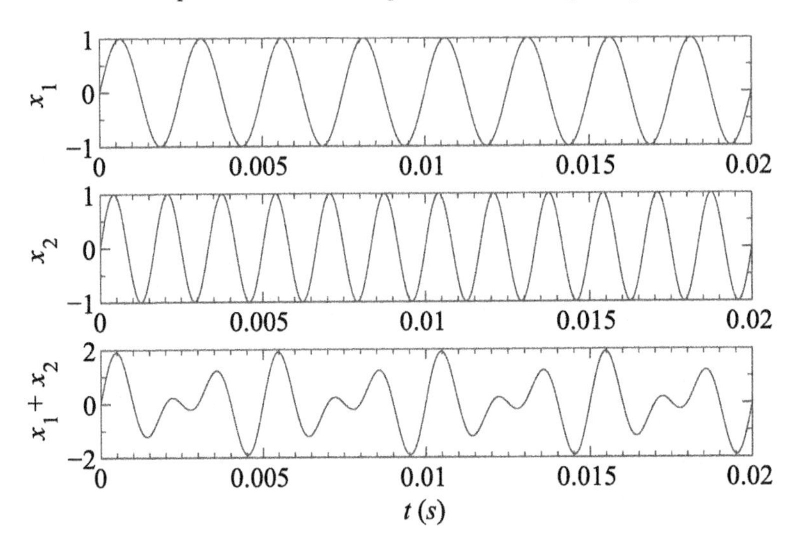

FIGURE 1.4: Superposition of two oscillations with commensurable periods (2:3) with zero at $t = 0$.

$\pi/4$, we can obtain the resultant motion by following the procedure outlined in Fig. 1.6. It is again an ellipse but with major and minor axes not along x and y directions.

Perpendicular motions with distinct frequencies

The procedure outlined above can be used to depict the motion in this case. In Fig. 1.7 we show the resultant motion for $\omega_2 = 2\omega_1$ as well as $\delta = 0$, $\pi/4$, $\pi/2$, $3\pi/4$ and π (from left to right, respectively). Such curves are sometimes referred to as Lissajous figures (after J. A. Lissajous, 1822–1880). For example, Fig. 1.7 can be reproduced using the diagrammatic approach as in Fig. 1.6. The curve depicted in the leftmost panel can be obtained by dividing the reference circle for the motion at frequency ω_2 into eight equal time intervals, i.e., into arcs subtending $\pi/4$ each, and by remembering that for $\omega_2 = 2\omega_1$, one complete cycle of ω_2 corresponds to only one half-cycle of ω_1.

1.3 Free oscillations

The restoring forces in any actual system are linear in the displacement only as an approximation. Nevertheless, in a vast majority of cases the

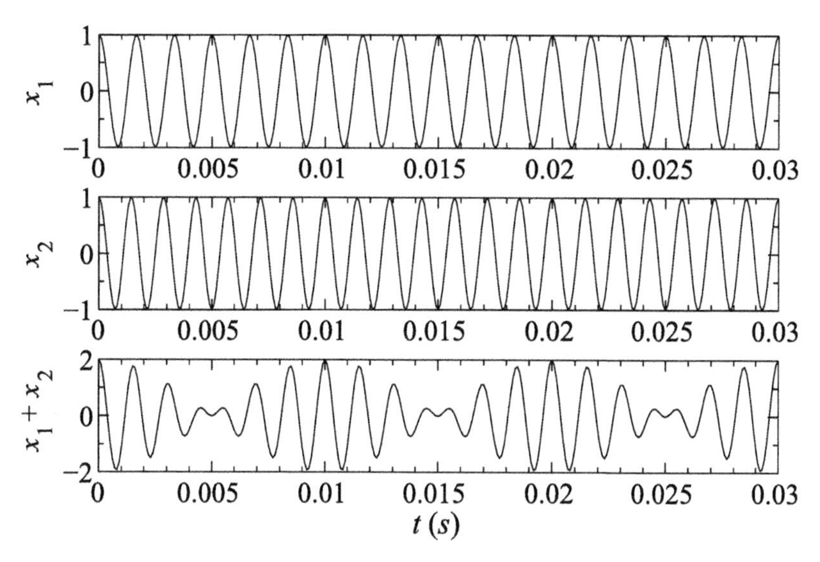

FIGURE 1.5: Beating of two oscillations with commensurable periods (6:7).

deformation produced results in restoring forces proportional to displacement and hence leads to simple harmonic motions. We consider several such cases by picking examples from various areas. But first we will carry out a deeper analysis of a basic mass-spring system, which serves as a prototype of many oscillatory systems. Specifically we consider a point mass attached to an ideal spring undergoing one-dimensional oscillatory motion. Two essential features necessary for oscillatory motion can immediately be identified. They are

- An inertial component capable of carrying kinetic energy and
- An elastic component capable of storing potential energy.

Assuming Hooke's law to be valid, we can obtain the potential energy as proportional to the square of the displacement, while the kinetic energy is $mv^2/2$. We can also write the equation of motion for the mass in either of two ways:

1. By Newton's law ($\mathbf{F} = m\mathbf{a}$),

$$-kx = ma, \quad \text{or} \tag{1.27}$$

2. By conservation of total mechanical energy,

$$\frac{1}{2}mv^2 + \frac{1}{2}kx^2 = E. \tag{1.28}$$

Note that Eq. (1.28) can be obtained from Eq. (1.27) by writing the first in the differential form, multiplying both sides by $(\frac{dx}{dt})$ and integrating both sides.

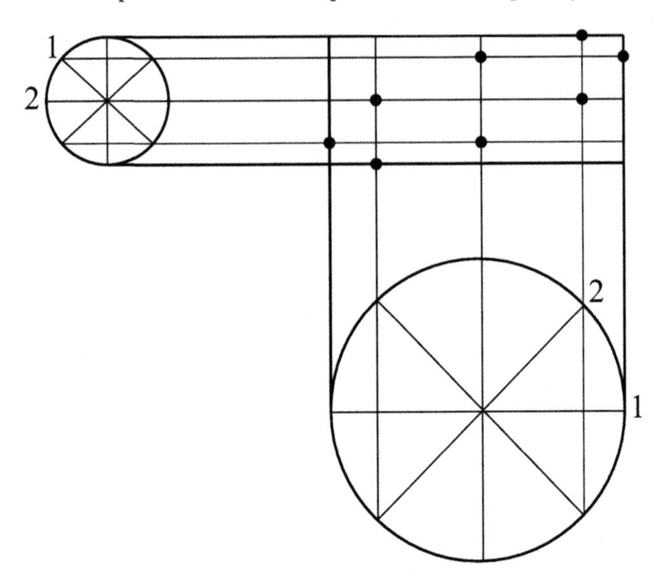

FIGURE 1.6: Superposition of two perpendicular oscillations with the same frequency, but with phase difference $\pi/4$.

Indeed,

$$m\frac{d^2x}{dt^2}\frac{dx}{dt} + kx\frac{dx}{dt} = 0, \tag{1.29}$$

$$\frac{1}{2}m\left(\frac{dx}{dt}\right)^2 + \frac{1}{2}kx^2 = E. \tag{1.30}$$

The solution to these equations can be written as

$$x = A\cos(\omega t + \alpha), \tag{1.31}$$

where $\omega^2 = k/m$ and the unknown constants A and α are to be determined from the initial conditions $x(t = 0) = x_0$ and $(\frac{dx}{dt})_{t=0} = v(t = 0) = v_0$.

1.3.1 General solution of the harmonic oscillator equation

Consider the equation for SHM given by

$$\frac{d^2x}{dt^2} + \omega^2 x = 0. \tag{1.32}$$

We seek the solution in the form

$$x = Ce^{pt}, \tag{1.33}$$

which after substitution in Eq. (1.32) yields the characteristic equation for p as

$$p^2 + \omega^2 = 0. \tag{1.34}$$

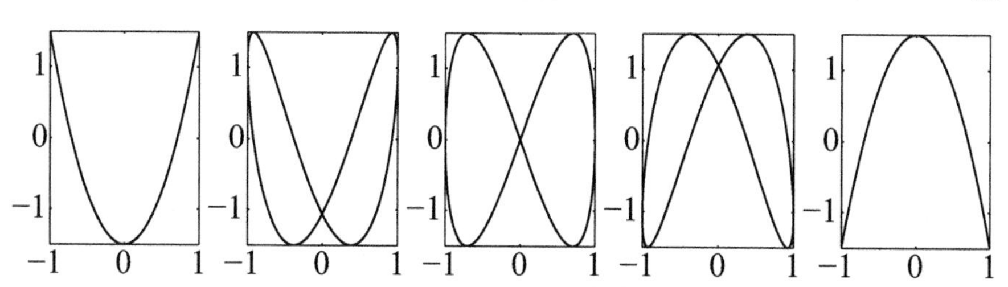

FIGURE 1.7: Superposition of two perpendicular oscillations with distinct frequencies with phase difference $\delta = 0$, $\pi/4$, $\pi/2$, $3\pi/4$ and π (from left to right).

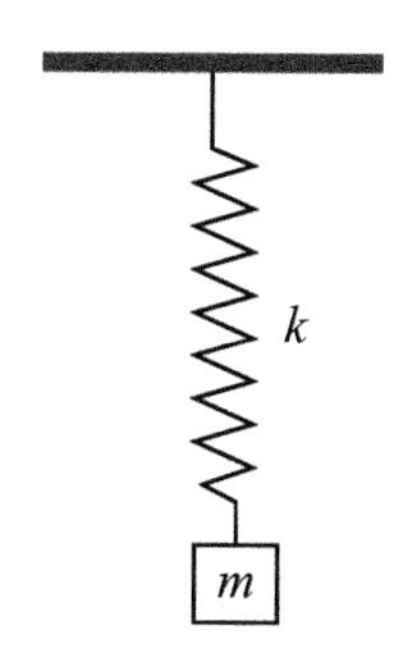

FIGURE 1.8: Mass-spring system.

Eq. (1.34) immediately leads to a pair of purely imaginary solutions, namely, $p = \pm i\omega$ and the general solution that can then be written as

$$x(t) = C_1 e^{i\omega t} + C_2 e^{-i\omega t}. \tag{1.35}$$

Note that the general solution of a second-order ordinary differential equation has two constants figuring in it. The physical solution will correspond to the real part of Eq. (1.35), which can be written as

$$x = (C_1' + C_2')\cos(\omega t) - (C_1'' - C_2'')\sin(\omega t), \tag{1.36}$$

where primes and double primes denote real and imaginary parts, respectively. Introducing notations as $C_1' + C_2' = A\cos\alpha$ and $C_1'' - C_2'' = A\sin\alpha$, Eq. (1.36) can be cast in the form

$$x = A\cos(\omega t + \alpha). \tag{1.37}$$

The same result can be obtained using the rotating vector representation. The first (last) term in Eq. (1.35) corresponds to the vector C_1 (C_2) rotating in the counterclockwise (clockwise) direction. These combine to give a harmonic oscillation along the x-axis if the lengths are the same. C_2 is rotated through

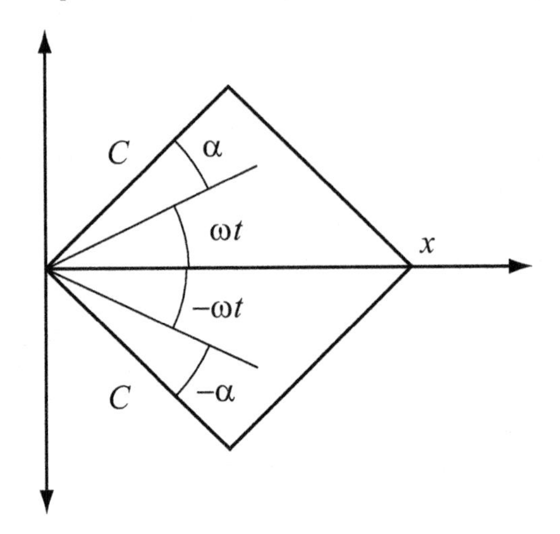

FIGURE 1.9: Superposition of complex solutions.

some angle α clockwise from $-\omega t$, provided that C_1 is rotated through α with respect to ωt (see Fig. 1.9). This analysis clearly reveals that linear motion can be obtained as a superposition of circular motions, which is just the opposite of the case of Lissajous figures, where we superposed linear motions to get 'circular' motion. The last statement has deep meaning in the context of the interchangeability of linear and circular polarizations.

1.3.2 Elasticity, Hooke's law and Young's modulus

Stretching a rod or a wire provides the simplest example amenable to easy analysis. We assume the system to be in static equilibrium.

1. For a given material with a given cross-sectional area A, the elongation Δl under a given force is proportional to the original length l_0. The dimensionless ratio $\Delta l/l_o$ is called the strain.

2. It is an experimental observation that for rods of a given material, but of different A, the same strain is caused by forces proportional to A. The ratio $\Delta F/A$ is called the stress and has the dimension of force per unit area, or pressure.

3. For small strains ($\leq 0.1\%$), the relation between stress and strain is linear in accordance with Hooke's law. The value of this constant for any given material is called Young's modulus of elasticity Y.

We thus have

$$\frac{dF/A}{dl/l_0} = -Y. \tag{1.38}$$

If we choose a different notation (x for displacement, and F for force) Eq. (1.38) can be recast in the standard form of SHM,

$$F = -\left(\frac{AY}{l_0}\right)x. \tag{1.39}$$

Here the spring constant can be identified as $k = AY/l_0$.

1.3.3 Pendulums

A conventional simple pendulum is shown in Fig. 1.10 and we must note that the motion of the bob is essentially two-dimensional in contrast to the problems discussed earlier. Indeed, though the motion is predominantly horizontal (along x), there is a vertical displacement (along y) associated with a change in the gravitational potential energy. This situation is well suited for a discussion based on the formulas for the conservation of energy

$$\frac{1}{2}mv^2 + mgy = E, \quad \text{where} \quad v^2 = \left(\frac{dx}{dt}\right)^2 + \left(\frac{dy}{dt}\right)^2. \tag{1.40}$$

It is clear from Fig. 1.10 that $l^2 = (l-y)^2 + x^2$ or $x^2 = 2ly - y^2$ and for small θ, $y \ll x$ and we have $x^2 = 2ly$ so that

$$y \approx \frac{x^2}{2l}. \tag{1.41}$$

Using this approximate relation and the other consequence $\left(\frac{dx}{dt}\right) \gg \left(\frac{dy}{dt}\right)$, we can recast the energy conservation relation Eq. (1.40) in the form

$$\frac{1}{2}m\left(\frac{dx}{dt}\right)^2 + \frac{1}{2}\frac{mg}{l}x^2 = E. \tag{1.42}$$

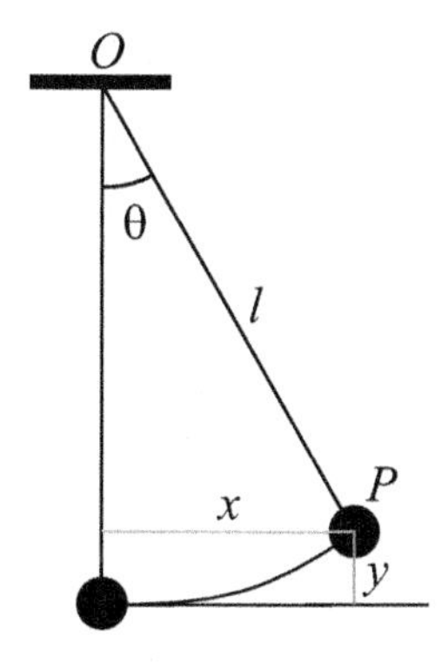

FIGURE 1.10: Schematic view of a simple pendulum.

Eq. (1.42) can easily be recognized as one describing SHM with frequency $\omega = \sqrt{g/l}$.

Damping of free oscillations

All physical systems are subjected to dissipative processes. For example, the motion of the bob of the pendulum is subjected to air resistance (frictional force). In a general case, in the lowest approximation we can write the damping force as

$$F_{damping} = -bv. \tag{1.43}$$

Note that this force is proportional to the magnitude of velocity and acts in the opposite direction of velocity. Accounting for this force, Newton's equation can be reduced to

$$\frac{d^2x}{dt^2} + 2\gamma\frac{dx}{dt} + \omega_0^2 x = 0, \tag{1.44}$$

where $2\gamma = b/m$, $\omega_0^2 = k/m$ (in case of a mass on a spring). Assuming a solution of the form $\exp(i\beta t)$, the characteristic equation can be written as

$$\omega_0^2 - \beta^2 + 2i\gamma\beta = 0, \tag{1.45}$$

which can be easily solved for β, yielding the pair of roots given by

$$\beta_{1,2} = i\gamma \pm \omega, \quad \omega^2 = \omega_0^2 - \gamma^2. \tag{1.46}$$

It is clear that damping modifies the oscillation frequency. In terms of system parameters, ω can be expressed as

$$\omega = \sqrt{\frac{k}{m} - \left(\frac{b}{2m}\right)^2}. \tag{1.47}$$

Using Eq. (1.46) one can write the general solution as

$$x = C_1 e^{i\beta_1 t} + C_2 e^{i\beta_2 t}, \tag{1.48}$$

$$= e^{-\gamma t}\left(C_1 e^{i\omega t} + C_2 e^{-i\omega t}\right). \tag{1.49}$$

Finally, taking the real part, Eq. (1.49) can be reduced to the following:

$$x = Ae^{-\gamma t}\cos(\omega t + \alpha). \tag{1.50}$$

Thus, the solution represents 'damped harmonic' oscillations with frequency ω distinct from the natural frequency ω_0 (modification due to damping) with an amplitude that decays in time in an exponential fashion. As expected this damping is determined by γ characterized by the damping force constant b. This is shown in Fig. 1.11. From this graph as well as Eq. (1.50), we can recognize γ as the reciprocal of the time required for the amplitude to decay to $1/e$ of its initial value.

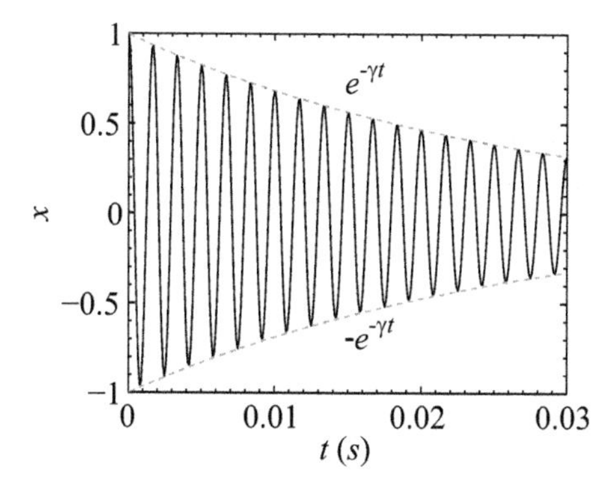

FIGURE 1.11: Damped harmonic oscillation with $f = 600$ Hz, $\gamma = 2\pi f/100$ and $\alpha = 0$.

Depending on how ω_0 and γ compare with each other, we can distinguish three different regimes, namely, (i) under-damped, (ii) over-damped and (iii) critically damped. In fact, Eqs. (1.46)–(1.50) describe the *under-damped* regime when $\gamma < \omega_0$ leads to real values for ω. In many cases γ is much smaller than the natural frequency ω_0 so that we almost have 'harmonic' oscillation, albeit with exponential damping as shown in Fig. 1.11. Keeping in mind that the total energy of harmonic oscillation is given by $kA^2/2$, we can find the temporal evolution of total energy as

$$E(t) = \frac{1}{2}kA_0^2e^{-2\gamma t} = E_0e^{-2\gamma t}, \tag{1.51}$$

where A_0 and E_0 are the initial amplitude and energy, respectively, at time $t = 0$. In the context of damped oscillatory systems, we often talk about a universal figure of merit (dimensionless), otherwise known as the quality factor or simply the Q-factor. It is defined as

$$Q = \frac{\omega_0}{2\gamma}. \tag{1.52}$$

It is clear that low values of γ imply large Q-factors, meaning thereby that the system is likely to sustain oscillations longer. The modified oscillation frequency can be expressed in terms of the Q-factor as

$$\omega^2 = \omega_0^2\left(1 - \frac{1}{4Q^2}\right). \tag{1.53}$$

Thus if $Q \gg 1$ it follows that $\omega \approx \omega_0$ and Eq. (1.50) can be written as

$$x = Ae^{-\omega_0 t/(2Q)}\cos(\omega_0 t + \alpha). \tag{1.54}$$

In the *over-damped* case, $\gamma > \omega_0$, and instead of Eq. (1.46) we have

$$\beta_{1,2} = i(\gamma \pm \xi), \quad \xi^2 = \gamma^2 - \omega_0^2, \tag{1.55}$$

and the general solution can be written as

$$x = c_1 e^{-(\gamma+\xi)t} + c_2 e^{-(\gamma-\xi)t}. \tag{1.56}$$

The *critically damped* case corresponds to the equality $\gamma = \omega_0$ and in that case the solution is written as

$$x = (A + Bt)e^{-\gamma t}. \tag{1.57}$$

1.4 Forced oscillations and resonance

In contrast to the previous section, we now consider a physical oscillatory system driven by a periodic force and try to explain the important phenomenon of 'resonance.' Resonances are encountered in a variety of seemingly different physical situations, ranging from atoms driven by laser light to a swing pushed periodically. Let us try to understand the phenomenon of resonance in layperson's terms. Any oscillatory system has a natural frequency of oscillation ω_0. If the driving field frequency ω is close to this natural frequency, then the amplitude of oscillation can be made very large by quite a small force. Away from the natural frequency (to both positive and negative sides), the effect of the same force is not very prominent, i.e., the amplitude produced can be very small. The farther away the frequency is from the resonance frequency, the smaller is the amplitude. The enhanced response of the system near the natural frequency is termed resonance.

For a model system we again pick the usual mass m on a spring with spring constant k. Let a sinusoidal driving force $F = F_0 \cos(\omega t)$ be applied to the system so that Newton's equation in absence of damping reads as

$$m\frac{d^2x}{dt^2} = -kx + F_0 \cos(\omega t) \tag{1.58}$$

or

$$\frac{d^2x}{dt^2} + \omega_0^2 = \frac{F_0}{m}\cos(\omega t). \tag{1.59}$$

A qualitative analysis of Eq. (1.59) reveals that driven from equilibrium, the oscillator has a tendency to oscillate at its natural frequency ω_0, while the driving force tries to leave its imprint at the driving frequency ω. Thus the resultant motion will be a superposition of oscillations at these two frequencies. However, due to inevitable damping (missing in the above equations), the natural oscillations will die out, leaving only the forced part. The initial stage when both the oscillations are present is termed 'transient,' while the long-time behavior is dictated by the forced oscillations. Mathematically, Eq. (1.59) is inhomogeneous since it has a right-hand side function of time not involving the dependent variable x or its derivative. The general solution of such an equation is a superposition of (i) a general solution of the homogeneous equation (when the right-hand side is equated to zero) and (ii) a particular solution of the inhomogeneous equation (in this case the forced part, not involving any constants). The general solution of the homogeneous equation, of course, contains two arbitrary constants (see, for example, Eq. (1.35)) and this is the part that dies out in the presence of damping (see Eq. (1.49)).

1.4.1 Solution for the forced undamped oscillator

Consider the complex equivalent of Eq. (1.59):

$$\frac{d^2x}{dt^2} + \omega_0^2 x = \frac{F_0}{m} e^{-i\omega t}. \tag{1.60}$$

In Eq. (1.60) we used the same notation for the complex dependent variable x. We have to remember to project it onto the real axis for the physical solution. Assuming a solution of the form

$$x = Ae^{-i(\omega t - \alpha)}, \tag{1.61}$$

and substituting in Eq. (1.60), we have

$$(\omega_0^2 - \omega^2)A = \frac{F_0}{m} e^{-i\alpha} = \frac{F_0}{m}(\cos\alpha - i\sin\alpha). \tag{1.62}$$

Equating the real and imaginary parts on both sides, we have

$$A = \frac{F_0/m}{(\omega_0^2 - \omega^2)} \cos\alpha, \tag{1.63}$$

$$0 = \frac{F_0}{m} \sin\alpha. \tag{1.64}$$

In order to ensure that A is positive on both sides of the resonance frequency ω_0, we pick $\alpha = 0$ for $\omega < \omega_0$ and $\alpha = \pi$ for $\omega > \omega_0$. Thus the complete particular solution to Eq. (1.59) can be written as

$$x = \frac{F_0/m}{(\omega_0^2 - \omega^2)} \cos\alpha \cos(\omega t - \alpha), \tag{1.65}$$

where $\alpha = 0$ for $\omega < \omega_0$ and $\alpha = \pi$ for $\omega > \omega_0$. Thus transition through resonance ($\omega = \omega_0$) is associated with a jump in phase by π. The behavior of amplitude and phase is shown in Fig. 1.12.

1.4.2 Forced damped oscillations

In Section A.3.2 of Appendix A, we looked at the mathematics of forced damped oscillations in order to highlight the advantages of complex notations.

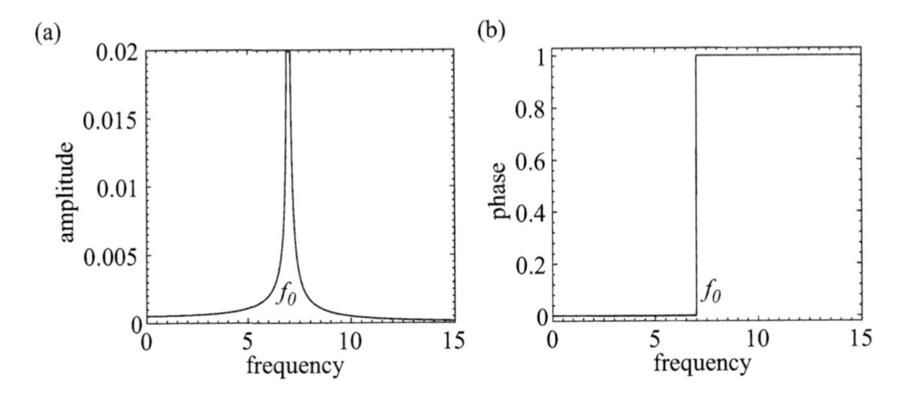

FIGURE 1.12: Undamped forced harmonic oscillation: (a) shows the amplitude while (b) shows the phase in units of π.

However, we redo the exercise again due to notational changes and for maintaining the consistency. Including damping, Eq. (1.60) can be rewritten as

$$\frac{d^2x}{dt^2} + 2\gamma\frac{dx}{dt} + \omega_0^2 x = \frac{F_0}{m}e^{-i\omega t}. \tag{1.66}$$

Substituting a solution of the form (1.61) and equating the real and imaginary parts on both the sides, we have

$$A = \frac{F_0/m}{(\omega_0^2 - \omega^2)}\cos\alpha, \tag{1.67}$$

$$2\gamma\omega A = \frac{F_0}{m}\sin\alpha. \tag{1.68}$$

Squaring and adding both sides of Eqs. (1.67) and (1.68), we can deduce the expression for A as

$$A = \frac{F_0/m}{\left[(\omega_0^2 - \omega^2)^2 + (2\gamma\omega)^2\right]^{1/2}}. \tag{1.69}$$

Dividing each side of Eq. (1.68) by those of Eq. (1.67), we can derive the expression for the phase

$$\tan(\alpha) = \frac{2\gamma\omega}{\omega_0^2 - \omega^2}. \tag{1.70}$$

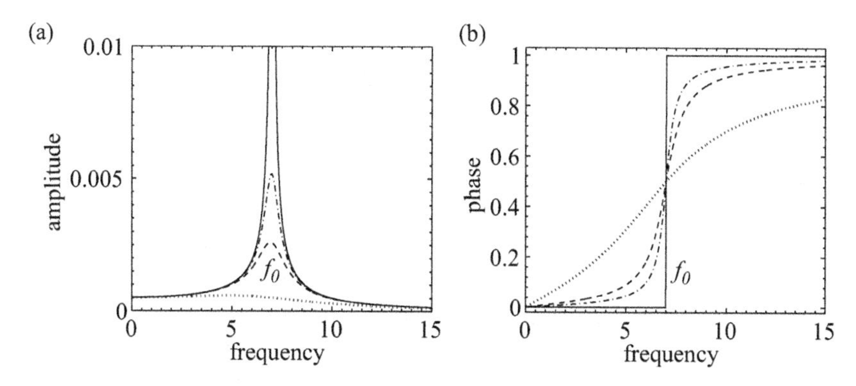

FIGURE 1.13: Damped forced harmonic oscillation with resonance frequency $f_0 = \omega_0/2\pi$: (a) shows the amplitude while (b) shows the phase in units of π.

In terms of the quality factor $Q = \omega_0/(2\gamma)$, the expressions for the amplitude and phase can be rewritten as

$$A = \frac{F_0/m}{\left[(\omega_0^2 - \omega^2)^2 + (\frac{\omega\omega_0}{Q})^2\right]^{1/2}}, \tag{1.71}$$

$$\tan(\alpha) = \frac{\frac{\omega\omega_0}{Q}}{\omega_0^2 - \omega^2}. \tag{1.72}$$

The results for the amplitude and phase for this case are shown in Fig. 1.13, where we have plotted these quantities for three different values of Q, namely, $Q = 10$ (dash-dotted), 5 (dashed) and 1 (dotted), respectively. For comparison we have also shown the case when there is no damping (solid lines). Note that finite damping removes the singularity at $\omega = \omega_0$. Besides, a higher-quality factor leads to a higher response with larger amplitudes.

Transients

In the transient regime, the solution, as mentioned earlier, can be written as

$$x = Be^{-\gamma t}\cos(\omega_1 t + \alpha_1) + A\cos(\omega t - \alpha), \tag{1.73}$$

where B and α_1 are arbitrary constants, $\omega_1^2 = \omega_0^2 - \gamma^2$ and A and α are given by Eqs. (1.69) and (1.70) (or Eqs. (1.71) and (1.72)). It is clear from Eq. (1.73) that in the absence of damping and for nearby frequencies, the solution represents the beating of two sinusoids. In the presence of damping in off-resonant cases, the beating persists for some time and finally the oscillations settle down to constant amplitude A. In the resonant case there is no beating, and the final amplitude is reached in a monotonic fashion. These features are shown in Fig. 1.14.

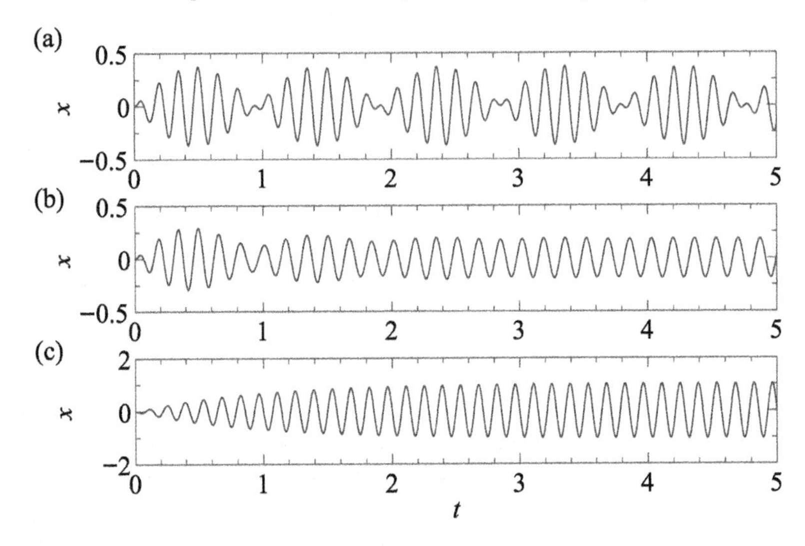

FIGURE 1.14: Transients in forced harmonic oscillation (temporal evolution of the amplitudes). (a) Corresponds to very large Q ($\sim$ 20000). Almost nonexistent damping leads to beating of the natural frequency ω_0 and the driving frequency ω. (b) and (c) are when $Q = 20$ and show the off-resonant ($\omega = 0.85\ \omega_0$) and the resonant ($\omega \approx \omega_0$) behavior, respectively.

1.5 Coupled oscillations and normal modes

In most of our earlier discussions, we concentrated on the type of oscillations that have mostly the same frequency. In reality the system may have components that oscillate with different frequencies. Each oscillating component has specific effects on the others and vice versa. For example, a solid body is composed of many atoms or molecules. Every atom may behave like an oscillator, vibrating about the equilibrium position. Motion of each atom affects the neighbors. Thus all the atoms are coupled together. A question results: How does the coupling affect the behavior of individual oscillators?

1.5.1 Two coupled pendulums

Consider the system of two identical pendulums A and B joined by a spring of rest length equal to the distance between the bobs (see Fig. 1.15). This system serves as a prototype toward understanding more complicated phenomena involving many oscillators. Let pendulum A be pulled to a distance, keeping B held at equilibrium, and then let both be released. Oscillations of A will decrease while those of B will gain in amplitude. Finally, the motion of A will

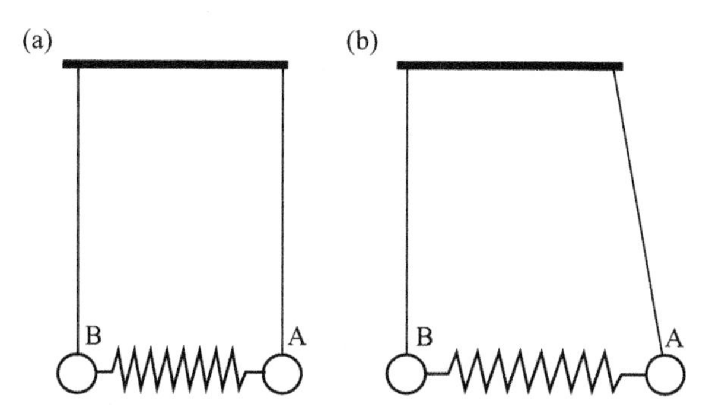

FIGURE 1.15: Two coupled pendulums, (a) at rest and (b) when B is kept at equilibrium with A displaced and both then released.

be transferred to B and then this pattern of exchange of motion will continue. Individual motion of A or B resembles that of beating with two frequencies. Indeed, there are two characteristic frequencies of the coupled system and they are termed the normal modes. Any oscillation of the coupled system can always be written as a superposition of these two normal modes.

A change in the initial conditions makes it easy to recognize the normal modes (see Fig. 1.16). Suppose we draw both A and B to one side by equal amounts and release them. Let the distance between them be fixed and equal to the relaxed length of the spring. Both A and B will oscillate in phase. Since the spring is not stretched it does not affect oscillation frequency, which is just the individual oscillation frequency $\omega_0 = \sqrt{(g/l)}$ of the pendulums. In the absence of damping, these oscillations will continue forever. This is one of the two normal modes and ω_0 is one normal mode frequency. The solutions in this case are given by

$$x_A = C\cos(\omega_0 t), \qquad x_B = C\cos(\omega_0 t). \tag{1.74}$$

If A and B are drawn to opposite sides by equal amounts and then released, such oscillations can persist forever. This is the other normal mode with a higher frequency. We calculate this frequency as follows. If the pendulums were free, a displacement of x would correspond to a restoring force of $m\omega_0^2 x$. In the presence of coupling spring, either it is stretched or compressed by an amount $2x$ and hence the additional force is $2kx$ (k is the spring constant). Thus the equation of motion for A is

$$m\frac{d^2x_A}{dt^2} + m\omega_0^2 x_A + 2kx_A = 0, \tag{1.75}$$

or

$$\frac{d^2x_A}{dt^2} + \omega_0^2 x_A + 2\omega_c^2 x_A = 0, \tag{1.76}$$

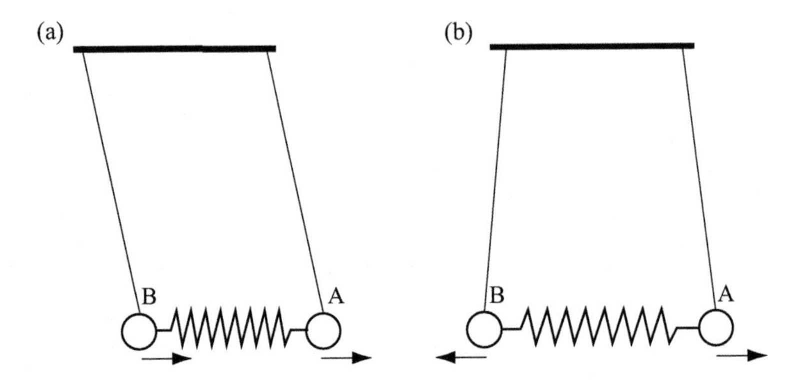

FIGURE 1.16: (a) Symmetric and (b) antisymmetric normal modes.

where $\omega_c^2 = k/m$. We can easily recognize the resonance frequency ω' from Eq. (1.76) as

$$\omega'^2 = \omega_0^2 + 2\omega_c^2 = \left(\frac{g}{l} + \frac{2k}{m}\right). \tag{1.77}$$

For the said initial conditions, the solutions are

$$x_A = D\cos(\omega' t), \qquad x_B = -D\cos(\omega' t). \tag{1.78}$$

Note that at any given moment, the motion of B is the mirror image of motion of A. The motions of A and B are π out of phase.

It is important to observe that if any pendulum is clamped, the angular frequency of the other has two contributions, one from the gravity and the other from the spring. Thus the net frequency is $\sqrt{(\omega_0^2 + \omega_c^2)}$. If this frequency (characteristic of one oscillator) is taken as the reference, then the two normal mode frequencies lie on the two sides of it (i.e., one greater and the other smaller).

1.5.2 Superposition of normal modes

We mentioned earlier that once excited in any of the normal modes (of course for suitable initial conditions), the system continues in the same oscillatory state. For any other initial conditions resulting in more complicated oscillatory pattern, the resultant oscillations can always be perceived as a superposition of the normal modes. In mathematical terms these normal modes form a suitable basis for representing any free motion of the coupled system. Here we show how this is done.

Pick any arbitrary moment when A is displaced by x_A and B is displaced by x_B resulting in a stretching of the spring by an amount $x_A - x_B$ (see Fig. 1.17). Thus the spring pulls on A and B with a force with magnitude

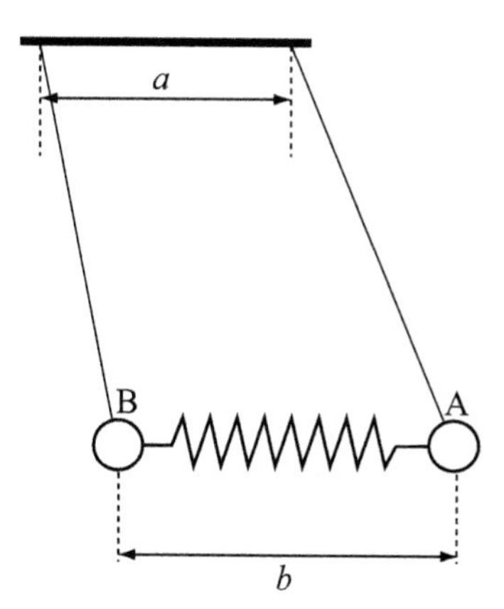

FIGURE 1.17: Displacement of the two pendulums at any arbitrary moment.

proportional to $k(x_A - x_B)$. Since the direction of this force is opposite for A and B, the restoring forces on A and B, respectively, are given by

$$m\omega_0^2 x_A + k(x_A - x_B) \quad \text{and} \quad m\omega_0^2 x_B - k(x_A - x_B) \tag{1.79}$$

so that the equations of motion are written as

$$m\frac{d^2x_A}{dt^2} + m\omega_0^2 x_A + k(x_A - x_B) = 0, \tag{1.80}$$

$$m\frac{d^2x_B}{dt^2} + m\omega_0^2 x_B - k(x_A - x_B) = 0. \tag{1.81}$$

Using the notations introduced earlier, the equations above can be reduced to

$$\frac{d^2x_A}{dt^2} + (\omega_0^2 + \omega_c^2)x_A - \omega_c^2 x_B = 0, \tag{1.82}$$

$$\frac{d^2x_B}{dt^2} + (\omega_0^2 + \omega_c^2)x_B - \omega_c^2 x_A = 0. \tag{1.83}$$

Adding and subtracting the above two equations, we have the equations for the sum $(x_A + x_B = q_1)$ and difference $(x_A - x_B = q_2)$ displacements,

$$\frac{d^2q_1}{dt^2} + \omega_0^2 q_1 = 0, \tag{1.84}$$

$$\frac{d^2q_2}{dt^2} + (\omega_0^2 + 2\omega_c^2)q_2 = 0. \tag{1.85}$$

Using the notation $\omega' = \sqrt{\omega_0^2 + 2\omega_c^2}$, we can write one solution (not the general one) as

$$q_1 = C\cos(\omega_0 t), \qquad q_2 = D\cos(\omega' t), \tag{1.86}$$

where C and D are to be evaluated from initial conditions. The frequencies ω_0 and ω' are known as the normal frequencies.

It is important to note the differences between the pairs of equations (1.82), (1.83) and (1.84), (1.85), though they describe the same system. While the first two are coupled (in the sense that one cannot be solved without the other), in the second pair the equations are independent of each other, each one representing a normal mode of the system. This is why if any of the normal modes is excited, it persists forever, not being affected by the other. In other words if through some algebraic manipulations we have reduced a coupled system into its independent components, then we have reduced the system to its normal modes. The dependent variables q_1 and q_2 are sometimes referred to as normal coordinates, and this procedure is termed normal mode decomposition.

Going back to the original displacements x_A and x_B, the solutions can be written as

$$x_A = \frac{1}{2}(q_1 + q_2) = \frac{1}{2}(C\cos(\omega_0 t) + D\cos(\omega' t)), \tag{1.87}$$

$$x_B = \frac{1}{2}(q_1 - q_2) = \frac{1}{2}(C\cos(\omega_0 t) - D\cos(\omega' t)). \tag{1.88}$$

It is clear that if $C = 0$, both pendulums oscillate at one normal frequency ω', while $D = 0$ implies oscillation at the other frequency ω_0. Thus one important characteristic of the normal frequency is that both the bobs can oscillate at that frequency.

For initial conditions given by

$$x_A = A_0, \quad \frac{dx_A}{dt} = 0, \quad x_B = 0, \quad \frac{dx_B}{dt} = 0, \tag{1.89}$$

which correspond to B at the equilibrium position with null velocity while A is moved to A_0 and released (initial null velocity), we can solve for the unknown constants to obtain $C = A_0,\ D = A_0$. Hence

$$x_A = \frac{1}{2}A_0(\cos(\omega_0 t) + \cos(\omega' t)), \tag{1.90}$$

$$x_B = \frac{1}{2}A_0(\cos(\omega_0 t) - \cos(\omega' t)). \tag{1.91}$$

Eqs. (1.90) and (1.91) can be reduced to the following form:

$$x_A = A_0 \cos\left(\frac{\omega' - \omega_0}{2}t\right)\cos\left(\frac{\omega' + \omega_0}{2}t\right), \tag{1.92}$$

$$x_B = A_0 \sin\left(\frac{\omega' - \omega_0}{2}t\right)\sin\left(\frac{\omega' + \omega_0}{2}t\right). \tag{1.93}$$

Both of these represent oscillation at the average frequency $(\omega' + \omega_0)/2$ with a low frequency modulation. The amplitude of one goes to the peak value while that of the other goes to zero.

1.5.3 Coupled oscillations as an eigenproblem: Exact analysis

Let us now address the coupled mode problem from a different angle, namely, as an eigenvalue problem. We first reduce the two second-order equations, Eqs. (1.82) and (1.83), to a set of four coupled first-order equations by writing the dependent variables as

$$x_1 = x_A, \quad x_2 = \frac{dx_A}{dt}, \quad x_3 = x_B, \quad x_4 = \frac{dx_B}{dt}. \tag{1.94}$$

With these definitions Eqs. (1.82) and (1.83) take the following compact matrix form

$$\frac{d}{dt}\begin{pmatrix} x_1 \\ x_2 \\ x_3 \\ x_4 \end{pmatrix} = \begin{pmatrix} 0 & 1 & 0 & 0 \\ -\omega_{oc}^2 & 0 & \omega_c^2 & 0 \\ 0 & 0 & 0 & 1 \\ \omega_c^2 & 0 & -\omega_{oc}^2 & 0 \end{pmatrix} \begin{pmatrix} x_1 \\ x_2 \\ x_3 \\ x_4 \end{pmatrix}. \tag{1.95}$$

Here we have defined $\omega_0^2 + \omega_c^2 = \omega_{oc}^2$. The solution of Eq. (1.95) is determined by the eigenvalues and the corresponding eigenvectors of the 4×4 matrix on the right-hand side. The eigenvalues of the above 4×4 matrix are given by the roots of the characteristic equation

$$\lambda^4 + 2\lambda^2\omega_{oc}^2 + \omega_{oc}^4 - \omega_c^4 = 0. \tag{1.96}$$

The roots of this equation are given by

$$\lambda^2 = \omega_{oc}^2 \pm \omega_c^2. \tag{1.97}$$

One pair of roots is given as $\lambda = \pm i\omega_0$, while the other pair is $\pm i\omega'$, where $\omega' = \sqrt{\omega_0^2 + 2\omega_c^2}$ as before. We can thus recover the oscillation frequencies of the normal modes expressed in Eqs. (1.84) and (1.85). Further, calculation of the eigenvectors will lead to the normal modes q_1 and q_2 described earlier.

1.5.4 Coupled oscillations as an eigenproblem: Approximate analysis

In the literature, an approximation, referred to as the *slowly varying envelope approximation* (SVEA), is often used. The purpose SVEA serves is to get rid of the higher-order derivatives on physical grounds and thus reduce the complexity of the problem. Here we demostrate how SVEA works in the context of normal modes discussed earlier. We will make use of the complex notations in order to solve Eqs. (1.82) and (1.83). We write the solutions as

$$x_A = x_a(t)e^{-i\omega_0 t}, \tag{1.98}$$

$$x_B = x_b(t)e^{-i\omega_0 t}, \tag{1.99}$$

and treat $x_a(t)$ and $x_b(t)$ as slowly varying so that the change in $x(t)$ (for both the subscripts) over one high-frequency period ($= 2\pi/\omega_0$) is negligible compared to the function itself,

$$\left|\frac{d^2x(t)}{dt^2}\right| \ll \omega_0 \left|\frac{dx(t)}{dt}\right| \ll \omega_0^2 \left|x(t)\right|. \tag{1.100}$$

Making use of the approximation (1.100) and Eqs. (1.98) and (1.99), the set of coupled Eqs. (1.82) and (1.83) can be written in matrix form as

$$\begin{pmatrix} \dot{x}_a \\ \dot{x}_b \end{pmatrix} = \frac{i}{2\omega_0}\begin{pmatrix} -\omega_c^2 & \omega_c^2 \\ \omega_c^2 & -\omega_c^2 \end{pmatrix}\begin{pmatrix} x_a \\ x_b \end{pmatrix} = A\begin{pmatrix} x_a \\ x_b \end{pmatrix}. \tag{1.101}$$

The eigenvalues of the matrix A determine the frequencies for the normal modes. The eigenvalues λ_1 and λ_2 are found to be

$$\lambda_1 = 0, \tag{1.102}$$

$$\lambda_2 = -i\frac{\omega_c^2}{\omega_0}. \tag{1.103}$$

Thus, one of the normal modes oscillates at ω_0, and the other at $\omega_0 + (\omega_c^2/\omega_0)$. Note that the latter can easily be recognized as $\omega' = \sqrt{\omega_0^2 + 2\omega_c^2} \approx \omega_0 + (\omega_c^2/\omega_0)$ when the coupling is weak ($\omega_c \ll \omega_0$). In fact, SVEA is applicable in this weak coupling situation. Further, we can find out the corresponding eigenvectors that will correspond to the normal modes mentioned earlier. We can refer to Goldstein [2] for a detailed analysis of similar cases.

Chapter 2

Scalar and vector waves

2.1 Plane waves

The plane wave is the simplest example of a three-dimensional wave. They are characterized by plane wavefronts (see below for a definition), which are perpendicular to the direction of propagation. In other words, they refer to the case when all the surfaces, upon which the disturbance has a constant phase, form a set of planes, each generally perpendicular to the propagation direction.

We first derive the equation of a plane that is perpendicular to a given vector $\mathbf{k} = (k_x, k_y, k_z)$ (in our case the propagation vector) and that passes through some point (x_0, y_0, z_0). The position vector $\mathbf{r} = (x, y, z)$ for such a plane satisfies

$$(\mathbf{r} - \mathbf{r}_0) \cdot \mathbf{k} = 0, \tag{2.1}$$

$$k_x(x - x_0) + k_y(y - y_0) + k_z(z - z_0) = 0, \tag{2.2}$$

$$k_x x + k_y y + k_z z = a = k_x x_0 + k_y y_0 + k_z z_0, \tag{2.3}$$

$$\mathbf{k} \cdot \mathbf{r} = \text{constant} = a. \tag{2.4}$$

The plane given by Eq. (2.4) is the locus of all points whose projection on the $\mathbf{k}$ direction is a constant. Let $\psi(\mathbf{r})$ vary sinusoidally in space as $\psi(\mathbf{r}) = Ae^{i\mathbf{k}\cdot\mathbf{r}}$ or $\psi(\mathbf{r}) = A\cos(\mathbf{k}\cdot\mathbf{r})$. For each of these expressions, ψ is constant over every plane defined by the plane $\mathbf{k} \cdot \mathbf{r} =$ constant. Since these functions are harmonic, they must repeat themselves in space after λ in the direction of $\mathbf{k}$:

$$\psi(\mathbf{r}) = \psi(\mathbf{r} + \lambda\mathbf{k}/k). \tag{2.5}$$

In terms of the exponents, Eq. (2.5) reduces to

$$Ae^{i\mathbf{k}\cdot\mathbf{r}} = Ae^{i\mathbf{k}\cdot\mathbf{r}}\, e^{i\lambda k}. \tag{2.6}$$

Therefore $\lambda k = 2\pi$ and we get

$$k = \frac{2\pi}{\lambda}. \tag{2.7}$$

At a fixed point in space, both $\mathbf{r}$ and $\psi(\mathbf{r})$ are constants, and the planes are motionless. For things to move, $\psi(\mathbf{r})$ has to vary in time:

$$\psi(\mathbf{r}, t) = Ae^{i(\mathbf{k}\cdot\mathbf{r}-\omega t)}. \tag{2.8}$$

For waves propagating along the z direction, Eq. (2.8) ($\mathbf{k}$ having the only nonzero component k along z) can be written in the form

$$\psi(z, t) = Ae^{i(kz-\omega t)}. \tag{2.9}$$

It is easy to verify that Eq. (2.9) satisfies the scalar wave equation given by

$$\frac{\partial^2\psi}{\partial z^2} - \frac{1}{v^2}\frac{\partial^2\psi}{\partial t^2} = 0, \tag{2.10}$$

where the *phase velocity* v is given by

$$v = \omega/k. \tag{2.11}$$

We can arrive at Eq. (2.11) by making use of another simple argument. Phase velocity is defined by the space-time invariance of the phase of the disturbance (see Eq. (2.9)), i.e., by demanding

$$kz - \omega t = \text{constant}. \tag{2.12}$$

Taking the time derivative of Eq. (2.12), we arrive at $v = \frac{dz}{dt} = \frac{\omega}{k}$. We shall be discussing another type of velocity, namely, the group velocity when the notion of a wave packet is introduced.

The scalar plane wave given by Eq. (2.9) is only a particular solution of the scalar wave equation. In fact, any arbitrary function having the form $f(z-vt)$ or $g(z+vt)$ or their superpositions can be a solution. In order to verify this, we take the second partial derivatives of, say, $f(z-vt)$ with respect to z and t:

$$\frac{\partial^2 f}{\partial z^2} = f'', \quad \frac{\partial^2 f}{\partial t^2} = f''v^2, \tag{2.13}$$

where the primes denote derivatives with respect to the argument. Eq. (2.13) substituted into Eq. (2.10) leads to an identity.

2.2 Maxwell's equations and vector waves

In this section we develop the notion of a vector wave where the disturbance is represented no longer by a scalar but by a vector. Thus we must worry about the polarization of the wave that sheds light about the direction of oscillation. As a typical example we consider the case of the electromagnetic (EM) waves. The understanding of the EM waves requires an in-depth knowledge of the basic laws of electromagnetics, namely, Maxwell's equations. Due to the lack of proper mathematical base, we will start from these equations, try to understand the notations, and finally derive the necessary relations for plane monochromatic waves. The detailed coverage of Maxwell's equations with the associated physics can be found in standard textbooks on electromagnetic theory [3, 4]. In many places we will cite the results without deriving them.

In the absence of any sources, Maxwell's equations for a homogeneous isotropic medium with dielectric permittivity ε and magnetic permeability μ can be written as

$$\nabla \cdot \mathbf{E} = 0, \tag{2.14}$$

$$\nabla \cdot \mathbf{B} = 0, \tag{2.15}$$

$$\nabla \times \mathbf{E} = -\frac{\partial \mathbf{B}}{\partial t}, \tag{2.16}$$

$$\nabla \times \mathbf{B} = \mu\varepsilon \frac{\partial \mathbf{E}}{\partial t}. \tag{2.17}$$

The material equations are given by

$$\mathbf{B} = \mu \mathbf{H}, \quad \mathbf{D} = \varepsilon \mathbf{E}. \tag{2.18}$$

In Eqs. (2.14)–(2.17) the vector operator ∇ is given by

$$\nabla = \hat{\mathbf{i}}\frac{\partial}{\partial x} + \hat{\mathbf{j}}\frac{\partial}{\partial y} + \hat{\mathbf{k}}\frac{\partial}{\partial z}. \tag{2.19}$$

From Eqs. (2.14)–(2.17) we can eliminate $\mathbf{B}$ to arrive at the vector wave equation

$$\nabla^2 \mathbf{E} - \frac{1}{v^2}\frac{\partial^2 \mathbf{E}}{\partial t^2} = 0, \tag{2.20}$$

with

$$v = 1/\sqrt{\varepsilon\mu}, \quad \nabla^2 = \frac{\partial^2}{\partial x^2} + \frac{\partial^2}{\partial y^2} + \frac{\partial^2}{\partial z^2}. \tag{2.21}$$

We have an identical wave equation for $\mathbf{B}$. Note that in vacuum $\varepsilon = \varepsilon_0$ and $\mu = \mu_0$, consequently v is replaced by $c = 1/\sqrt{\varepsilon_0\mu_0}$. The refractive index of the medium is defined as

$$n = \frac{c}{v}. \tag{2.22}$$

Since c is a universal constant, the denser the medium (n is larger) the smaller is the phase velocity. Using Eq. (2.11) we can rewrite the expression for the wave-vector magnitude k as

$$k = \frac{\omega}{v} = \frac{\omega n}{c} = k_0 n, \tag{2.23}$$

where k_0 is the magnitude of the wave-vector in vacuum.

For plane waves with exponential factor like $\exp[i(\mathbf{k}\cdot\mathbf{r} - \omega t)]$, the operators $\nabla\cdot$ and $\nabla\times$ reduce to

$$\nabla\cdot = i\mathbf{k}\cdot, \quad \nabla\times = i\mathbf{k}\times, \tag{2.24}$$

and hence for the complex vector amplitudes, Eqs. (2.14)–(2.17) reduce to

$$i\mathbf{k}\cdot\mathbf{E} = 0, \tag{2.25}$$

$$i\mathbf{k}\cdot\mathbf{B} = 0, \tag{2.26}$$

$$i\mathbf{k}\times\mathbf{E} = i\omega\mathbf{B}, \tag{2.27}$$

$$i\mathbf{k}\times\mathbf{B} = -i\varepsilon\mu\omega\mathbf{E}. \tag{2.28}$$

The first two equations imply that both $\mathbf{E}$ and $\mathbf{B}$ are perpendicular to $\mathbf{k}$, while the third equation implies the orthogonality of $\mathbf{B}$ and $\mathbf{E}$. Thus $\mathbf{k}$, $\mathbf{E}$ and $\mathbf{B}$ form a right-handed mutually orthogonal triplet. This also confirms the transverse nature of electromagnetic plane waves whereby both the electric and magnetic field vectors oscillate in a plane perpendicular to the direction of propagation. For future use introducing the unit vector along the direction of propagation $\hat{\mathbf{k}}$, we rewrite Eq. (2.27) as

$$\hat{\mathbf{k}}\times\mathbf{E} = v\mathbf{B}. \tag{2.29}$$

A negative refractive index medium (see Chapter 5) is defined as having both ε and μ negative. It can be shown that for such a medium, $\mathbf{k}$, $\mathbf{E}$ and $\mathbf{H}$ form a left-handed triplet, while for a standard medium they correspond to a right-handed triplet.

2.3 Wave propagation in dispersive media

In this section we will assume the medium to be nonmagnetic (i.e., $\mu = \mu_0$). A medium is said to be dispersive if the corresponding refractive index depends on the frequency, i.e.,

$$n = c/v = c/v(\omega) = n(\omega). \tag{2.30}$$

It is clear from Eq. (2.30) that different frequency components of a wave 'packet' propagate with different velocities and for specific conditions may come closer to each other leading to a spread of the pulse as a function of time (hence the name dispersion). Further, the dispersion is said to be normal if

$$\frac{\partial n}{\partial \omega} > 0. \tag{2.31}$$

Otherwise the medium is said to have anomalous dispersion.

In order to assess the origin, nature and consequences of dispersion, it is necessary to incorporate the oscillatory nature of atoms (forming the dielectric medium) under the action of the incident electromagnetic field. This model was developed by Lorentz, and it is referred to as the Lorentz model of dielectrics.

2.3.1 Lorentz model for dispersion in a dielectric

As mentioned earlier, the uniform dielectric medium is assumed to be a collection of oscillators with natural frequency ω_0 and decay constant γ. Let the number density of such oscillators be given by N (i.e., N oscillators per unit volume). The applied EM field forces any of these oscillators to execute SHM and this leads leads to a dipole moment (microscopic polarization),

$$\mathbf{p} = -e\mathbf{x}, \tag{2.32}$$

where e is the electron charge and $\mathbf{x}$ is the displacement from the equilibrium position. Henceforth we will ignore the vector nature of the relevant quantities, assuming the same direction for all of them. The macroscopic resultant dipole moment is called the polarization P and it is given by

$$P = -Nex. \tag{2.33}$$

Thus the problem of finding the polarization is reduced to finding the displacement x under the action of the EM field, leading to a driving force

$$F = -eE(z,t) = -eE_0 \cos(kz - \omega t). \tag{2.34}$$

In Eq. (2.34) we only need to worry about the temporal dependence since time is the only independent variable governing the motion of the oscillator. Thus in complex notation, the equation for the driven oscillator can be written as

$$\frac{d^2x}{dt^2} + 2\gamma\frac{dx}{dt} + \omega_0^2 x = -\frac{eE_0}{m}e^{-i\omega t}. \tag{2.35}$$

We have already dealt with equations like Eq. (2.35), and the particular solution that survives in the long run can be written as

$$x(t) = \frac{-e/m}{[(\omega_0^2 - \omega^2) - (2i\gamma\omega)]}E_0 e^{-i\omega t} = \frac{-e/m}{[(\omega_0^2 - \omega^2) - (2i\gamma\omega)]}E(t). \tag{2.36}$$

And for polarization P we have

$$P(t) = \frac{Ne^2/m}{[(\omega_0^2 - \omega^2) - (2i\gamma\omega)]} E(t). \tag{2.37}$$

The only step left is to write the expression for P in terms of the medium response or the susceptibility χ as

$$P(t) = \varepsilon_0 \chi E(t). \tag{2.38}$$

Indeed, for most of the dielectric media at low intensities, the relation between the 'effect' polarization and 'cause' electric field is linear. Comparing Eqs. (2.37) and (2.38), we obtain the expression for the linear susceptibility χ:

$$\chi(\omega) = \frac{Ne^2/(m\varepsilon_0)}{[(\omega_0^2 - \omega^2) - (2i\gamma\omega)]}. \tag{2.39}$$

For induction $\mathbf{D}$ we can write the following relation:

$$\mathbf{D} = \varepsilon_0 \mathbf{E} + \mathbf{P} = \varepsilon_0 (1 + \chi) \mathbf{E} = \varepsilon \mathbf{E}. \tag{2.40}$$

Thus the relation between the dielectric constant ε, refractive index n and susceptibility χ of a nonmagnetic medium is given by

$$\varepsilon/\varepsilon_0 = n^2 = (1 + \chi) \tag{2.41}$$

Using the above results, it can shown that for $\gamma = 0$, n satisfies the equation $(n^2 - 1)^{-1} = -C\lambda^{-2} + C\lambda_0^{-2}$. Evaluation of the expression for C is left as an exercise. We often introduce the plasma frequency ω_p and rewrite n as follows:

$$n = \sqrt{1 + \chi(\omega)} = \sqrt{1 + \frac{\omega_p^2}{[(\omega_0^2 - \omega^2) - (2i\gamma\omega)]}}, \quad \omega_p^2 = Ne^2/(m\varepsilon_0). \tag{2.42}$$

The complex nature of the susceptibility leads to a complex refractive index $n = n' + in''$. The physical meaning becomes clear if we look at the spatial part of a plane wave:

$$e^{ikz} = e^{ik_0 nz} = e^{ik_0(n' + in'')z} = e^{ik_0 n' z} e^{-k_0 n'' z}. \tag{2.43}$$

The first exponential factor on the right-hand side of Eq. (2.43) corresponds to spatial oscillation, while the second leads to damping as the wave propagates in the positive z direction. Thus the real (imaginary) part of n defines phase propagation (absorption). In Fig. 2.1 we show the real and imaginary parts of n as functions of normalized frequency ω/ω_0 for $\omega_p/\omega_0 = 0.1$ and $\gamma/\omega_0 = 0.005$. A close inspection of Fig. 2.1 reveals that close to the resonance ($\omega/\omega_0 = 1$) we have anomalous dispersion, while away from the resonance the character of dispersion is normal.

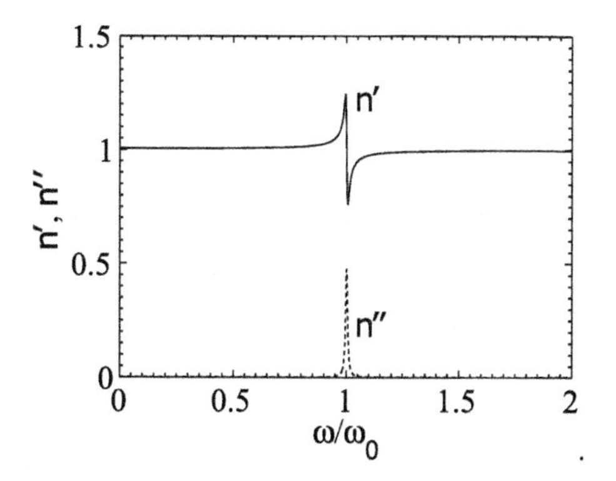

FIGURE 2.1: Real and imaginary parts of the refractive index.

2.4 Phase and group velocities: Sub- and superluminal light

In Section 2.1 the concept of phase velocity was discussed briefly. Here we introduce the notion of group velocity by considering first the beat wave, which results as the superposition of two co-propagating plane waves. Let the constituent waves have the same amplitude E_0 but with slightly different frequencies ω_1 and ω_2 as well as wave vectors k_1 and k_2 so that the resultant wave can be written as

$$E(z,t) = E_0 \left[\cos(k_1 z - \omega_1 t) + \cos(k_2 z - \omega_2 t)\right]. \tag{2.44}$$

We can rewrite Eq. (2.44) as

$$E(z,t) = 2E_0 \left[\cos\frac{1}{2}[(k_1 - k_2)z - (\omega_1 - \omega_2)t] \ \cos\frac{1}{2}[(k_1 + k_2)z - (\omega_1 + \omega_2)t]\right]. \tag{2.45}$$

Introducing new variables for the average frequencies and wave-vectors as well as the corresponding differences, we have

$$\bar{k} = (k_1 + k_2)/2, \quad \bar{\omega} = (\omega_1 + \omega_2)/2. \ \ \Delta k = (k_1 - k_2)/2, \quad \Delta\omega = (\omega_1 - \omega_2)/2, \tag{2.46}$$

or

$$E(z,t) = 2E_0 \cos(\Delta k z - \Delta\omega t) \ \cos(\bar{k}z - \bar{\omega}t). \tag{2.47}$$

In Eq. (2.47) we can easily recognize the quickly oscillating phase term that defines the phase velocity as $v = \bar{\omega}/\bar{k}$, while the low frequency ($\Delta\omega$) envelope is determined by the first cosine term. Thus the envelope moves with a velocity determined by

$$\Delta k z - \Delta\omega t = \text{constant}. \tag{2.48}$$

This velocity is termed as the group velocity v_g and, in the limiting case of vanishing Δk, takes the form

$$v_g = \frac{\partial \omega}{\partial k}. \tag{2.49}$$

The above concepts can easily be generalized to the case of a spatio-temporal pulse. Such a pulse is characterized by a spread of frequency about a mean frequency and a spread of wave-vector about an 'average' wave-vector. If the dispersion-induced distortions are not too much, we can still talk about a group velocity defined by Eq. (2.49) evaluated at the mean frequency.

We now demonstrate how both phase and group velocities can exceed the velocity of light in vacuum. Light is referred to as sub- or superluminal depending on whether the corresponding velocities can exceed that of light in vacuum. Thus the case of $v < c$ is referred to as subluminal while the opposite case bears the name of superluminal. The fact that phase velocity can exceed c has been known for quite some time [5], since the experiments of Wood on D_2 lines of sodium. Close to the resonance, due to anomalous dispersion, the refractive index can become less than one, resulting in $v = c/n > c$. Initially, it was believed that group velocity could not exceed c. Since the beautiful experiment of Chu and Wong [6], it has been demonstrated that we can achieve both sub- and superluminal group velocities and make use of positive and negative slopes of the dispersion curves. We now show how this can be achieved. Using Eq. (2.23) or $k = \omega n(\omega)/c$, we can calculate the group velocity as

$$v_g = \frac{\partial \omega}{\partial k} = \frac{1}{\frac{\partial k}{\partial \omega}} = \frac{c}{n + \omega \frac{\partial n}{\partial \omega}}. \tag{2.50}$$

One introduces the so-called group index n_g to rewrite Eq. (2.50) as follows:

$$v_g = \frac{c}{n_g}, \quad n_g = n + \omega \frac{\partial n}{\partial \omega}. \tag{2.51}$$

It is clear from Eq. (2.50) that in the absence of dispersion ($\frac{\partial n}{\partial \omega} = 0$ or $n_g = n$), group and phase velocities coincide. For positive slope of the dispersion curve, ($\frac{\partial n}{\partial \omega} > 0$) we usually have $n_g > 1$ and hence subluminal light. In the case of anomalous dispersion or negative slope of the dispersion curve, n_g can be smaller than unity, zero or even negative. Thus the group velocity can be larger than c or even negative. Negative group velocity has the very interesting consequence that the pulse peak can arrive earlier than the time when it enters the medium. Both sub- and superluminal group velocities have been observed in recent experiments. Crawling speed of light as small as 17 m/sec has been reported using the so-called phenomenon of electromagnetically induced transparency [7]. Finally, we conclude this discussion by saying (without proof) that the superluminality of group velocity does not violate Einstein's postulates of special theory of relativity or causality principle [7].

2.5 Energy and momentum of electromagnetic waves

Traveling waves carry energy and momentum [3, 8]. In this section we try to understand these aspects. The physical quantity that quantifies the flux of electromagnetic energy is the Poynting vector. We start with a discussion of the Poynting vector.

2.5.1 Poynting vector

Electromagnetic fields can store energy. Let u be the measure of this radiant energy density (per unit volume), which we simply label as energy density. For example, for a plane parallel plate capacitor, the energy density of the field in between the plates is given by

$$u_E = \frac{\varepsilon_0}{2}E^2. \tag{2.52}$$

Similarly, the energy density of the B field (e.g., for a current-carrying toroid) is given by

$$u_B = \frac{1}{2\mu_0}B^2. \tag{2.53}$$

For plane waves we have $E = cB$ in vacuum. Using the relation $c = 1/\sqrt{\varepsilon_0\mu_0}$ we can easily show that

$$u_E = u_B. \tag{2.54}$$

Thus the total energy density is given by

$$u = u_E + u_B = \varepsilon_0 E^2. \tag{2.55}$$

Let the electromagnetic wave travel with velocity c through an area A. Let S also represent the transport of energy per unit time (power) per unit area. The unit of S in SI will be W/m^2. During a very small interval Δt, the energy contained in the cylindrical volume $u(c\Delta tA)$ will cross A. Thus

$$S = \frac{cu\Delta tA}{\Delta tA} = uc \quad or \quad S = \frac{1}{\mu_0}EB. \tag{2.56}$$

For isotropic media we make a reasonable approximation that energy propagates in the direction of propagation of wave. This leads to the expression for the Poynting vector $\mathbf{S}$ as follows:

$$\mathbf{S} = \frac{1}{\mu_0}\mathbf{E}\times\mathbf{B} \quad \text{or} \quad \mathbf{S} = c^2\varepsilon_0\mathbf{E}\times\mathbf{B}. \tag{2.57}$$

For a harmonic plane wave propagating along $\mathbf{k}$ given by

$$\mathbf{E} = \mathbf{E}_0\cos(\mathbf{k}\cdot\mathbf{r} - \omega t), \tag{2.58}$$

$$\mathbf{B} = \mathbf{B}_0\cos(\mathbf{k}\cdot\mathbf{r} - \omega t), \tag{2.59}$$

$\mathbf{S}$ takes the form

$$\mathbf{S} = c^2\varepsilon_0\mathbf{E_0} \times \mathbf{B_0}\cos^2(\mathbf{k}\cdot\mathbf{r} - \omega t). \tag{2.60}$$

It is clear from Eq. (2.60) that at optical frequencies ($\sim 10^{15-16}$ Hz), $\mathbf{S}$ is a rapidly oscillating function. It is impractical to follow the instantaneous value of the Poynting vector, which suggests the need for averaging over many optical cycles. This is also consistent with the fact that we absorb the radiant energy during some finite interval of time by, for example, a photocell, film plate or the retina of the human eye. The averaging procedure for $\cos^2$ leads to a factor 0.5. Thus the time-averaged Poynting vector has the expression

$$\langle\mathbf{S}\rangle = \frac{c^2\varepsilon_0}{2}|\mathbf{E_0} \times \mathbf{B_0}| = \frac{c\varepsilon_0}{2}E_0^2, \tag{2.61}$$

which is also known as the irradiance I. These were derived for a vacuum. For a linear homogeneous and isotropic dielectric, the irradiance can be written as

$$I = v\varepsilon\langle E^2\rangle. \tag{2.62}$$

2.5.2 Photons

Light is absorbed and emitted in tiny portions or quanta. An elementary portion of electromagnetic 'stuff' is called a photon. Each photon has the energy

$$\mathcal{E} = \hbar\omega. \tag{2.63}$$

The corresponding momentum is given by

$$\mathbf{p} = \hbar\mathbf{k}. \tag{2.64}$$

We refer to Section 3.3.3 of the book by Hecht for a detailed description of the nature and properties of photons [8].

Chapter 3

Reflection and refraction

3.1 Huygens-Fresnel principle

We first define the notion of a wavefront. The wavefront is a surface over which an optical disturbance has a constant phase. A plane wave has a planar wavefront perpendicular to the direction of propagation of the wave, while a spherical wave has a wavefront in the shape of a spherical surface. The Huygens-Fresnel principle states that every point on a primary wavefront serves as the source of spherical secondary wavelets such that the wavefront at some later time is the envelope of these wavelets. According to Fresnel these secondary wavelets can interfere. The wavelets advance with a speed and frequency equal to that of the primary wave at each point in space. Fig. 3.1, for example, explains the refraction of light using the Huygens principle.

In optics it is often useful to exploit the notion of a light ray. A ray is a line drawn in space corresponding to the direction of flow of radiant energy. We can derive the laws of reflection and refraction using the Huygens-Fresnel principle. These laws (referred to as Snell's laws) govern the way light rays

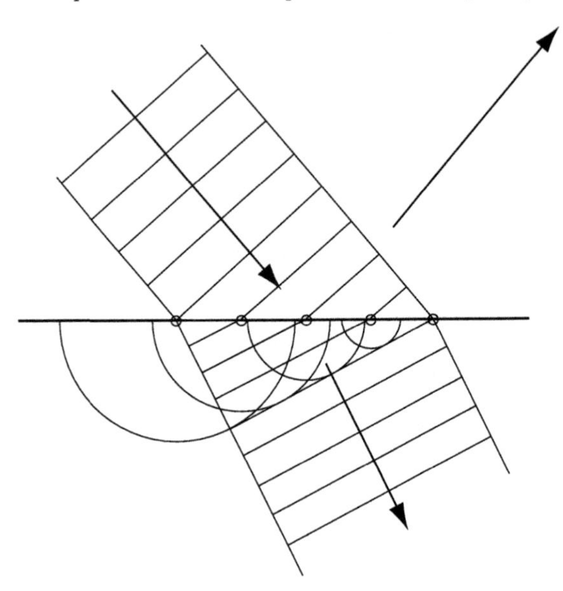

FIGURE 3.1: Refraction using the Huygens-Fresnel principle.

reflect or refract at the interface between two media. We shall derive these laws in the next section.

3.2 Laws of reflection and refraction

Consider the interface between two dielectric media with refractive indices n_1 and n_2, respectively. Let the incident plane wave be given by the expression

$$\mathbf{E}_i = \mathbf{E}_{0i} \cos(\mathbf{k} \cdot \mathbf{r} - \omega_i t). \tag{3.1}$$

Without any loss of generality, we can write the expressions for the reflected and transmitted waves as

$$\mathbf{E}_r = \mathbf{E}_{0r} \cos(\mathbf{k}_r \cdot \mathbf{r} - \omega_r t + \phi_r), \tag{3.2}$$

$$\mathbf{E}_t = \mathbf{E}_{0t} \cos(\mathbf{k}_t \cdot \mathbf{r} - \omega_t t + \phi_t). \tag{3.3}$$

The boundary conditions state that the tangential components of the field must be continuous across the interface. Let the unit normal to the interface be given by $\mathbf{u_n}$. Thus at the interface we require the following equality:

$$\mathbf{u}_n \times \mathbf{E}_i + \mathbf{u}_n \times \mathbf{E}_r = \mathbf{u}_n \times \mathbf{E}_t. \tag{3.4}$$

Eq. (3.4) with Eqs. (3.1)–(3.3) can be satisfied for arbitrary t and $\mathbf{r}$ at the interface plane if and only if the following relation holds:

$$\mathbf{k}_i \cdot \mathbf{r} - \omega_i t = \mathbf{k}_r \cdot \mathbf{r} - \omega_r t + \phi_r = \mathbf{k}_t \cdot \mathbf{r} - \omega_t t + \phi_t. \tag{3.5}$$

Eq. (3.5) has to hold for all t and hence the coefficients of t must match:

$$\omega_i = \omega_r = \omega_t. \tag{3.6}$$

Eq. (3.5) easily leads to the following relations at the interface:

$$(\mathbf{k}_i - \mathbf{k}_r) \cdot \mathbf{r} = \phi_r, \tag{3.7}$$

$$(\mathbf{k}_i - \mathbf{k}_t) \cdot \mathbf{r} = \phi_t. \tag{3.8}$$

Eq. (3.7) implies that the tip of $\mathbf{r}$ sweeps out a plane (our interface) and this plane is perpendicular to the difference vector $\mathbf{k}_i - \mathbf{k}_r$. Since $\mathbf{u}_n$ is also normal to this plane, $\mathbf{u}_n$ and $\mathbf{k}_i - \mathbf{k}_r$ are collinear and their cross product, $\mathbf{u}_n \times (\mathbf{k}_i - \mathbf{k}_r)$, must vanish. We thus have

$$\frac{\omega n_1}{c} \sin\theta_i = \frac{\omega n_1}{c} \sin\theta_r \quad \mathit{or} \quad \theta_i = \theta_r. \tag{3.9}$$

Analogously, Eq. (3.8) leads to

$$\frac{\omega n_1}{c} \sin\theta_i = \frac{\omega n_2}{c} \sin\theta_t \quad \mathit{or} \quad \frac{\sin\theta_i}{\sin\theta_t} = \frac{n_2}{n_1}. \tag{3.10}$$

3.3 Fermat's principle and laws of reflection and refraction

Fermat's Principle: Light follows the path of least time. The laws of reflection and refraction can be derived from this principle, as we explain in this section.

3.3.1 Reflection

The path L from A to B (see Fig. 3.2) is given by

$$L = \sqrt{a^2 + x^2} + \sqrt{b^2 + (d - x)^2}. \tag{3.11}$$

Note that x (still unknown) determines the angles of incidence and reflection. Since the incident and reflected light travel in the same medium, their velocities are the same. Hence the 'least time' implies least path. The least

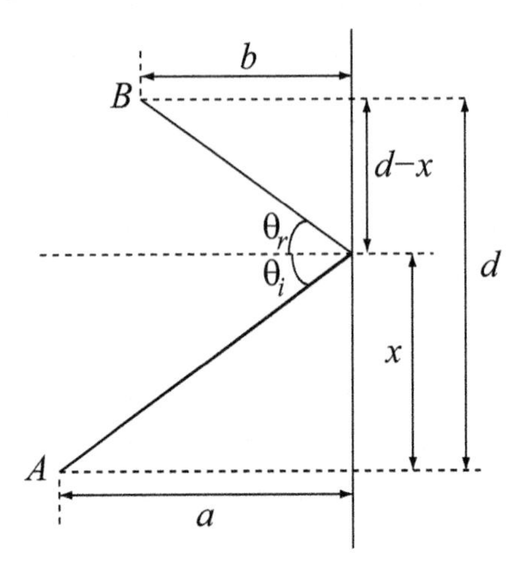

FIGURE 3.2: Schematics of reflection.

path results when $\frac{dL}{dx} = 0$ or, in other words, when x satisfies the following equation:

$$\frac{dL}{dx} = \frac{1}{2}\frac{2x}{\sqrt{a^2+x^2}} + \frac{1}{2}\frac{2(d-x)(-1)}{\sqrt{b^2+(d-x)^2}} = 0. \tag{3.12}$$

Note that with

$$\sin(\theta_i) = \frac{x}{\sqrt{a^2+x^2}}\ , \quad \sin(\theta_r) = \frac{(d-x)}{\sqrt{b^2+(d-x)^2}}, \tag{3.13}$$

we have

$$\theta_i = \theta_r. \tag{3.14}$$

3.3.2 Refraction

Consider the interface between two media as shown in Fig. 3.3. Since the velocities in the two media are different, the time t taken to cover the path from A to B (see Fig. 3.3) is given by

$$t = \frac{\sqrt{a^2+x^2}}{v} + \frac{\sqrt{b^2+(d-x)^2}}{v'}\ , \quad v = c/n\ , \quad v' = c/n'\ . \tag{3.15}$$

Applying Fermat's principle we require that this time is minimal, i.e.,

$$\frac{dt}{dx} = \frac{1}{v}\frac{x}{\sqrt{a^2+x^2}} - \frac{1}{v'}\frac{(d-x)}{\sqrt{b^2+(d-x)^2}} = 0. \tag{3.16}$$

As per the figure, the angle of incidence θ_i and refraction θ_t are given by

$$\sin(\theta_i) = \frac{x}{\sqrt{a^2+x^2}}\ , \quad \sin(\theta_t) = \frac{(d-x)}{\sqrt{b^2+(d-x)^2}}. \tag{3.17}$$

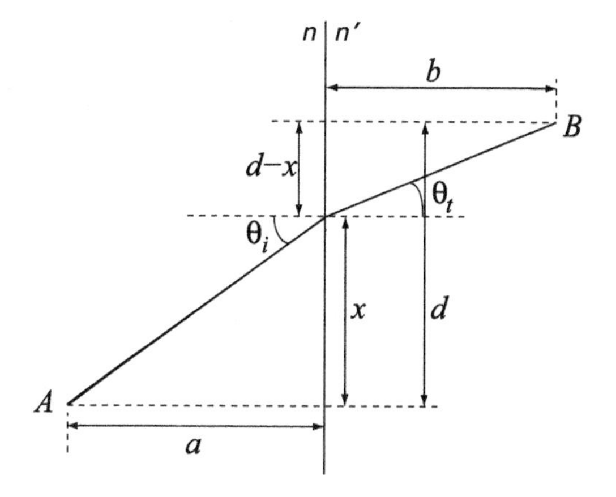

FIGURE 3.3: Schematics of refraction.

Eq. (3.17) thus leads to the relation

$$\frac{\sin(\theta_i)}{\sin(\theta_t)} = \frac{v}{v'} = \frac{n'}{n}. \tag{3.18}$$

3.4 Fresnel formulas

Earlier we derived the relationships connecting the various angles of reflection and refraction based on the analysis of the phases of the incident, reflected and refracted light. We now try to derive the relationships between the complex amplitudes. Whatever the polarization of the waves, we shall try to resolve its **E** and **B** into the components parallel and perpendicular to the plane of incidence. Depending on whether **E** is parallel or perpendicular, these are usually referred to as p (or TM) and s (TE) polarizations. Any arbitrary polarization can always be thought of as a superposition of these two polarizations. Let also the interface be given by the plane $z = 0$.

3.4.1 s-Polarization (electric field perpendicular to the plane of incidence)

Again making use of the continuity of the tangential components of the electric fields at the boundary, we have

$$\mathbf{E}_{0i} + \mathbf{E}_{0r} = \mathbf{E}_{0t}. \tag{3.19}$$

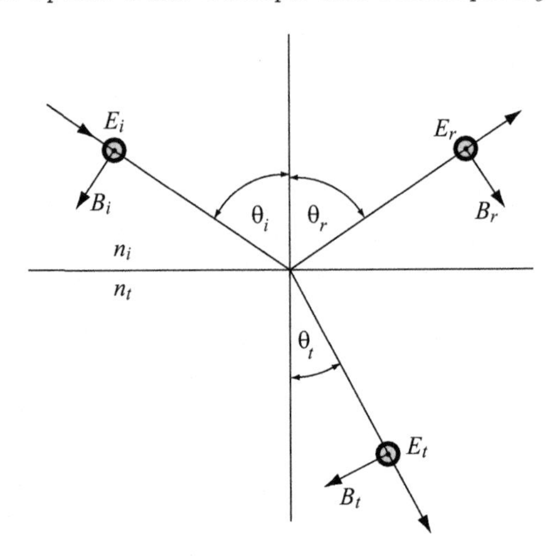

FIGURE 3.4: Schematics of reflection and refraction for s-polarization.

We have taken the directions of the electric fields as shown in Fig. 3.4. The directions of the B-fields then follow from Eq. (2.27) and they are also shown in Fig. 3.4. Note that the boundary conditions demand that the tangential components of $\mathbf{B}/\mu$ must be continuous at the interface. This leads to

$$-\frac{B_i}{\mu_i}\cos\theta_i + \frac{B_r}{\mu_i}\cos\theta_r = -\frac{B_t}{\mu_t}\cos\theta_t. \tag{3.20}$$

Henceforth we will assume that both media are nonmagnetic, i.e., $\mu_i = \mu_t = 1$. Moreover, making use of the relations $B_i = E_i/v_i$, $B_r = E_r/v_r$, $B_t = E_t/v_t$ and also $v_i = v_r$ and $\theta_i = \theta_r$, Eq. (3.20) reduces to

$$\frac{1}{v_i}(E_i - E_r)\cos\theta_i = \frac{1}{v_t}E_t\cos\theta_t. \tag{3.21}$$

Making use of the relation $v = c/n$ and the fact that the phases are the same at the interface, we have

$$n_i(E_{0i} - E_{0r})\cos\theta_i = n_t E_{0t}\cos\theta_t, \tag{3.22}$$

$$(E_{0i} + E_{0r}) = E_{0t}. \tag{3.23}$$

The two equations above can be solved to yield the amplitude reflection and transmission coefficients

$$r_\perp = \left(\frac{E_{0r}}{E_{0i}}\right)_\perp = \frac{n_i\cos\theta_i - n_t\cos\theta_t}{n_i\cos\theta_i + n_t\cos\theta_t}, \tag{3.24}$$

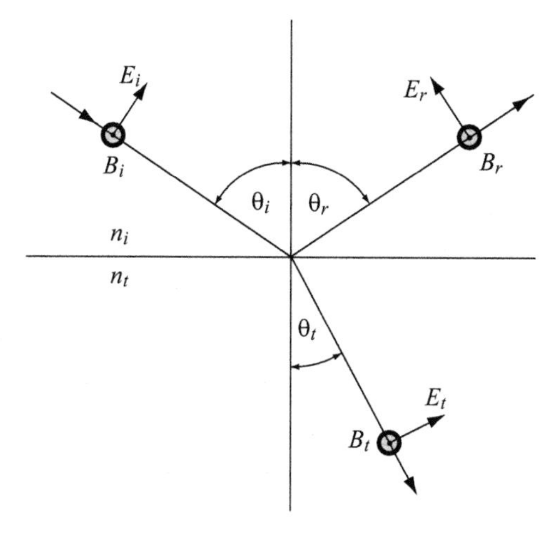

FIGURE 3.5: Schematics of reflection and refraction for p-polarization.

$$t_\perp = \left(\frac{E_{0t}}{E_{0i}}\right)_\perp = \frac{2n_i\cos\theta_i}{n_i\cos\theta_i + n_t\cos\theta_t}. \tag{3.25}$$

3.4.2 p-Polarization (electric field parallel to the plane of incidence)

We can derive a similar pair of equations as in the case of the s-polarization. Referring to Fig. 3.5 for the continuity of the tangential component of $\mathbf{E}$, we now have

$$E_{0i}\cos\theta_i - E_{0r}\cos\theta_r = E_{0t}\cos\theta_t. \tag{3.26}$$

The continuity of the tangential component of $\mathbf{B}/\mu$ across the interface leads to

$$n_i(E_{0i} + E_{0r}) = n_t E_{0t}. \tag{3.27}$$

Eqs. (3.26) and (3.27) lead to the expressions for the amplitude reflection and transmission coefficients for the parallel components:

$$r_\parallel = \left(\frac{E_{0r}}{E_{0i}}\right)_\parallel = \frac{n_t\cos\theta_i - n_i\cos\theta_t}{n_t\cos\theta_i + n_i\cos\theta_t}, \tag{3.28}$$

$$t_\parallel = \left(\frac{E_{0t}}{E_{0i}}\right)_\parallel = \frac{2n_t\cos\theta_i}{n_t\cos\theta_i + n_i\cos\theta_t}. \tag{3.29}$$

Finally, a drastic simplification takes place if we make use of Snell's law:

$$r_\perp = -\frac{\sin(\theta_i - \theta_t)}{\sin(\theta_i + \theta_t)}, \tag{3.30}$$

$$r_\parallel = +\frac{\tan(\theta_i - \theta_t)}{\tan(\theta_i + \theta_t)}, \tag{3.31}$$

$$t_\perp = +\frac{2\sin\theta_t \cos\theta_i}{\sin(\theta_i + \theta_t)}, \tag{3.32}$$

$$t_\parallel = +\frac{2\sin\theta_t \cos\theta_i}{\sin(\theta_i + \theta_t)\cos(\theta_i - \theta_t)}. \tag{3.33}$$

3.5 Consequences of Fresnel equations

In this section we consider the various implications of Fresnel equations both in terms of amplitudes and phases. Note that the complex amplitude reflection and transmission coefficients given by Eqs. (3.30)–(3.33) have both amplitudes and phases. The amplitude tells us how much will be reflected or transmitted, while the phase carries information about the phase shift in the wave for each act of reflection or transmission. We first look at the amplitudes, concentrating later on the phases.

3.5.1 Amplitude relations

For normal incidence ($\theta_i = 0$), Eqs. (3.24) and (3.28) can be reduced to the following form:

$$[r_\parallel]_{\theta_i=0} = -[r_\perp]_{\theta_i=0} = \frac{n_t - n_i}{n_t + n_i}. \tag{3.34}$$

The magnitude of amplitude reflection for normal incidence, for example, for air ($n_i = 1.0$) and glass ($n_t = 1.5$) interface is 0.2. The equality of the reflection coefficients is a consequence of the fact that for normal incidence, we cannot distinguish between s- and p-polarized light (since the unique plane of incidence cannot be defined). For $n_i < n_t$, it follows from Snell's law that $\theta_i > \theta_t$ and $r_\perp$ is negative for all values of the angle of incidence (see Eq. (3.30)). In contrast, $r_\parallel$ shows a different behavior: Starting from a positive value, it becomes negative, passing through zero when $\theta_i + \theta_t = \pi/2$ since $\tan(\pi/2)$ is infinite (see Eq. (3.31)). The angle θ_i at which this occurs is labeled θ_B and it is known as the polarization angle or the Brewster angle. Indeed, for incidence of arbitrarily polarized light at this angle, the reflectivity of the parallel component will be zero and hence the reflected light will be polarized only

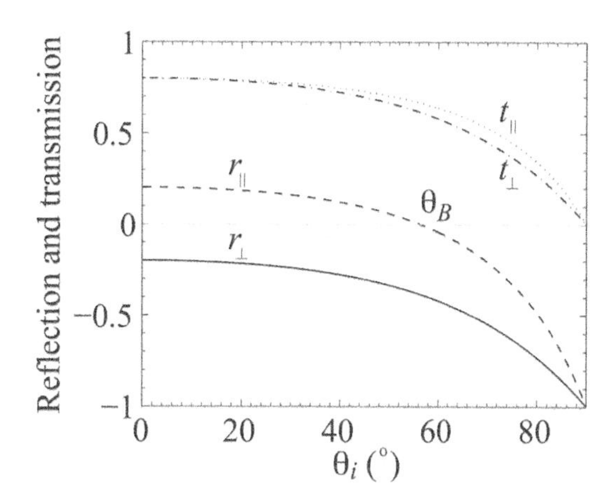

FIGURE 3.6: Amplitude reflection and transmission coefficients $r_\perp$, $r_\parallel$, $t_\perp$ and $t_\parallel$, respectively (from bottom to top), for $n_i = 1.0$ and $n_t = 1.5$.

with $\perp$ polarization. In order to have a comprehensive idea of the behavior of all the amplitude reflection and transmission coefficients, we have plotted them in Fig. 3.6 for $n_i = 1.0$ and $n_t = 1.5$. The Brewster angle for this case is $\theta_B = 56.3°$.

The situation changes drastically when $n_i > n_t$, resulting in $\theta_t > \theta_i$. In this case $r_\perp$ is always positive while $r_\parallel$ passes through zero at the Brewster angle (see Fig (3.7)). However, both the perpendicular and parallel amplitude reflection coefficients become complex beyond a critical angle θ_c given by

$$\theta_c = \sin^{-1}(n_t/n_i). \tag{3.35}$$

It is easy to verify that $\theta_i = \theta_c$ corresponds to an angle of refraction equal to $\pi/2$. For angles of incidence larger than the critical angle the magnitude of both the amplitude reflection coefficients equals one, implying thereby the return of all the energy to the medium of incidence. Hence the critical angle is also referred to as the angle of total internal reflection (TIR). The Brewster and the TIR angles for this case are given by $\theta_B = 33.7°$ and $\theta_c = 41.8°$. Note the complementary nature of the Brewster angles for the cases of Figs. (3.6) and (3.7).

3.5.2 Phase shifts

We first look at the phase change in reflection for the case $\theta_i > \theta_t$, i.e., when light enters a denser medium from a lighter one ($n_i < n_t$). It is clear from Eq. (3.30) that, for any angle of incidence, $r_\perp$ is negative. Thus the perpendicular (to plane of incidence) component of the electric field undergoes a phase shift of π under reflection when the incident medium has a lower refractive index than the transmitting medium.

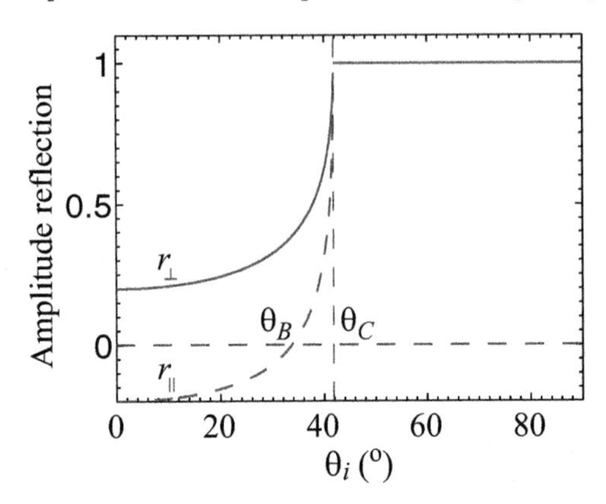

FIGURE 3.7: Amplitude reflection coefficients $r_{\|} and r_{\perp}$, respectively (from bottom to top), for $n_i = 1.5$ and $n_t = 1.0$.

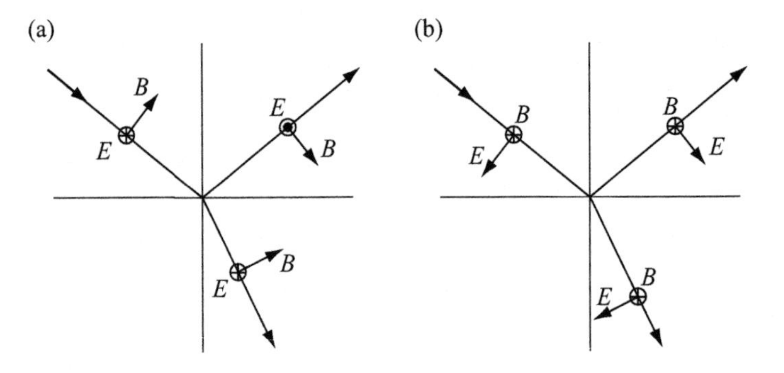

FIGURE 3.8: Explanation of (a) out-of-phase and (b) in-phase components.

The situation is slightly more complicated for the components on the plane of incidence, since we need to define the terms *in-phase* and *out-of-phase*. Fig. 3.8 is handy for understanding these notions. Two fields in the incidence plane are *in-phase* if their components along the unit normal to the surface are parallel and are *out-of-phase* if these components are antiparallel. If two **E** fields are antiparallel, then the same holds for the corresponding two **B** fields.

It is clear from Eq. (3.28) that $r_{\|}$ is positive or the phase difference between the reflected and incident components is zero ($\Delta\phi_{\|} = 0$) so long as the numerator is positive. The inequality that the numerator is positive can be rewritten in the form

$$\sin(\theta_i - \theta_t)\cos(\theta_i + \theta_t) > 0. \tag{3.36}$$

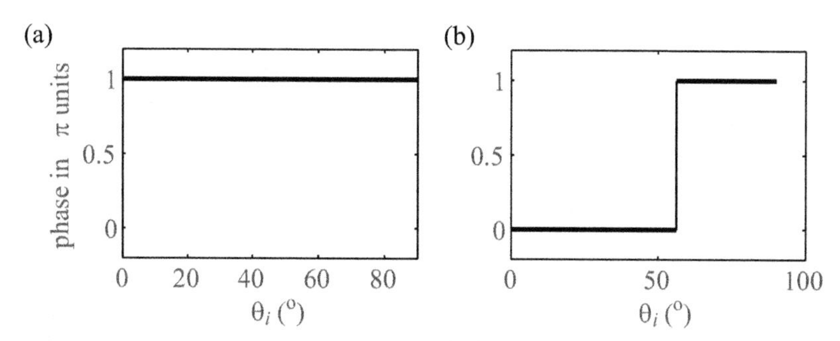

FIGURE 3.9: Phase angles (a) $\Delta\phi_\perp$ and (b) $\Delta\phi_\parallel$ components in units of π for $n_i < n_t$.

For $n_i < n_t$, this translates into

$$\theta_i + \theta_t < \pi/2, \tag{3.37}$$

while for $n_i > n_t$, we have

$$\theta_i + \theta_t > \pi/2. \tag{3.38}$$

Thus for $n_i < n_t$ there will be zero phase lag between the reflected and incident components in the range ($\theta_i = 0 - \theta_B$). Thereafter there will be a phase difference of π (see Fig. 3.9). In contrast, for a TIR case (for $n_i > n_t$) $r_\parallel$ is negative until θ reaches the Brewster angle, which implies that $\Delta\phi_\parallel = \pi$. From θ_B to θ_c, $\Delta\phi_\parallel = 0$. Beyond the critical angle the reflection coefficient is complex and $\Delta\phi_\parallel$ increases gradually to π at 90°. The results for this case are shown in Fig. 3.10, where the last plot gives the difference between the parallel and perpendicular phases. These results will be used in order to understand the change of polarization state under total internal reflection and their use in various polarization devices.

3.5.3 Reflectance and transmittance

Let a light beam of circular cross-section be incident on the surface of a dielectric at an angle θ_i. Since we will be dealing with the reflected and transmitted intensities, it is better to recall the meaning and expressions of the Poynting vector $\mathbf{S}$ and irradiance I. The Poynting vector gives the power per unit area crossing a normal surface and is given by

$$\mathbf{S} = c^2\varepsilon_0 \mathbf{E} \times \mathbf{B}. \tag{3.39}$$

The radiant flux density or the irradiance is the time-averaged Poynting vector and has the expression

$$< S >= \frac{c\varepsilon_0}{2} E_0^2. \tag{3.40}$$

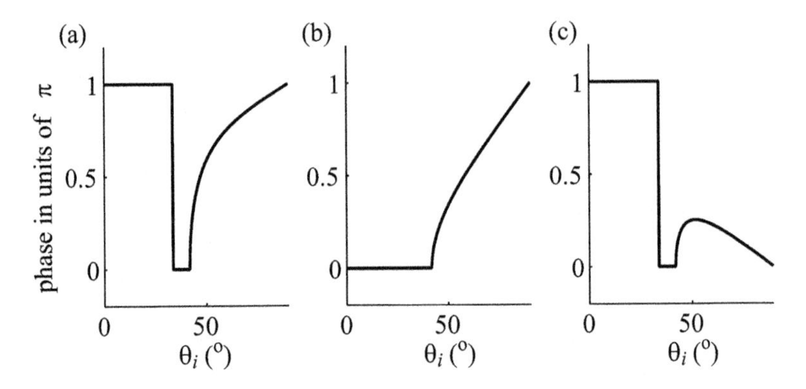

FIGURE 3.10: Phase angles (a) $\Delta\phi_{\parallel}$ and (b) $\Delta\phi_{\perp}$ components in units of π for $n_t < n_i$. (c) gives the difference $\Delta\phi_{\parallel} - \Delta\phi_{\perp}$.

The irradiance has the unit W/m^2 and it is the average energy crossing a unit area normal to **S**.

Let I_i, I_r and I_t be the incident, reflected and transmitted irradiances, respectively. The corresponding cross-sectional areas are $A\cos\theta_i$, $A\cos\theta_r$ and $A\cos\theta_t$, respectively. The incident, reflected and transmitted powers are $I_i A\cos\theta_i$, $I_r A\cos\theta_r$ and $I_t A\cos\theta_t$, respectively. The first quantity is the incident energy per unit time associated with the incident beam, and hence it is the incident power that falls on A. We define reflectance R as the ratio of reflected and incident power (flux) as

$$R = \frac{I_r A\cos\theta_r}{I_i A\cos\theta_i} = \frac{I_r}{I_i}. \tag{3.41}$$

Similarly, transmittance T can be defined as

$$T = \frac{I_t \cos\theta_t}{I_i \cos\theta_i}, \tag{3.42}$$

where

$$I_j = v_j \varepsilon_j E_{0j}^2, \quad j = i, r, t. \tag{3.43}$$

Noting that both the medium of incidence and reflection are the same, Eq. (3.41) can be rewritten as

$$R = \left(\frac{E_{0r}}{E_{0i}}\right)^2 = r^2. \tag{3.44}$$

Likewise for T, we have

$$T = \frac{n_t\cos\theta_t}{n_i\cos\theta_i}\left(\frac{E_{0t}}{E_{0i}}\right)^2 = \left(\frac{n_t\cos\theta_t}{n_i\cos\theta_i}\right)t^2. \tag{3.45}$$

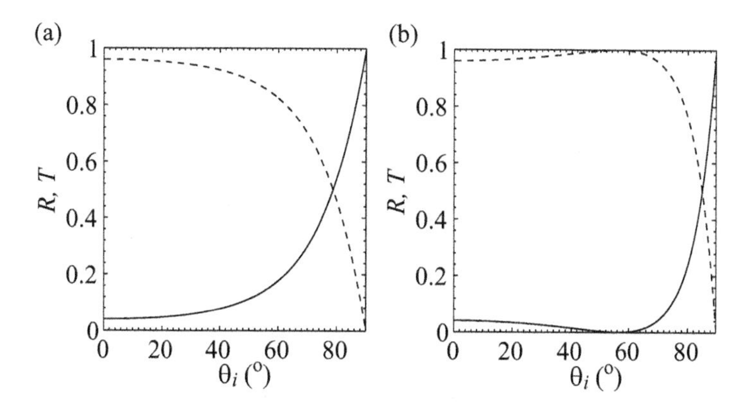

FIGURE 3.11: Reflectance (solid line) and transmittance (dashed line) for (a) $\perp$ and (b) $\|$ orientations, respectively, for $n_i = 1.0$ and $n_t = 1.5$.

In Eq. (3.45) we made use of the relation $\mu_0\varepsilon_j = \mu_0\varepsilon_0 n_j^2 = n_j^2/c^2 = 1/v_j^2$ $(j = i, t)$. For normal incidence when $\theta_i = \theta_r = \theta_t = 0$, reflectance and transmittance are given by the ratios of the corresponding irradiances. Note that T is not simply equal to t^2 for two reasons. First, the velocity of light in the two media is not the same. Hence the ratio of the refractive indices must appear in T. Second, the cross-sectional areas for incident and transmitted beams are not the same. The energy flow per unit area will be affected accordingly.

3.5.4 Energy conservation

The energy flowing into the area A per unit time must be the same as that flowing out of it. Thus

$$I_i A\cos\theta_i = I_r A\cos\theta_r + I_t A\cos\theta_t. \tag{3.46}$$

Multiplying both sides by c, Eq. (3.46) can be written as

$$n_i E_{0i}^2 \cos\theta_i = n_r E_{0r}^2 \cos\theta_r + n_t E_{0t}^2 \cos\theta_t, \tag{3.47}$$

expressed differently as

$$1 = \left(\frac{E_{0r}}{E_{0i}}\right)^2 + \left(\frac{n_t\cos\theta_t}{n_i\cos\theta_i}\right)\left(\frac{E_{0t}}{E_{0i}}\right)^2, \quad \text{or} \quad R + T = 1. \tag{3.48}$$

Eq. (3.48) just states that total reflectance and transmittance in passing through an interface add up to unity in absence of losses. The part that is not transmitted is bound to be reflected. In component form these relations read as

$$R_\perp = r_\perp^2, \tag{3.49}$$

$$R_\parallel = r_\parallel^2, \tag{3.50}$$

$$T_\perp = \left(\frac{n_t \cos\theta_t}{n_i \cos\theta_i}\right) t_\perp^2, \tag{3.51}$$

$$T_\parallel = \left(\frac{n_t \cos\theta_t}{n_i \cos\theta_i}\right) t_\parallel^2, \tag{3.52}$$

$$R_\perp + T_\perp = 1, \tag{3.53}$$

$$R_\parallel + T_\parallel = 1. \tag{3.54}$$

The results for the reflectance and transmittance are shown in Fig. 3.11. It is easy to identify the Brewster angle at which there is null reflectance and unity transmittance for the p-polarized light. For normal incidence ($\theta_i = 0$), the expressions for reflectance and transmittance reduce to

$$R = R_\parallel = R_\perp = \left(\frac{n_1 - n_2}{n_1 + n_2}\right)^2 \tag{3.55}$$

and

$$T = T_\parallel = T_\perp = \frac{4n_1 n_2}{(n_1 + n_2)^2}. \tag{3.56}$$

3.5.5 Evanescent waves

It can be easily verified that it is impossible to satisfy the boundary conditions if we assume that there is no transmitted wave in case of total internal reflection. In order to understand this, we rewrite Eqs. (3.24) and (3.28) in the form

$$r_\perp = \frac{n_i \cos\theta_i - n_t \cos\theta_t}{n_i \cos\theta_i + n_t \cos\theta_t} = \frac{\cos\theta_i - (n_{ti}^2 - \sin^2\theta_i)^{1/2}}{\cos\theta_i + (n_{ti}^2 - \sin^2\theta_i)^{1/2}}, \tag{3.57}$$

$$r_\parallel = \frac{n_t \cos\theta_i - n_i \cos\theta_t}{n_t \cos\theta_i + n_i \cos\theta_t} = \frac{n_{ti}^2 \cos\theta_i - (n_{ti}^2 - \sin^2\theta_i)^{1/2}}{n_{ti}^2 \cos\theta_i + (n_{ti}^2 - \sin^2\theta_i)^{1/2}}, \tag{3.58}$$

where $n_{ti} = n_t/n_i < 1$ for a TIR situation. For $\theta_i > \theta_c$, $\sin(\theta_i) > \sin(\theta_c) = n_{ti}$ and both $r_\perp$ and $r_\parallel$ are complex. Even then, $r_\perp r_\perp^* = r_\parallel r_\parallel^* = 1$ and the reflectance R equals unity. This means that though there is a transmitted wave, it cannot carry any energy across the boundary. In order to have a deeper understanding, let us write the expression for the transmitted wave as follows (we assume that the xz plane forms the plane of incidence with the rarer medium occupying the half-space $z > 0$):

$$\mathbf{E}_t = \mathbf{E}_{0t} \exp i(\mathbf{k}_t \cdot \mathbf{r} - \omega t). \tag{3.59}$$

For the chosen geometry,

$$\mathbf{k}_t \cdot \mathbf{r} = k_{tx}x + k_{tz}z = k_t \sin(\theta_t)x + k_t \cos(\theta_t)z. \tag{3.60}$$

Using Snell's law we have

$$k_{tz} = k_t \cos(\theta_t) = \pm\, k_t \left(1 - \frac{\sin^2\theta_i}{n_{ti}^2}\right)^{1/2} = \pm i\; k_t \left(\frac{\sin^2\theta_i}{n_{ti}^2} - 1\right)^{1/2} = \pm i\beta. \tag{3.61}$$

In writing Eq. (3.61), we used the fact that when $\theta_i > \theta_c$, $\sin\theta_i > n_{ti}$. For the surface component we have

$$k_{tx} = (k_t/n_{ti})\sin(\theta_i). \tag{3.62}$$

Thus for the transmitted wave we have the expression

$$\mathbf{E}_t = \mathbf{E}_{0t}e^{\mp\beta z}e^{i[(k_t/n_{ti})\sin(\theta_i)x - \omega t]}. \tag{3.63}$$

Here we neglect the positive exponential because it is unphysical. We thus have a wave whose amplitude decays as we see deeper in the rarer medium. The wave advances along the surface as a surface or evanescent wave. The amplitude decays very fast (over a few wavelengths) as one moves away from the interface. Thus this represents an inhomogeneous wave. For this case the surfaces of constant phase (parallel to the yz plane) are perpendicular to surfaces of constant amplitude (parallel to the xy plane).

One of the areas where TIR is used extensively is to design beam-splitters with precise control of the transmission in two or more arms. Another important area involves the use of phase change beyond the critical angle to design various polarizers and for conversion of polarization from linear to circular and vice versa [8].

Chapter 4

Elements of polarization, anisotropy and birefringence

Before Einstein's theory of relativity and the famous Michelson-Morley experiment, scientists used to believe that light propagates through a 'medium,' the so-called ether, as longitudinal waves just like sound waves. The transverse character of light was first comprehended in the experiments carried out by Fresnel and Young with birefringent materials. The origin of the hypothesis was based on the interference of polarized light conducted by Fresnel. It was observed that light waves polarized in mutually orthogonal directions cannot interfere. In order to explain this phenomenon, Young put forth his theory of transverse nature of light waves. Despite the fact that this contradicted earlier theories of longitudinal character of light, Fresnel used this successfully to derive many useful results, including the Fresnel formulas, which were discussed in Chapter 3. In this chapter we define the basic states of polarization and also discuss how they can be changed as light propagates through different kinds of media.

4.1 Basic types of polarization: Linear and elliptically polarized waves

Let a plane monochromatic wave propagate in the positive z direction. Hence the **E** and the **B** fields will oscillate in the xy plane, while the propagation vector **k** will be directed along z (see Fig. 4.1). Since the triplet of vectors **E**, **B** and **k** are mutually orthogonal, it suffices to look at only the behavior of vector **E**. The wave may be considered as the superposition of two waves with **E** having components along, say, the x and y directions, propagating along **k**:

$$E_x(z,t) = E_{0x}\ exp[i(kz - \omega t)] = a_x \exp[-i\phi_x]\ \exp[i(kz - \omega t)], \tag{4.1}$$
$$E_y(z,t) = E_{0y}\ exp[i(kz - \omega t)] = a_y \exp[-i\phi_y]\ \exp[i(kz - \omega t)]. \tag{4.2}$$

The ratio of the complex amplitudes can be written as

$$\frac{E_{0y}}{E_{0x}} = \frac{a_y}{a_x}\ \exp[-i\delta], \tag{4.3}$$

where $\delta = \phi_y - \phi_x$ is the phase difference between the orthogonal components. Depending on the value of δ, different polarization states are realized. In order to appreciate this, start with real fields at $z = 0$ written as

$$E_x(0,t) = a_x \cos(\omega t + \phi_x), \tag{4.4}$$
$$E_y(0,t) = a_y \cos(\omega t + \phi_y). \tag{4.5}$$

We note the following (recalling superposition of mutually perpendicular oscillations with the same frequency from Chapter 1) for $a_x = a_y$:

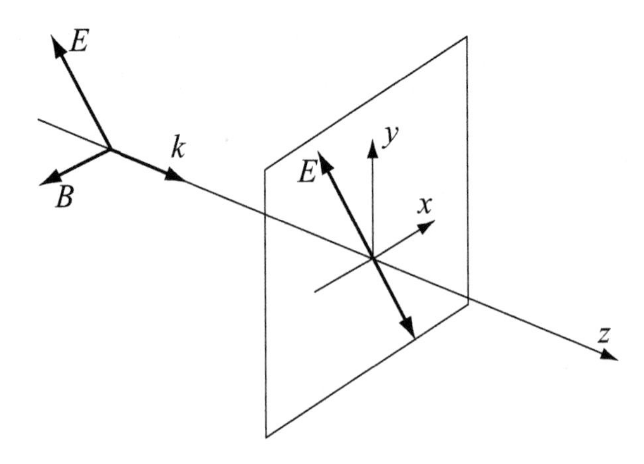

FIGURE 4.1: Schematics of a plane monochromatic wave with the orientations of the various vectors.

1. If E_{0y}/E_{0x} is real (i.e., $\delta = n\pi$, $n = 0, \pm1, \pm2 \cdots$), then the resulting electric field is linearly polarized, oscillating along a straight line. The ratio of the components a_y/a_x (in case they are different), along with the sign of $\exp(-i\delta)$, will determine the slope of this straight line.

2. If $\delta = \pi/2$, then we have right circularly polarized light, while $\delta = -\pi/2$ corresponds to the left circularly polarized light. Right-handed polarization is defined by clockwise rotation of the electric field vector as the wave travels toward the observer.

3. If the phase difference is not a multiple of $\pi/2$, then in general we have elliptical polarization. In this case $E_{0y}/E_{0x} = \exp(-i\delta)$ has a complex value and $\mathbf{E}$ rotates around the z-axis with the tip describing an ellipse.

We just demonstrated how superposition of linearly polarized light can generate circularly polarized light. We now show how the reverse can be achieved, i.e., how superposition of circularly polarized light leads to linear polarization. The general results can be extended to elliptically polarized light as well. We consider a left circularly (denoted by superscript L) polarized light given by

$$E_x^L(0,t) = a\cos(\omega t), \tag{4.6}$$

$$E_y^L(0,t) = a\cos(\omega t - \pi/2) = a\sin(\omega t). \tag{4.7}$$

Consider also a right circular wave propagating in the same direction but with additional phase ϕ:

$$E_x^R(0,t) = a\cos(\omega t + \phi), \tag{4.8}$$

$$E_y^R(0,t) = a\cos(\omega t + \phi + \pi/2) = -a\sin(\omega t + \phi). \tag{4.9}$$

We now show that the superposition of the left and right circular components leads to a linearly polarized light. Indeed, for the x and y components we have

$$E_x(0,t) = \quad E_x^L(0,t) + E_x^R(0,t) = 2a\cos(\phi/2)\cos(\omega t + \phi/2), \tag{4.10}$$

$$E_y(0,t) = \quad E_y^L(0,t) + E_y^R(0,t) = -2a\sin(\phi/2)\cos(\omega t + \phi/2). \tag{4.11}$$

It is clear from these equations that for $\phi = 0$ or π, oscillations are along the x- or y-axis, respectively, while $\phi = \pm\pi/2$ corresponds to oscillation angles $\mp\pi/4$. For the arbitrary value of ϕ, the fact that the resulting wave is linearly polarized emerges from the relation

$$E_y(0,t) = -\tan(\phi/2)E_x(0,t). \tag{4.12}$$

Based on the discussions above and the general definition (see Eq. (4.3)), we can now state the conditions for having elliptically polarized light. Elliptical polarization will occur

1. for $a_x \neq a_y$ for $\delta \neq 0, \pm\pi, \pm2\pi, \cdots$,

2. for $a_x = a_y$ for $\delta \neq 0, \pm\pi/2, \pm\pi, \pm3\pi/2, \pm2\pi, \cdots$.

4.1.1 Polarizers and analyzers

Pure polarization states are never emitted by real sources of light. It is thus necessary to use optical elements in order to have plane polarized light. Such elements are called *polarizers*. Polarizers can operate on different physical ground. We have already encountered some of them. For example, the Brewster angle phenomenon can be used to get polarized light. Phase change in total internal reflection can be used to convert plane polarized light to circularly polarized and vice versa.

One of the most convenient methods to produce plane polarized light is the use of polaroid films, based on the effect of dichorism. Certain materials have different absorption coefficients for lights polarized in different directions. PVA (polyvinyl alcohol) films doped with iodine transmit about 80% of light in one plane while only 1% is transmitted in the plane at right angles.

A helium-neon laser with Brewster windows generates plane polarized light. If a polarizer is inserted in the beam, we can have maximum transmission if the plane of polarization of the polarizer coincides with the polarization plane of the laser light. The transmission is minimal if these planes are crossed (at right angles to each other). This way we can determine the principal direction of the polarizer. Unpolarized light from a source like mercury lamps can be converted into plane polarized light by means of a polarizer. Another polarizer (often called the analyzer) can be used to diagnose the polarization direction and the degree of polarization (described below).

4.1.2 Degree of polarization

A beam of natural light can be considered a composition of wavelets with two mutually orthogonal linear polarizations in 50/50 ratio. With simple geometrical considerations we estimate the intensity of light passing through a system of polarizer and analyzer, the estimate being a function of the angle α between the polarizer and analyzer. The electric field vector oscillates along the principal direction of the polarizer after passing through it. Let its amplitude be E. The projection of this vector on the principal direction of the analyzer is $E_\alpha = E\cos\alpha$. The intensity of light passing through the analyzer is proportional to the square of the amplitude and thus we have

$$I_\alpha = I\cos^2\alpha, \tag{4.13}$$

where I is the intensity of light after passing through the polarizer. If the light source is a natural one (with indeterminate state of polarization) a maximum of one-half of the incident light can pass through the polarizer. Thus the maximum half of the incident intensity will be available after the polarizer. However, there is always a small component of the orthogonal polarization that can pass through the polarizer as well. Denoting the former and the latter intensities by $I_\parallel$ and $I_\perp$, respectively, one can define the degree of polarization

P as

$$P = \frac{I_{\parallel} - I_{\perp}}{I_{\parallel} + I_{\perp}}. \tag{4.14}$$

The value of P less than unity corresponds to partially polarized light while $P = 1$ implies perfectly polarized light. P is always less than unity for any realistic polarizer.

4.2 Stokes parameters and Jones vectors

4.2.1 Representation of polarization states of a monochromatic wave

It is clear that the amplitudes a_x and a_y in Eqs. (4.1) and (4.2) cannot be measured directly in any experiment. Only intensities proportional to the squares of these amplitudes can be recorded by a detector. For characterizing the polarization states, Stokes proposed the parameters

$$s_0 = a_x^2 + a_y^2, \tag{4.15}$$

$$s_1 = a_x^2 - a_y^2, \tag{4.16}$$

$$s_2 = 2a_x a_y \cos\delta, \tag{4.17}$$

$$s_3 = 2a_x a_y \sin\delta, \tag{4.18}$$

where $\delta = \phi_x - \phi_y$ gives the phase difference between the orthogonal components. Only three Stokes parameters will be independent since only three independent quantities, namely, a_x, a_y and δ, are involved in describing them. Indeed, it is easy to verify that

$$s_0^2 = s_1^2 + s_2^2 + s_3^2. \tag{4.19}$$

Use of an analyzer makes it possible to measure the intensities of the two orthogonal polarizations a_x^2 and a_y^2. Thus s_0 is proportional to the intensity of light. Later we describe how to measure the various Stokes parameters.

Another useful way to express the polarization state of light is through the Jones vector, which expresses the electric field as a column vector:

$$\mathbf{E} = \begin{bmatrix} E_x \\ E_y \end{bmatrix} = \begin{bmatrix} E_{0x}\exp(-i\phi_x) \\ E_{0y}\exp(-i\phi_y) \end{bmatrix}. \tag{4.20}$$

In such notation right-handed polarization can be expressed as

$$\mathbf{E}_R = \begin{bmatrix} E_{0x}\exp(-i\phi_x) \\ E_{0x}\exp(-i\phi_x - i\pi/2) \end{bmatrix}, \tag{4.21}$$

where we use $E_{0y} = E_{0x}$ and $\phi_y = \phi_x + \pi/2$. Dividing both sides by the length of the vector $E_R = \sqrt{E_{0x}^2 + E_{0y}^2} = \sqrt{2}E_{0x}$, we have

$$\mathbf{E}_R/E_R = \frac{1}{\sqrt{2}} \exp(-i\phi_x) \begin{bmatrix} 1 \\ \exp(-i\pi/2) \end{bmatrix}. \tag{4.22}$$

Dropping the arbitrary phase, the expression for a right circularly polarized light takes the form

$$\mathbf{E}_R = \frac{1}{\sqrt{2}} \begin{bmatrix} 1 \\ -i \end{bmatrix}. \tag{4.23}$$

In an analogous fashion, left circularly polarized light can be expressed by

$$\mathbf{E}_L = \frac{1}{\sqrt{2}} \begin{bmatrix} 1 \\ i \end{bmatrix}. \tag{4.24}$$

It is easy to verify that a superposition of right and left circular light leads to linear polarization:

$$\frac{1}{\sqrt{2}} \begin{bmatrix} 1 \\ -i \end{bmatrix} + \frac{1}{\sqrt{2}} \begin{bmatrix} 1 \\ i \end{bmatrix} = \sqrt{2} \begin{bmatrix} 1 \\ 0 \end{bmatrix}. \tag{4.25}$$

It is also easy to verify that the left- and right-handed polarizations are mutually orthogonal in the sense that their scalar product $\mathbf{E}_R \cdot \mathbf{E}_L^*$ vanishes:

$$\mathbf{E}_R \cdot \mathbf{E}_L^* = \frac{1}{\sqrt{2}} \begin{bmatrix} 1 \\ -i \end{bmatrix} \cdot \frac{1}{\sqrt{2}} \begin{bmatrix} 1 \\ -i \end{bmatrix} = \frac{1}{2}(1 + i^2) = 0. \tag{4.26}$$

Analogous relations hold for any orthogonally polarized light pair.

4.2.2 Measurement of Stokes parameters

It is now clear that pure polarization states do not exist. Nor do the purely unpolarized states exist. There is always some residual polarization due to reflection and scattering. It thus follows that Stokes parameters must be expressed in terms of mean intensities. To phrase it differently, Stokes parameters must be described in terms of partially polarized light.

Let the two orthogonal components be given by

$$E_x^{(r)} = A_x \cos(\omega t + \phi_x), \tag{4.27}$$

$$E_y^{(r)} = A_y \cos(\omega t + \phi_y), \tag{4.28}$$

where A_x, ϕ_x, A_y and ϕ_y are slowly varying functions of time and the superscript r denotes the real field components. Defining the phase difference δ as $\delta = \phi_x - \phi_y$, the Stokes parameters can be written as

$$s_0 = \langle A_x^2 \rangle + \langle A_y^2 \rangle, \tag{4.29}$$

$$s_1 = \langle A_x^2 \rangle - \langle A_y^2 \rangle, \tag{4.30}$$

$$s_2 = 2\langle A_x A_y \cos\delta \rangle, \tag{4.31}$$

$$s_3 = 2\langle A_x A_y \sin\delta \rangle. \tag{4.32}$$

The various Stokes parameters can be measured in terms of various intensities. In the following we show how this can be done. We use the complex notations

$$E_x = A_x \exp[-i(\omega t + \phi_x)], \tag{4.33}$$
$$E_y = A_y \exp[-i(\omega t + \phi_y)]. \tag{4.34}$$

Let a retardation plate producing retardation ϕ and a polarizer be put in the path of the beam. Let the principal direction of the polarizer be at an angle θ with the x-axis. The projection of the E-field on the principal direction can be written as

$$E(\theta, \phi) = E_x \cos\theta + E_y e^{-i\phi} \sin\theta. \tag{4.35}$$

The intensity of light described by the amplitude above can be expressed as

$$\begin{aligned} I(\theta, \phi) = \langle EE^* \rangle &= \langle E_x E_x^* \rangle \cos^2\theta + \langle E_y E_y^* \rangle \sin^2\theta \\ &+ (\langle E_x E_y^* \rangle e^{i\phi} + \langle E_y E_x^* \rangle e^{-i\phi}) \cos\theta \sin\theta, \end{aligned} \tag{4.36}$$

or

$$\begin{aligned} I &= I_x \cos^2\theta + I_y \sin^2\theta \\ &+ (\langle A_x A_y \rangle e^{-i(\delta-\phi)} + \langle A_y A_x \rangle e^{+i(\delta-\phi)}) \cos\theta \sin\theta, \\ &= I_x \cos^2\theta + I_y \sin^2\theta + \langle A_x A_y \rangle \cos(\delta - \phi) \sin 2\theta. \end{aligned} \tag{4.37}$$

Here $I_{x,y} = \langle A_{x,y}^2 \rangle$. It is clear from Eq. (4.37) that a set of six measurements yielding $I(0,0)$, $I(\pi/2,0)$, $I(\pi/4,0)$, $I(3\pi/4,0)$, $I(\pi/4,\pi/2)$ and $I(3\pi/4,\pi/2)$ is adequate to determine the Stokes parameters, since

$$s_0 = I(0,0) + I(\pi/2,0), \tag{4.38}$$
$$s_1 = I(0,0) - I(\pi/2,0), \tag{4.39}$$
$$s_2 = I(\pi/4,0) - I(3\pi/4,0), \tag{4.40}$$
$$s_3 = I(\pi/4,\pi/2) - I(3\pi/4,\pi/2). \tag{4.41}$$

When any of the parameters s_1, s_2 or s_3 has a nonzero value, the light is polarized (at least partially) and $\sqrt{s_1^2 + s_2^2 + s_3^2}$ describes the intensity of the polarized portion of the beam, while the unpolarized portion is given by $s_0 - \sqrt{s_1^2 + s_2^2 + s_3^2}$. Thus the degree of polarization is given by

$$P = \frac{\sqrt{s_1^2 + s_2^2 + s_3^2}}{s_0}. \tag{4.42}$$

It is obvious that for a partially polarized light, $P \leq 1$.

4.3 Anisotropy and birefringence

Light propagation through crystalline media is often associated with interesting physical effects. One of them is double refraction or birefringence.

Birefringence occurs when light entering a uniaxial (with one optic axis) crystal splits into two beams. These two beams have mutually orthogonal polarization, usually referred to as the ordinary and the extraordinary waves. The ordinary wave behaves just like a wave in an isotropic medium, while the extraordinary wave has some peculiar characteristics. Most important is the dependence of the magnitude of the refractive index on the direction of propagation of light. Note that the ordinary wave has the same refractive index in all directions. Along one direction both the waves have the same refractive index. This particular direction is identified as the optic axis of the crystal. There can be uniaxial or biaxial crystals depending on whether there exists one or two optic axes.

It is clear that the dependence of the refractive index increases due to the anisotropic nature of the crystal through which the light propagates. It was mentioned earlier that all the information about the optical properties of the material is carried by a pair of physical quantities, namely, the dielectric permittivity ε and the magnetic permeability μ. For a nonmagnetic medium, the anisotropic nature of the dielectric function ε reflects the directional dependence. In particular ε is a second-rank tensor (denoted by $\bar{\bar{\varepsilon}}$ below) with 3×3 elements, while it is a scalar for an isotropic medium. In general for an anisotropic nonmagnetic medium, the relation between $\mathbf{D}$ and $\mathbf{E}$ can be written as

$$\mathbf{D} = \bar{\bar{\varepsilon}}\mathbf{E}. \tag{4.43}$$

In terms of components

$$D_x = \varepsilon_{xx}E_x + \varepsilon_{xy}E_y + \varepsilon_{xz}E_z, \tag{4.44}$$

$$D_y = \varepsilon_{yx}E_x + \varepsilon_{yy}E_y + \varepsilon_{yz}E_z, \tag{4.45}$$

$$D_z = \varepsilon_{zx}E_x + \varepsilon_{zy}E_y + \varepsilon_{zz}E_z. \tag{4.46}$$

Thus fields along y can cause induction along the x direction, provided ε_{xy} is nonzero. Note that in an isotropic medium, E_y can lead to induction only along the y direction. By means of coordinate transformation, the 3×3 matrix ε_{ij} can be diagonalized.

4.3.1 Birefringence in crystals like calcite

Calcite ($CaCO_3$) can be found in nature as large crystals, which can be made into a cube, pressed slightly along one great diagonal. It can also be polished along the crystal faces. Objects seen through this cube form two images, which explains the origin of the name: double refraction or birefringence. If a narrow beam is directed along the normal to a natural face of the crystal, two beams exit from the opposite side parallel to the incident beam. The ordinary beam passes along the direction of the incident beam while the extraordinary beam is shifted with respect to the direction of the original beam. To put it differently, the angle of refraction of the extraordinary beam is not zero. A rotation of the crystal about the axis of the incident beam, the spot due to the extraordinary beam rotates in a circular path about the same axis. For unpolarized light the two spots from ordinary and extraordinary beams have the same intensity. Since the two rays have mutually orthogonal polarization, an analyzer placed after the crystal can filter them out. The rotation of the analyzer results in a periodic decrease in the ordinary ray intensity and an increase of the same for the extraordinary wave. This can be understood from a different angle. Let a plane polarized light be incident on the crystal having optic axis along OO' perpendicular to $\mathbf{k}$. Let the incident field vector oscillate along AA', which makes an angle α with the normal to the optic axis BB'. The ordinary wave $\mathbf{E}$ oscillates along BB', while that of the extraordinary wave oscillates along OO'. Thus amplitudes of oscillations in the ordinary and extraordinary waves are given by

$$(E_0)_o = (E_0)_{inc} \cos\alpha, \tag{4.47}$$

$$(E_0)_e = (E_0)_{inc} \sin\alpha. \tag{4.48}$$

For intensities we have

$$I_o = I_{inc} \cos^2\alpha, \tag{4.49}$$

$$I_e = I_{inc} \sin^2\alpha. \tag{4.50}$$

Hence

$$\frac{I_e}{I_o} = \tan^2\alpha, \tag{4.51}$$

$$I_{inc} = I_o + I_e. \tag{4.52}$$

It is clear from the above equations that the ratio of intensities vary in a periodic fashion, and in the region of overlap of the spots, the intensity is constant. In the literature this is often referred to as Malus' law.

4.3.2 Polarizers based on birefringence: Nicol and Wollaston prisms

The Nicol prism is cut from a calcite crystal as shown in Fig. 4.2. This crystal is split into two and joined using glue with refractive index $n_{glue} = 1.549$,

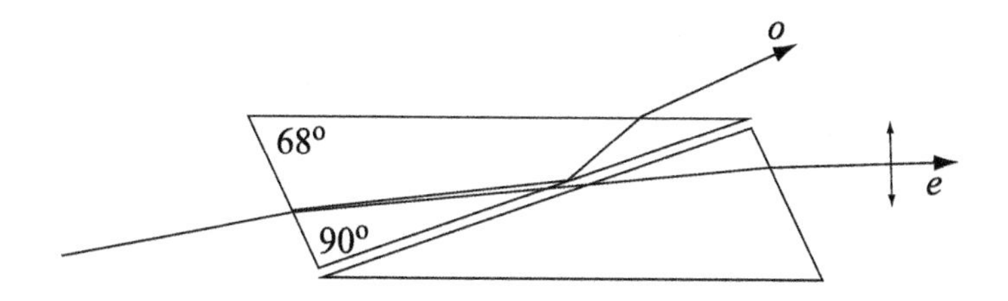

FIGURE 4.2: Schematics of a Nicol prism.

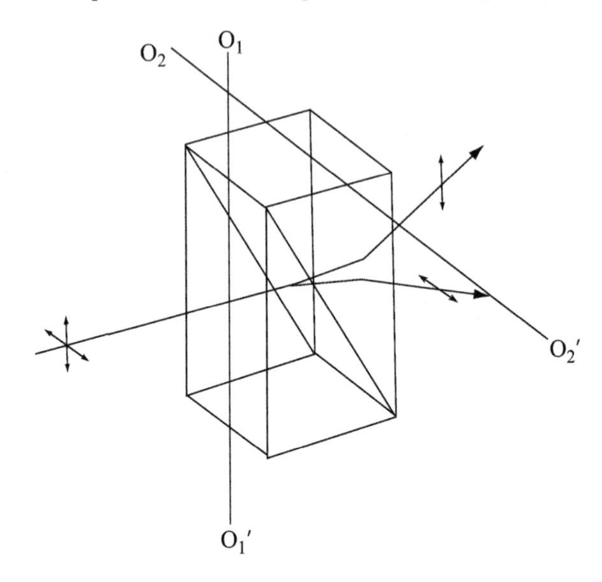

FIGURE 4.3: Schematics of a Wollaston prism.

which is in between the ordinary and extraordinary refractive indices $n_o = 1.66$ and $n_e = 1.49$. For the specified geometry, the ordinary wave undergoes total internal reflection while the extraordinary ray passes through. The two outgoing waves are linearly polarized with mutually orthogonal polarization.

The Wollaston prism gives rise to two orthogonally polarized light beams as shown in Fig. 4.3. It is made of two triangular prisms glued along the hypotenuse in such a way that its optic axes O_1O_1' and O_2O_2' are orthogonal to each other. Both ordinary and extraordinary rays travel in the same direction in the first prism. An extraordinary wave leaving the first prism travels in the second as an ordinary wave and vice versa. The refractive index of the initially extraordinary beam in the second prism is higher and hence it bends upward, closer to the normal to the interface, while the second beam (initially ordinary) sees a higher to lower refractive index and bends downward, away from the normal.

4.3.3 Retardation plates

Consider a thin plate of a uniaxial material. Assume the material to be cut in such a way that the optic axis is parallel to the surfaces of the plate. Let a plane polarized monochromatic wave be incident on the plate normally. Let also the direction of oscillation of the field vector $\mathbf{E}$ make some angle with the optic axis. Then E_x (the projection onto the normal to the optic axis) and E_y (the projection along the optic axis) determine the ordinary and extraordinary components within the plate. Both propagate in the same direction albeit with different phase velocities since $n_e \neq n_o$. At the input face both are in phase, while the phase difference at the output face of a crystal with width h is given by

$$\delta = k_0(n_e - n_o)h, \tag{4.53}$$

where $k_0 = \omega/c = 2\pi/\lambda$ is the vacuum wave-vector magnitude. In general this will lead to elliptic polarization. In particular, for $E_x = E_y$ (when the angle is 45°) and $\delta = \pi/2$, the superposition of the orthogonal components will lead to circular polarization. For example, for a negative crystal, when $n_e < n_o$, the phase of the ordinary wave is retarded by δ with respect to that of the extraordinary ray. The field at the output face can be written as

$$\mathbf{E}(t) = \hat{x}\frac{E_0}{\sqrt{2}}\cos(\omega t) + \hat{y}\frac{E_0}{\sqrt{2}}\cos(\omega t + \delta). \tag{4.54}$$

For

$$\delta = (2m+1)\pi/2 \quad \text{or} \quad h(n_e - n_o) = (2m+1)\lambda/4, \tag{4.55}$$

the tip of the vector **E** undergoes a rotation in a circular path and for this case, the total field can be written as

$$\mathbf{E}(t) = \hat{x}\frac{E_0}{\sqrt{2}}\cos(\omega t) - \hat{y}\frac{E_0}{\sqrt{2}}\sin(\omega t). \tag{4.56}$$

It is clear that a retardation plate, also called a $\lambda/4$ plate, converts linear polarization to circular polarization under certain conditions (Eq. (4.55)). We can devise a matrix tool based on Jones vectors in order to calculate the conversion between various states of polarization.

4.4 Artificial birefringence

There are several artificial ways by which an otherwise isotropic transparent medium can be rendered anisotropic. We discuss here two of them based on stress-induced anisotropy and the Kerr effect.

4.4.1 Stress-induced anisotropy

A transparent plate of plastic material or glass exerted to a mechanical force of compression (rarefaction) along a given direction can lead to induced anisotropy. The nature of the anisotropic material is uniaxial, with the optic axis coinciding with the direction of the force. A typical setup and the necessary optics to analyze the structure is shown in Fig. 4.4. The polarizer and the analyzer are crossed and the principal axis of the polarizer is at 45° with the optic axis (direction of force). Using this setup, we can measure $\Delta n = n_e - n_0$, which turns out to be proportional to F. Thus similar setups can be used for pressure sensing.

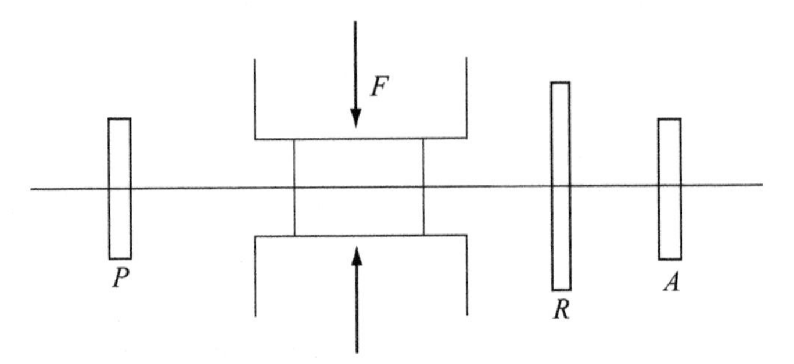

FIGURE 4.4: Schematics for observing artificial mechanical anisotropy. P-polarizer, A-analyzer, R-retarder plate ($\lambda/4$ plate).

4.4.2 Kerr effect

Under the action of an electric field, some liquids like nitrobenzene exhibit induced anisotropy. Like in mechanically induced anisotropy, here also the optic axis lies along the direction of the applied field. The liquid is kept in a cell in between the plates of a capacitor and a similar detection setup is used. An empirical law shows that the phase difference δ resulting from the difference of the refractive indices $\Delta n = n_e - n_0$ is proportional to the square of the electric field amplitude E^2. In fact, it can be written as

$$\delta = \beta d E^2, \tag{4.57}$$

where β depends on the characteristics of the liquid and d is the thickness of the liquid layer between the capacitor plates. The glass cell used for the observation of the Kerr effect is called a Kerr cell. Kerr cells can be very effective for laser light modulation because of the short time response of nitrobenzene ($\sim 10^{-9}$ sec). An arrangement like in Fig. 4.4 without the retarder can be used for such purposes. In the absence of the electric field, light will be blocked by the system since the polarizer and the analyzer are crossed and incident light is polarized along the principal axis of the polarizer (at 45° with the optic axis). The applied electric field creates elliptical polarization that cannot be blocked by the analyzer.

4.5 Optical activity and rotation of plane of polarization

Materials that can rotate the plane of polarization of incident plane polarized light are said to be optically active. Such materials can be crystalline as well as amorphous. The classical example of the setup leading to rotation

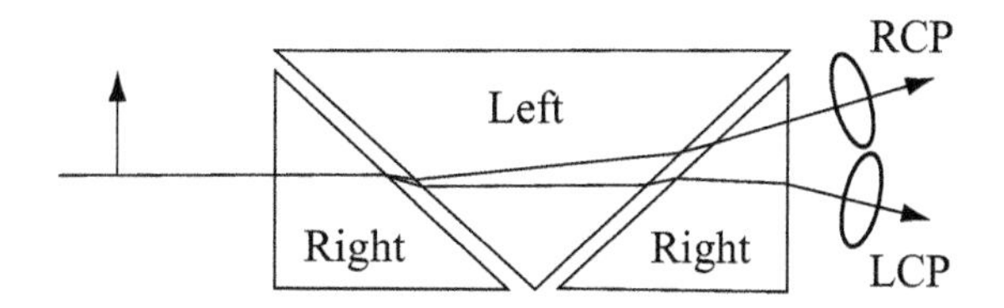

FIGURE 4.5: Fresnel prism.

of the plane of polarization is as follows. A uniaxial material like quartz, with the optic axis along the direction of propagation, is placed between a polarizer and a crossed analyzer. In absence of the quartz crystal, light is blocked by the polarizer-analyzer system. Insertion of the quartz leads to some finite transmission. A rotation of the analyzer by an angle ϕ again extinguishes transmission. This implies that after exiting the quartz crystal, the polarization plane has undergone a rotation by an angle ϕ. Experiments show that different samples of quartz exhibit different types of rotation, either clockwise (right-handed) or counterclockwise (left-handed). One type happens to be the mirror image of the other. Experiment also shows that $\phi = \alpha d$ with $\alpha \sim 1/\lambda^2$. For example, for $d = 1$ mm, $\phi_{yellow} = 20°$ and $\phi_{violet} = 50°$. Thus there is a significant dependence of optical activity on the wavelength.

Optical activity is also observed in some materials like sugar, nicotine, camphor, etc. In this case $\phi = \alpha dc'$, where $\alpha \sim 1/\lambda^2$ is the rotation constant, d is the length of the cell and c' is the concentration. Similar experiments form the foundation of how to determine the concentration of the optically active materials. It has many chemical and biological applications. The explanation of the rotation of plane of polarization in optically active media was first given by Fresnel, who first showed the similarity between double refraction and optical activity. The explanation is based on the fact that any linearly polarized light can be thought of as a superposition of left and right circular waves as in Fig. 4.6. Fresnel further assumed that in an optically active medium, the right and left waves propagate with different velocities. On this basis all the optically active media can be divided in two classes, namely, *right* ($v_R > v_L$) and *left* ($v_R < v_L$). For experimental demonstration of the validity of this assumption, Fresnel prepared a special prism as in Fig. 4.5. The inequality $n_R < n_L$ holds for the first and third prisms, while for the second $n_R > n_L$. The angle of refraction of the RCP is smaller than that of LCP when the wave leaves the first prism. While passing through the second prism the RCP component is refracted to a higher degree than the LCP and the angle between the two directions increases. Finally, the spatial separation between the rays increases after passing the last prism.

We now explain the rotation of plane of polarization. Due to different velocities of the right and left components, the time taken to traverse the same length of the material will be different for these components. Hence the

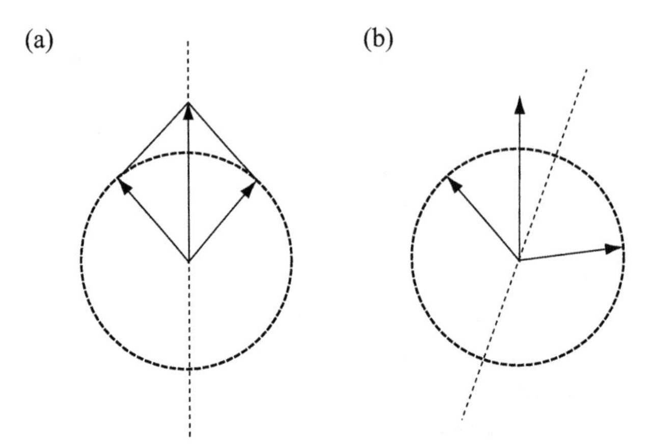

FIGURE 4.6: Explanation of the rotation of plane of polarization. The dashed line gives the direction of oscillation of the electric field. (a) and (b) represent the polarizations at the input and the output faces, respectively.

angle spanned by the right component will be different from that of the left component. The symmetry direction is then given by the arithmetic average of these two angles (see the dashed line in Fig. 4.6). Thus the angle of rotation will be given by

$$\phi = \phi_R - \frac{\phi_R + \phi_L}{2} = \frac{\phi_R - \phi_L}{2}. \tag{4.58}$$

The right and left circular components for a plane polarized light at the input face of the uniaxial material can be written as

$$E_R = E_0(\hat{x} - i\hat{y})e^{i(k_R z - \omega t)}, \tag{4.59}$$

$$E_L = E_0(\hat{x} + i\hat{y})e^{i(k_L z - \omega t)}. \tag{4.60}$$

Note that in the above equations we have incorporated Fresnel's postulate that inside the material the right and left circular waves have different wave vectors or different velocities. At the output face of the slab with thickness h, the x and y components of the electric field can be written as

$$E_x = 2E_0 \exp\left[i\left(\frac{k_R + k_L}{2}h - \omega t\right)\right] \cos\left(\frac{k_R - k_L}{2}h\right), \tag{4.61}$$

$$E_y = -2E_0 \exp\left[i\left(\frac{k_R + k_L}{2}h - \omega t\right)\right] \sin\left(\frac{k_R - k_L}{2}h\right), \tag{4.62}$$

so that

$$\frac{E_y}{E_x} = -\tan\left(\frac{k_R - k_L}{2}h\right). \tag{4.63}$$

It is clear that this ratio is a real number and hence for arbitrary thickness of the slab, light remains linearly polarized. The direction of E now makes an angle ϕ with the x-axis, which is given by

$$\phi = -\frac{k_R - k_L}{2} h = -k_0 \frac{n_R - n_L}{2} h = -\frac{\omega}{c} \frac{n_R - n_L}{2} h. \tag{4.64}$$

For a right (left) material, the rotation is in the clockwise (counterclockwise) direction.

Chapter 5

Optical properties of dielectric, metal and engineered materials

Material properties can be determined from the real-time response of the system. For example, for a dielectric, the dipoles constituting the medium take finite time to respond to the quickly oscillating radiation ($\sim 10^{15}$ Hz in optical domain) passing through the dielectric. We present a brief sketch of the linear response function theory in order to understand the frequency dependence of the dielectric function $\epsilon(\omega)$. We discuss this for $\epsilon(\omega)$, while similar arguments can be developed for magnetic response. While the details were worked out for the Lorentz model of the dielectric in Section 2.3.1, here we develop the response for metals (known as the Drude model). In dealing with the susceptibilities, we pay due attention to causality and its manifestation in the form of Kramers-Kronig relations.

Most of our attention is focused on some of the properties of composite materials, which are formed by two or more constituents. It so happens that the effective dielectric and material properties of the composites can turn out to be better than those of the constituents. There are several such examples from linear and nonlinear optics [9, 10], where the composites are engineered to have the desired optical response or to have larger nonlinear effective susceptibilities. Our goal will be to understand the mechanism of how the properties of the composites can be manipulated. There can be different kinds of composites from a geometrical viewpoint. For example, we can talk about metallic/dielectric sub-wavelength layers stacked together or we can have heterostructures with tiny metal (dielectric) inclusions embedded in a (metal) dielectric host. Of late such nano-composites are very much in focus because of their interesting properties and wide application potentials. We first look at the two-component layered composites to highlight the remarkable possibilities with such structures. Indeed, in Section 5.4 we show that a metal-dielectric composite can exhibit extremely large anisotropy though both the constituents are isotropic in nature. We then use the local field modification to obtain the effective dielectric function of the heterostructures. Note that some of these issues have been discussed in detail in textbooks [3] or in monographs [11]. We follow these sources to arrive at the Clausius-Mossotti relation and the effective dielectric function for two or multi-component heterogeneous media. Special attention is paid to metal inclusion in a dielectric host since this is used to a large extent for many other problems in other chapters. The interest in such metal nano-composites stems from the fact that they can support localized plasmon resonances. The excitation of the localized plasmons can lead to large local fields that are needed for low-threshold optical processes (see Chapter 10 and Section 10.2.3 for more details and applications).

5.1 Linear response theory and dielectric response

5.1.1 Time domain picture

Consider the propagation of light through a medium. In case of a dielectric, the medium can be thought of as a collection of charged particles, namely, electrons and positively charged cores. Under the action of the applied electromagnetic field, the positive charges move along the field, while the negative ones move in the other direction. In conductors some of the charged particles are free to move around. In dielectric materials the charges are bound together, though there is a certain flexibility of movement, resulting in induced dipole moments. Light at optical frequencies ($\sim 10^{15}$ Hz) interacting with matter leads to a very interesting situation. The electric component makes the elementary dipoles oscillate. The effect of optical magnetic fields is orders of magnitude smaller and we neglect that in our considerations. The positively charged nucleus (ion core) is heavier and thus in the visible–UV region what really matters is the motion of the electrons. A very important aspect emerges when we probe the response of the medium to the incident light that has a very high frequency. No realistic medium can react instantaneously to such fast oscillations and the response of the medium in real time can be described in the framework of a linear response theory.

The dielectric response embodied by polarization $\mathbf{P}(t)$ in optics is not instantaneous and can depend on the cause $\mathbf{E}(t)$ at any previous times:

$$\mathbf{P}(t) = \epsilon_0 \int_0^{\infty} \bar{\bar{\mathbf{R}}}(\tau) \cdot \mathbf{E}(t-\tau) d\tau, \tag{5.1}$$

where $\bar{\bar{\mathbf{R}}}(\tau)$ is the linear polarization response function, which is a second-rank tensor with the property

$$\bar{\bar{\mathbf{R}}}(\tau) = 0, \quad \text{for} \quad \tau < 0. \tag{5.2}$$

The vanishing of the response function for negative arguments (Eq. (5.2)) is the statement of causality. Indeed, for negative τ $(= -|\tau|)$, a field $\mathbf{E}(t + |\tau|)$ at a later time causes a polarization at an earlier time $\mathbf{P}(t)$. Fulfillment of Eq. (5.2) is necessary in order to ensure that such violation of causality does not happen. In writing Eq. (5.1) we have assumed that the response is local in space, meaning thereby that whatever happens at $\mathbf{r}$ is not affected by the fields at any other nearby point. Note that generalization to nonlocality in space can lead to spatial dispersion, which we do not discuss here. The time invariance principle was applied to arrive at Eq. (5.1). This principle stresses the importance of the duration of the physical process, irrespective of whether it took place in the remote past, now or in future. It is another way of saying that physical laws remain invariant in time. Mathematically, it leads to the fact that the linear response function $\bar{\bar{\mathbf{R}}}$ is a function of only one argument τ. In component notation

$$P_i(t) = \epsilon_0 \sum_j \int R_{ij}(\tau) E_j(t-\tau)\, d\tau. \tag{5.3}$$

Eq. (5.3) is linear and hence the principle of superposition remains valid. The theory can be extended to nonlinear response of the medium for intense fields and the related issues fall in the domain of nonlinear optics.

5.1.2 Frequency domain picture

We now move from time domain to frequency domain description by means of the Fourier transformations

$$\boldsymbol{\mathcal{E}}(\omega) = \int_{-\infty}^{\infty} \mathbf{E}(t) e^{i\omega t}\, dt, \tag{5.4}$$

$$\mathbf{E}(t) = \frac{1}{2\pi} \int_{-\infty}^{\infty} \boldsymbol{\mathcal{E}}(\omega) e^{-i\omega t}\, d\omega. \tag{5.5}$$

Similar integrals can be written for all other dynamical variables like $\mathbf{P}(t)$ and $\mathbf{D}(t)$. Substituting Eq. (5.5) in Eq. (5.1), we get

$$\mathbf{P}(t) = \epsilon_0 \int_0^\infty d\tau \; \bar{\bar{\mathbf{R}}}(\tau) \cdot \left[\int_{-\infty}^{\infty} \frac{d\omega}{2\pi} \mathbf{E}(\omega) e^{-i\omega(t-\tau)} \right], \tag{5.6}$$

$$= \epsilon_0 \int_{-\infty}^{\infty} \left\{ \int_0^\infty \bar{\bar{\mathbf{R}}}(\tau) e^{i\omega\tau} d\tau \right\} \cdot \mathbf{E}(\omega) \; e^{-i\omega t} \frac{d\omega}{2\pi}. \tag{5.7}$$

We introduce the linear electric susceptibility $\bar{\bar{\chi}}_e(\omega)$ as

$$\bar{\bar{\chi}}_e(\omega) = \int_0^\infty \bar{\bar{\mathbf{R}}}(\tau) e^{i\omega\tau} d\tau, \tag{5.8}$$

and using Eq. (5.8), $\mathbf{P}(t)$ can be written as

$$\mathbf{P}(t) = \epsilon_0 \int_{-\infty}^{\infty} \bar{\bar{\chi}}_e(\omega) \cdot \mathbf{E}(\omega) \; e^{-i\omega t} \frac{d\omega}{2\pi}. \tag{5.9}$$

In component form Eqs. (5.8) and (5.9) can be written as

$$P_i(t) = \epsilon_0 \sum_j \int_{-\infty}^{\infty} \chi_{ij}(\omega) \mathcal{E}_j(\omega) e^{-i\omega t} \frac{d\omega}{2\pi}, \tag{5.10}$$

$$\chi_{ij}(\omega) = \int_0^\infty R_{ij}(\tau) \; e^{i\omega\tau} \; d\tau. \tag{5.11}$$

In Eq. (5.11) we have dropped the subscript e (for the electric susceptibility) for χ_{ij}. In contrast to standard Fourier transform where ω is treated as a real variable, we take ω in the complex plane, making use of the analytic continuation. Choice of ω in the upper half complex plane (Im $\omega > 0$), along with the condition $R(\tau) = 0$ for $\tau < 0$, ensures the convergence of Eq. (5.11). Thus $\bar{\bar{\chi}}_e(\omega)$ is an analytic function in the upper half complex plane, which is another way to express causality albeit in the frequency domain. For a detailed treatment, readers are referred to Butcher and Cotter [12].

5.2 Kramers-Kronig relations

The analytic properties of $\bar{\bar{\chi}}_e(\omega)$ lead to a fundamental relation in optics, namely, the Kramers-Kronig relations. A complex frequency-dependent susceptibility implies that the medium has dispersion and absorption, both depending on frequency. These two important properties are not completely independent of each other. In fact, one can be expressed as a principal value integral depending on the other. In what follows, we show this by exploiting

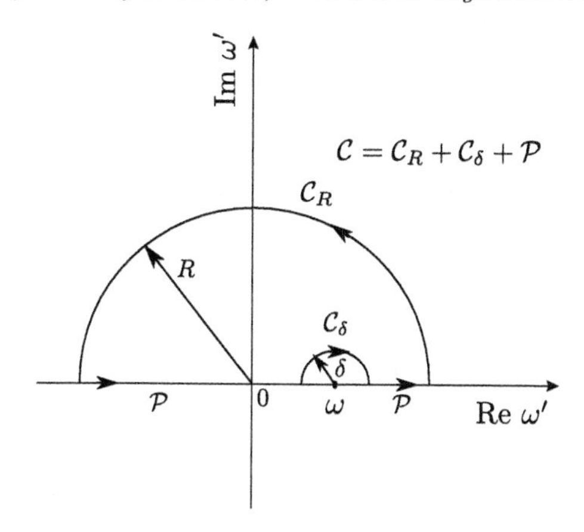

FIGURE 5.1: Contour $\mathcal{C}$ for evaluating the integral in Eq. (5.14).

the properties of complex analytic functions. In order to avoid notational complexity, we will adhere to a scalar susceptibility. Consider the integral given by

$$\mathcal{I} = \mathrm{P}\int_{-\infty}^{\infty} \frac{\chi(\omega')}{\omega' - \omega} d\omega', \tag{5.12}$$

where P denotes the principal value integral as follows:

$$\mathrm{P}\int_{-\infty}^{\infty} \frac{\chi(\omega')}{\omega' - \omega} d\omega' = \lim_{\delta \to 0} \left[\int_{-\infty}^{\omega-\delta} \frac{\chi(\omega')}{\omega' - \omega} d\omega' + \int_{\omega+\delta}^{+\infty} \frac{\chi(\omega')}{\omega' - \omega} d\omega' \right]. \tag{5.13}$$

We integrate in the complex plane along the contour shown in Fig. 5.1:

$$\int_{\mathcal{C}} \frac{\chi(\omega')}{\omega' - \omega} d\omega'. \tag{5.14}$$

In Eq. (5.14), $\mathcal{C} = \mathcal{C}_{\mathcal{R}} + \mathcal{C}_{\delta} + \mathcal{P}$, where $\mathcal{C}_{\mathcal{R}}$ ($\mathcal{C}_{\delta}$) is the part of the contour from the larger (smaller) arc of radius R (δ) and $\mathcal{P}$ gives the part of the contour along the real axis excluding the singular point $\omega' = \omega$. The complex susceptibility $\chi(\omega)$ is analytic with $\int_{\mathcal{C}} \frac{\chi(\omega')}{\omega'-\omega} d\omega' = 0$ and it is easily shown that $\int_{\mathcal{C}_{\mathcal{R}}} \frac{\chi(\omega')}{\omega'-\omega} d\omega' \to 0$ as $R \to \infty$ because of the finite response of the system for a finite input. We are left with

$$0 = \mathrm{P}\int_{-\infty}^{\infty} \frac{\chi(\omega')}{\omega' - \omega} d\omega' - i\pi\chi(\omega). \tag{5.15}$$

In writing Eq. (5.15), we have taken lim $\delta \to 0$ and used the residue theorem. From Eq. (5.15) we arrive at

$$\chi(\omega) = \chi' + i\chi'' = \frac{-i}{\pi}\,\mathrm{P}\int_{-\infty}^{\infty} \frac{\chi(\omega')}{\omega' - \omega} d\omega', \tag{5.16}$$

where χ' (χ'') is the real (imaginary) part of χ. Separating the real and the imaginary parts, we have

$$\chi'(\omega) = \frac{1}{\pi}\,\mathrm{P}\int_{-\infty}^{\infty}\frac{\chi''(\omega')}{\omega'-\omega}d\omega', \tag{5.17}$$

$$\chi''(\omega) = \frac{-1}{\pi}\,\mathrm{P}\int_{-\infty}^{\infty}\frac{\chi'(\omega')}{\omega'-\omega}d\omega'. \tag{5.18}$$

Note that integration includes the negative frequency components of $\chi(\omega)$, which has no transparent physical meaning. We invoke another important physical concept, namely, the reality principle in terms of the response function. It says that $\bar{\bar{\mathbf{R}}}$ must be real in order to ensure real $\mathbf{P}(t)$ for a real cause $\mathbf{E}(t)$. Taking the complex conjugate of Eq. (5.8) (in scalar form),

$$\chi^*(\omega) = \int_0^{\infty} R(\tau)e^{-i\omega^*\tau}d\tau. \tag{5.19}$$

A comparison of Eq. (5.19) with Eq. (5.8) establishes the relation

$$\chi^*(\omega) = \chi(-\omega^*). \tag{5.20}$$

Further, for real ω we have

$$\chi^*(\omega) = \chi(-\omega), \tag{5.21}$$

and for real and imaginary parts we have

$$\chi'(\omega) = \chi'(-\omega), \quad \chi''(\omega) = -\chi''(-\omega). \tag{5.22}$$

Thus, reality of the response and the frequency leads to the even and odd character of the dispersion ($\chi'(\omega)$) and the absorption ($\chi''(\omega)$), respectively.

Using Eq. (5.22) in the Kramers-Kronig relations, we can easily arrive at the following:

$$\chi'(\omega) = \frac{1}{\pi}\,\mathrm{P}\int_{-\infty}^{\infty}\frac{\chi''(\omega')(\omega'+\omega)}{\omega'^2-\omega^2}d\omega', \tag{5.23}$$

$$= \frac{1}{\pi}\,\mathrm{P}\left[\int_{-\infty}^{\infty}\frac{\omega'\,\chi''(\omega')}{\omega'^2-\omega^2}d\omega' + \int_{-\infty}^{\infty}\frac{\omega\,\chi''(\omega')}{\omega'^2-\omega^2}d\omega'\right], \tag{5.24}$$

$$= \frac{2}{\pi}\,\mathrm{P}\int_{0}^{\infty}\frac{\omega'\,\chi''(\omega')}{\omega'^2-\omega^2}d\omega'. \tag{5.25}$$

The second integral in Eq. (5.24) equates to zero since the integrand is an odd function. In the same way we can derive the following relation for $\chi''(\omega)$:

$$\chi''(\omega) = -\frac{2\omega}{\pi}\,\mathrm{P}\int_{0}^{\infty}\frac{\chi'(\omega')}{\omega'^2-\omega^2}d\omega'. \tag{5.26}$$

It is clear from Eqs. (5.25) and (5.26) that dispersion and absorption, embodied by real and imaginary parts of the susceptibility, respectively, are not independent of each other. Since measuring absorption is easier than measuring dispersion, we make use of these relations to obtain the refractive index at a given frequency from the absorption data over a range of frequencies.

5.3 Dispersion in metals: Drude model

As mentioned earlier, electrons are free to move around in metals, and accordingly, the term proportional to the restoring force $m\omega_0^2 x$ in Eq. (2.35) can be dropped. Thus for the displacement we have the equation of motion

$$\ddot{x} + 2\gamma\dot{x} = -\frac{e}{m} E_0 e^{-i\omega t}. \tag{5.27}$$

Proceeding as before we have

$$x(t) = \frac{e}{m} \frac{E_0 e^{-i\omega t}}{\omega(\omega + 2i\gamma)}, \tag{5.28}$$

$$P(t) = -Nex(t) = -\frac{Ne^2}{m} \frac{E_0 e^{-i\omega t}}{\omega(\omega + 2i\gamma)} = \epsilon_0 \chi_e E(t). \tag{5.29}$$

From Eq. (5.29) it follows that

$$\chi_e = -\frac{\omega_p^2}{\omega(\omega + 2i\gamma)}, \tag{5.30}$$

$$\epsilon(\omega) = 1 - \frac{\omega_p^2}{\omega(\omega + 2i\gamma)}, \tag{5.31}$$

where $\omega_p^2 = \frac{Ne^2}{m\epsilon_0}$ as before. Eq. (5.27) considers all the electrons to be free. But in reality it is not so and some electrons are bound to the lattice sites. In fact, the interband transitions due to these electrons give the shining yellow color of gold near 450 nm, since the bound electrons have resonance occurring at around 450 nm that eats away all the green light and reflects the yellow light (see for example, Shalaev [11]). These corrections can be introduced through ϵ_∞, which sums up all the contributions from interband transitions, and Eq. (5.31) can be rewritten as

$$\epsilon(\omega) = \epsilon_\infty - \frac{\omega_p^2}{\omega^2 + (2\gamma)^2} + i\frac{2\gamma\omega_p^2}{\omega(\omega^2 + (2\gamma)^2)}. \tag{5.32}$$

It is clear from Eq. (5.32) that for lower frequencies metals can possess large and negative values of Re $\epsilon(\omega)$, and this leads to the possibility of surface and localized modes at metal/dielectric interfaces (see Chapter 10).

A great deal of research has gone into the experimental studies of dispersion and the absorption in noble metals. They are tabulated in different sources [13, 14]. In fact, there can be significant differences in the effects using different data sources. One of the most valuable sources is the work of Johnson and Christie [13]. A nice comparison has been made by J. Dionne et al. [15, 16]. The results for ϵ' and ϵ'' ($\epsilon = \epsilon' + i\epsilon''$) for the real and imaginary parts are presented in Fig. 5.2 using the interpolated Johnson and Christie data [13].

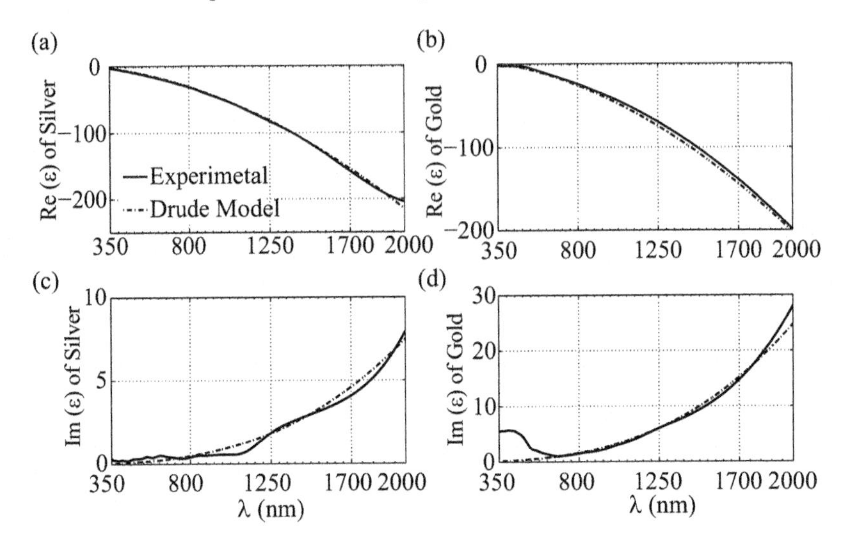

FIGURE 5.2: Complex dielectric functions of silver and gold as a function of wavelength. The solid line in (a) [(b)] depicts the real part and (c)[(d)] corresponds to the imaginary part of dielectric function of silver (gold) after interpolation [13]. The dash-dot line in each panel corresponds to a dielectric function calculated using the Drude model with $\epsilon_\infty = 5$ [$\epsilon_\infty = 9$], $2\gamma = 0.021$ eV [$2\gamma = 0.0072$ eV] and $\omega_p = 9.2$ eV [$\omega_p = 9.1$ eV] for silver [gold].

5.4 Planar composites and motivation for metal-dielectric structures

Consider the planar composite medium shown in Fig. 5.3, which consists of two nonmagnetic components with dielectric constants ε_1 and ε_2 and volume fractions f_1 and f_2, respectively, with $f_1 + f_2 = 1$. We assume each layer to be thin enough for quasi-statics to hold. We consider two representative field orientations that are parallel to the layers denoted by subscript $||$, and perpendicular to the layers denoted by subscript $\perp$. Note that both the constituent media satisfy the usual material relation $\mathbf{D} = \varepsilon\mathbf{E}$. For parallel orientation of the field, we have

$$E_1 = E_2 = E_{||}, \tag{5.33}$$

$$f_1\varepsilon_1 E_1 + f_2\varepsilon_2 E_2 = \varepsilon_{||}E_{||}. \tag{5.34}$$

Note that Eq. (5.33) reflects the continuity of the tangential components of the field at each interface, while Eq. (5.34) represents the approximation whereby the parallel component of the induction vector $D_{||}$ $(= \varepsilon_{||}E_{||})$ is taken as the

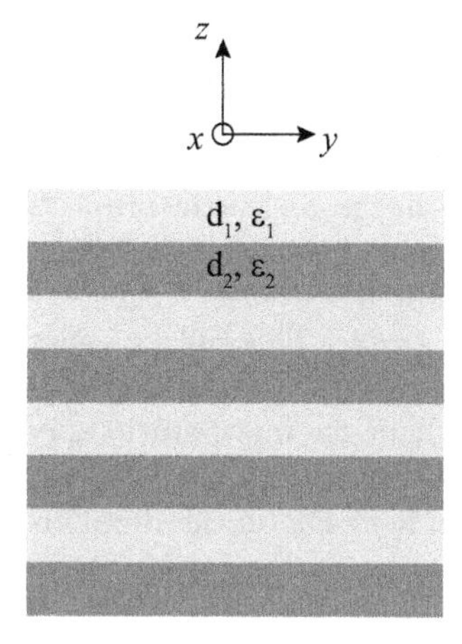

FIGURE 5.3: Schematic view of the layered medium with two components with volume fractions f_1 and f_2, each with sub-wavelength widths ($d_1, d_2 \ll \lambda$).

weighted average of similar components of D in media 1 and 2. For perpendicular orientation of the field, the normal to the surface component of D is continuous across the interface, while E for the effective medium is to be approximated by the weighted average. This leads to the equations

$$\varepsilon_1 E_1 = \varepsilon_2 E_2 = \varepsilon_\perp E_\perp, \tag{5.35}$$

$$f_1 E_1 + f_2 E_2 = E_\perp. \tag{5.36}$$

It may be noted that the field subscripts 1 and 2 in Eqs. (5.33)–(5.36) refer to the two adjacent layers. Eqs. (5.33)–(5.36) easily lead to the expression for the effective dielectric function of the composite for parallel and perpendicular excitation:

$$\varepsilon_{||} = f_1 \varepsilon_1 + f_2 \varepsilon_2, \tag{5.37}$$

$$\frac{1}{\varepsilon_\perp} = \frac{f_1}{\varepsilon_1} + \frac{f_2}{\varepsilon_2}. \tag{5.38}$$

Eqs. (5.37) and (5.38) reveal the remarkable possibilities for metal-dielectric composites. Since metals have a large negative real part of dielectric function in visible/IR range [13] with significant dispersion, we can have a very small effective $\varepsilon_{||}$ with a much larger $\varepsilon_\perp$. A simple two-parameter (λ, f_1) tuning is enough to reach such a regime, leading to a highly anisotropic material out of isotropic constituents. The significance of structured materials that brings the boundary conditions into play is easily seen from this simple example.

5.5 Metal-dielectric composites

We now discuss the linear dielectric properties of a composite medium constituting a host and some inclusions. For the sake of simplicity, we mostly deal with only a two-component system with two constituent materials, which are homogeneous and isotropic. This can be generalized to a multicomponent system very easily. We assume the average size of inclusions to be much smaller than the wavelength of light for quasi-static approximation to hold. We use the mean field approximation to calculate the effective dielectric response of the composite. The local field $\mathbf{E}_L$ on the dipole is given by

$$\mathbf{E}_L = \mathbf{E}_0 + \mathbf{E}_d + \mathbf{E}_S + \mathbf{E}_n, \tag{5.39}$$

with $\mathbf{E}_0$ as the applied electric field and $\mathbf{E}_d$ as the depolarizing field due to the charges induced on the surface of the sphere. The superposition $\mathbf{E}_0 + \mathbf{E}_d$ gives the homogenous macroscopic (Maxwell) field averaged over the entire volume of the material. In terms of the macroscopic polarization $\mathbf{P}$ the field $\mathbf{E}_d$ can be given by

$$\mathbf{E}_d = -\frac{\mathbf{P}}{\epsilon_0}. \tag{5.40}$$

In Eq. (5.39) $\mathbf{E}_S$ is the field due to the charges induced on the Lorentz sphere (of radius R) and $\mathbf{E}_n$ is the field caused by other dipoles inside the sphere. The field $\mathbf{E}_S$ can be evaluated following Ref. [3], and is given by

$$\mathbf{E}_S = \frac{\mathbf{P}}{3\epsilon_0}. \tag{5.41}$$

If now we choose the homogenous medium to have cubic crystal structure, then the contribution from E_n goes to zero due to inherent symmetry of the lattice structure. Thus for the local field we have

$$\mathbf{E}_L = \mathbf{E} + \frac{\mathbf{P}}{3\epsilon_0}. \tag{5.42}$$

The single molecule polarizability α can be related to the macroscopic polarization $\mathbf{P}$ as follows:

$$\mathbf{P} = N\alpha\mathbf{E}_L = N\alpha\left(\mathbf{E} + \frac{\mathbf{P}}{3\epsilon_0}\right), \tag{5.43}$$

where N denotes the average number of molecules per unit volume. Using Eq. (5.43), $\mathbf{D}$ can be written as

$$\mathbf{D} = \epsilon_0\left(1 + \frac{N\alpha/\epsilon_0}{1 - \frac{N\alpha}{3\epsilon_0}}\right)\mathbf{E} = \epsilon_0\epsilon_r\mathbf{E}. \tag{5.44}$$

Rewriting the relative permeability of the medium ϵ_r in terms of α yields the famous Clausius-Mossotti relation as

$$\frac{N\alpha}{3\epsilon_0} = \frac{\epsilon_r - 1}{\epsilon_r + 2}, \tag{5.45}$$

leading to the expression for α as

$$\alpha = \frac{3\epsilon_0}{N}\left(\frac{\epsilon_r - 1}{\epsilon_r + 2}\right). \tag{5.46}$$

Eq. (5.45) relates the microscopic quantity α to a macroscopic property ϵ. We will now extend this to spherical particle inclusions of relative permittivity ϵ_1 embedded in a host medium with relative permittivity ϵ_h. Let the effective dielectric function of the composite be denoted by ϵ_{eff}. The Clausius-Mossotti relation then reduces to

$$\frac{N\alpha}{3\epsilon_h\epsilon_0} = \frac{\epsilon_{eff} - \epsilon_h}{\epsilon_{eff} + 2\epsilon_h}, \tag{5.47}$$

and the polarizability α takes the form

$$\alpha = 3\epsilon_0\epsilon_h \frac{f}{N}\frac{\epsilon_1 - \epsilon_h}{\epsilon_1 + 2\epsilon_h}, \tag{5.48}$$

with f as the volume fraction of the inclusion. Note that $1/N$ in Eq. (5.48) is now replaced by f/N to account for the volume occupied by the inclusion. Replacing α in Eq. (5.48), we get

$$\frac{\epsilon_{eff} - \epsilon_h}{\epsilon_{eff} + 2\epsilon_h} = f\frac{\epsilon_1 - \epsilon_h}{\epsilon_1 + 2\epsilon_h}. \tag{5.49}$$

The medium ϵ_{eff} is then given by

$$\epsilon_{eff} = \epsilon_h \frac{1 + 2f\left(\frac{\epsilon_1 - \epsilon_h}{\epsilon_1 + 2\epsilon_h}\right)}{1 - f\left(\frac{\epsilon_1 - \epsilon_h}{\epsilon_1 + 2\epsilon_h}\right)}. \tag{5.50}$$

5.5.1 Maxwell-Garnett theory

For small f, Eq. (5.50) can be approximated by

$$\epsilon_{eff} = \epsilon_h + 3f\epsilon_h\frac{\epsilon_1 - \epsilon_h}{\epsilon_1 + 2\epsilon_h} + \mathcal{O}(f^2). \tag{5.51}$$

This is known as the Maxwell-Garnett (MG) formula. The same formula can be derived using a different method that is consistent with Eq. (5.51) for $f \ll 1$ [9]. When $\epsilon_1 + 2\epsilon_h = 0$ we have resonance, which is possible only when one of the components is a metal, since metals can have a large negative real part of the dielectric function (e.g., $\mathrm{Re}(\epsilon_1) < 0$). Such resonances are known as localized plasmon resonances. Note that ϵ_{eff} given by the MG formula cannot predict the percolation threshold, which is overcome in Bruggeman theory, discussed next.

5.5.2 Bruggeman theory for multicomponent composite medium

We will now derive the effective dielectric function for the n-component composite medium. Let ϵ_j denote the dielectric functions of the j-th component with volume fraction f_j in a host with dielectric function ϵ_h. Eq. (5.47) gets modified as

$$\frac{\epsilon_{eff} - \epsilon_h}{\epsilon_{eff} + 2\epsilon_h} = \frac{N_1 \alpha_1}{3\epsilon_0 \epsilon_h} + \frac{N_2 \alpha_2}{3\epsilon_0 \epsilon_h} + \cdots + \frac{N_j \alpha_j}{3\epsilon_0 \epsilon_h} + \cdots + \frac{N_n \alpha_n}{3\epsilon_0 \epsilon_h}, \tag{5.52}$$

where α_j, $(j = 1, 2, \cdots, n)$ is given by

$$\alpha_j = \frac{3\epsilon_0 \epsilon_h}{N_j} f_j \left(\frac{\epsilon_j - \epsilon_h}{\epsilon_j + 2\epsilon_h} \right), \tag{5.53}$$

and N_j denotes the average number of molecules of the j-th species per unit volume. After substituting α_j in Eq. (5.52), we have

$$\frac{\epsilon_{eff} - \epsilon_h}{\epsilon_{eff} + 2\epsilon_h} = f_1 \frac{\epsilon_1 - \epsilon_h}{\epsilon_1 + 2\epsilon_h} + f_2 \frac{\epsilon_2 - \epsilon_h}{\epsilon_2 + 2\epsilon_h} + \cdots + f_j \frac{\epsilon_j - \epsilon_h}{\epsilon_j + 2\epsilon_h} + \cdots + f_n \frac{\epsilon_n - \epsilon_h}{\epsilon_n + 2\epsilon_h}. \tag{5.54}$$

For an n-component effective composite medium, we have

$$f_1 + f_2 + \cdots + f_j + \cdots + f_n = 1, \tag{5.55}$$

and ϵ_h cannot be distinguished from the effective ϵ_{eff} ($\epsilon_{eff} = \epsilon_h$); the left-hand side of Eq. (5.54) reduces to zero and we arrive at the Bruggeman formula

$$\sum_{j=1}^{n} f_j \frac{\epsilon_j - \epsilon_{eff}}{\epsilon_j + 2\epsilon_{eff}} = 0, \quad \text{with} \quad \sum_{j=1}^{n} f_j = 1. \tag{5.56}$$

Note that the derivation for the n-component system presented here does not differentiate the inclusion and host as in the Maxwell-Garnett formula. The present approach treats both inclusions and host on an equal footing. The effective medium dielectric function ϵ_{eff} for a two-component ($n = 2$) composite simplifies to

$$f_1 \frac{\epsilon_1 - \epsilon_{eff}}{\epsilon_1 + 2\epsilon_{eff}} + f_2 \frac{\epsilon_2 - \epsilon_{eff}}{\epsilon_2 + 2\epsilon_{eff}} = 0, \quad \text{with} \quad f_1 = 1 - f_2. \tag{5.57}$$

Eq. (5.57) represents a quadratic equation in ϵ_{eff} having roots

$$\epsilon_{eff} = \frac{1}{4} \{ (3f_1 - 1)\,\epsilon_1 + (3f_2 - 1)\,\epsilon_2 \pm \sqrt{[(3f_1 - 1)\,\epsilon_1 + (3f_2 - 1)\,\epsilon_2]^2 + 8\epsilon_1 \epsilon_2} \}. \tag{5.58}$$

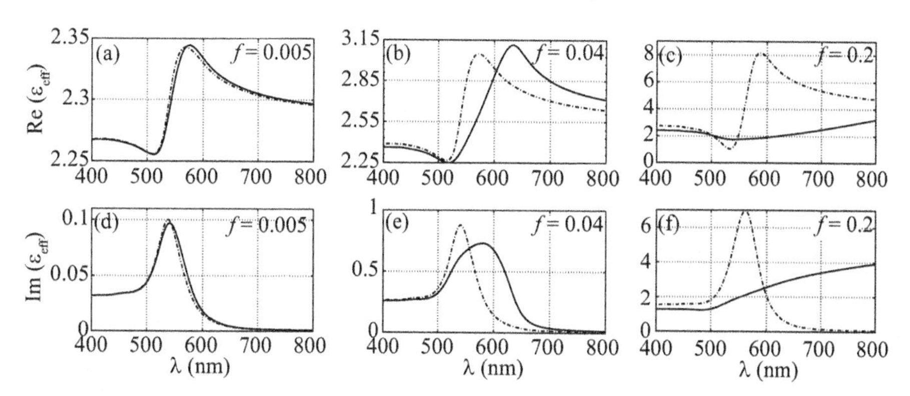

FIGURE 5.4: [(a), (b) and (c)] Real and [(d), (e) and (f)] imaginary parts of ϵ_{eff} as function of λ for a gold-silica composite evaluated using the Bruggeman (solid curve) and the Maxwell-Garnett (dash-dot) formula for three different volume fractions of gold inclusions, namely, $f = 0.005$ [(a) and (d)], $f = 0.04$ [(b) and (e)] and $f = 0.2$ [(c) and (f)]. The dielectric function of gold is taken from Johnson and Christie [13], while $\varepsilon_h = 2.25$.

The proper sign in Eq. (5.58) is to be chosen such that Im $(\epsilon_{eff}) > 0$ to ensure causality.

We now compare the estimates of ϵ_{eff} as predicted by the Maxwell-Garnett (MG) and Bruggeman formulas. As mentioned earlier, MG formula is valid only for lower values of f, as can be seen from Fig. 5.4. Fig. 5.4 shows the real and imaginary parts of ϵ_{eff} as a function of wavelength λ for different values of volume fraction of gold inclusions in silica. The solid (dash-dot) curve in Fig. 5.4 corresponds to the MG (Bruggeman) formula. It is evident that for lower values of f (for example, $f = 0.005$), both the formulas match very well, but for higher values of f they differ drastically. Thus due attention is to be paid when dealing with higher values of f of the metal inclusions since ϵ_{eff} can be completely different as estimated by the two approaches. Throughout the derivation we have considered the inclusions to be spherical, but this can be generalized to other geometries in a straightforward manner. In a similar vein, anisotropic character of the inclusions can be incorporated.

5.6 Metamaterials and negative index materials

It has been stressed now and again that all the material properties are contained in the dielectric permittivity ϵ and the permeability μ in the context of electromagnetics and optics. Recall that ϵ and μ are related to corresponding

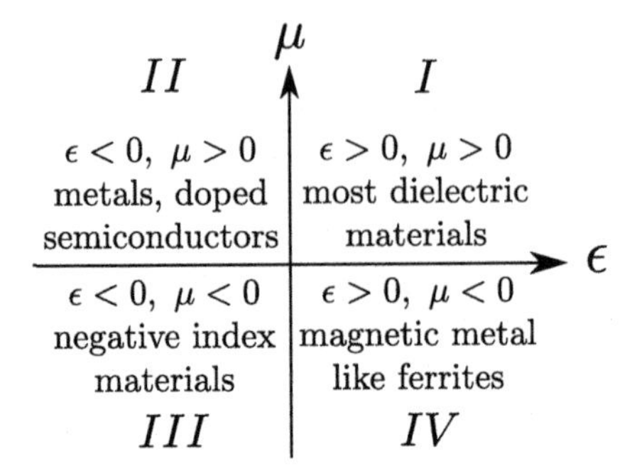

FIGURE 5.5: Characterization of materials based on the electric (ϵ) and magnetic (μ) response.

electric and magnetic susceptibilities through

$$\epsilon = \epsilon_0(1 + \chi_e) = \epsilon_0\epsilon_r, \tag{5.59}$$

$$\mu = \mu_0(1 + \chi_m) = \mu_0\mu_r. \tag{5.60}$$

Thus electric and magnetic properties of the medium are determined by the relative permittivity and permeability ϵ_r and μ_r, respectively. In 1968 Victor Veselago, in a seminal paper, posed a very interesting problem [17]: Does the standard electromagnetics hold when both ϵ_r and μ_r are simultaneously negative? It was shown by Veselago that there is no conceptual problem when both ϵ_r and μ_r are simultaneously negative over a certain frequency range. Of course, such materials do not occur in nature and, if at all, they are to be engineered and fabricated. Such materials were named as negative index materials (NIM) or left-handed materials, since we have to pick the negative sign for n ($n = -\sqrt{\epsilon_r\mu_r}$) and $\mathbf{k}$, $\mathbf{E}$ and $\mathbf{H}$ in such materials from a left-handed triplet. In contrast, for standard materials we pick the positive sign for the square root, and we have a right-handed triplet for $\mathbf{k}$, $\mathbf{E}$ and $\mathbf{H}$ (see Eqs. (2.25)–(2.28) in Section 2.2). Ignoring the imaginary part of ϵ and μ, the material parameter domain can be split into four quadrants as shown in Fig. 5.5. It is clear from Fig. 5.5 that quadrants *II* and *IV* correspond to metal-like behavior. Parameters corresponding to *II*, meanwhile, are for standard metals and doped semiconductors at low frequencies. Quadrant *IV* corresponds to 'magnetic' metals like some ferrites. Clearly *I* corresponds to most dielectric materials and *III* refers to negative index materials. The left-handed triplet character for NIMs leads to yet another counterintuitive result regarding the antiparallel nature of the phase velocity and the Poynting vector. Indeed, $\mathbf{S} = \frac{1}{\mu_0}\mathbf{E} \times \mathbf{H}$ is antiparallel to $\mathbf{k}$.

It is now clear that NIMs can lead to a host of unexpected and counterintuitive phenomena. For example, there can be negative refraction leading to the bending of the refracted ray on the same side of the normal. A few other effects are the negative Doppler effect, the anomalous Cherenkov effect, lensing, etc.

In the context of lensing, a real breakthrough was the theoretical paper by Sir Pendry. He showed that a thin slab of NIM can act as a perfect lens since it can focus not only the propagating waves but also the evanescent waves [18]. Such a perfect lens can offer superresolution beating the diffraction limit. In contrast, a standard lens can focus only the propagating part and thus cannot override the Rayleigh criterion. On the design front, the key suggestions again come from Sir Pendry and his coworkers [19, 20]. Designing materials with negative ϵ is not difficult, since we know that metals possess large negative ϵ at low frequencies. The job was to reduce the plasma frequency. With a 'dilute' metal in the form of array of wires, this was achieved in 1996 [19]. After about three years, in 1999, Sir Pendry theoretically showed how to achieve a magnetic material from nonmagnetic constituents [20]. Split-ring resonators (SRR) emerged as the core units in the artificial magnetic materials that could offer $\mu < 0$. A realization of the clever arrangement of wires and SRRs led to the first implementation of NIMs in the microwave range [21]. Fishnet structures emerged as the likely candidates to push the frequency range to near-IR and visible [22], since there were enormous difficulties (both conceptual and practical) in scaling down the SRRs to the optical domain.

The major difficulties in metamaterials (the Greek word *meta* means 'beyond,' these are materials beyond what is available in nature) and especially in NIM research have been the losses in such materials. In fact, some of them exploit the material resonances, which are associated with large losses. Most of the theoretical predictions on the prefect lensing, etc. assume low or no losses in the structure. It has been shown theoretically that inclusion of losses can wash out the sub-wavelength features, and it is difficult to achieve superresolution even with the parameters of the best NIM structures available to date [23]. Several schemes to overcome the devastating effects of losses have been proposed and most efforts have not yet achieved a 'perfect' metamaterial at optical frequencies.

Volumes have been written on such engineered materials and excellent reviews and monographs exist [11, 24, 25, 26]. In fact, one of the first metamaterials was conceived by Sir J. C. Bose, when he showed that a twisted medium can affect the polarization of light [27]. Here we introduced the readers to this fascinating area of engineered materials or metamaterials. Such metamaterials hold the key to many practical problems like perfect lensing, superresolution, invisibility cloaks, etc. Some of these effects are discussed briefly in Chapter 14.

Chapter 6

More on polarized light

In Chapter 4 it was noted that the experiments carried out by Fresnel and Young led to the discovery of transverse character of light that could satisfactorily describe the phenomenon of interference using polarized light. The basic definition of state of polarization of light via the polarization ellipse of the transverse EM field was briefly introduced in Chapter 4. When such polarized light passes through any anisotropic medium, its polarization state is transformed depending upon medium characteristics. The variations in the state of polarization of a wave thus enable us to characterize the system under consideration. A number of mathematical formalisms have been developed over the years to deal with the propagation of polarized light and its interaction with optical systems. Among these, the Jones calculus and the Stokes-Mueller calculus have been the most widely used. The former is a field-based model that assumes coherent addition of the phase and amplitude of EM waves, and the latter is an intensity-based model that instead utilizes the incoherent addition of wave intensities. In this chapter we define the various states of polarization of light waves and discuss the interaction of such polarized light with material media using the Jones and Stokes-Mueller calculus. We also briefly introduce the concepts of polarimetric measurements and touch upon representative applications of experimental polarimetry.

6.1 State of polarization of light waves

As introduced in Chapter 4, the classical concept of polarized light represents the state of polarization of a transverse electromagnetic (EM) wave by the evolution of transverse electric field vector **E** as a function of time at a given point of the space. If the vector extremity describes a stationary curve in their temporal evolution, the wave is *polarized.* Accordingly, the shape of the curve traced out by the **E** vector defines the polarization state of the wave in question. On the other hand, if the vector extremity follows random paths during the observation or measurement time, it is *unpolarized.* In the corresponding quantum mechanical description, it is assumed that each individual photon (energy quanta) is polarized, and its associated state vector corresponds to one of the classical polarization states. When a large number of photons are considered, their collective behavior is consistent with the classical limit (the wave solution to Maxwell's equations). In the case when all of the photons exhibit the same polarization, the light is said to be completely polarized. On the other hand, when there are photons of different polarizations but with a distribution favoring one particular state, the light is *partially polarized,* and when the photons are uniformly distributed over all possible polarization states, the light is said to be *unpolarized.* Nevertheless, the quantum polarization state vector for the photon is analogous to the Jones vector (described subsequently) in its classical description. Thus the quantum mechanical view of polarization and the corresponding classical formalisms are mutually consistent.

6.1.1 Jones vector representation of pure polarization states

In the classical description, the electric field vector of any transverse EM plane monochromatic wave of frequency ω, propagating along the z direction, can be expressed in terms of the two orthogonal components (x and y; note that other orthonormal coordinates are possible) in the right-handed Cartesian coordinate system as [28, 29, 30, 31, 32]

$$\mathbf{E}(z,t) = \begin{bmatrix} E_{0x}\cos(kz-\omega t-\delta_x) \\ E_{0y}\cos(kz-\omega t-\delta_y) \end{bmatrix}, \tag{6.1}$$

with

$$k = (n' + in'')\frac{\omega}{c}. \tag{6.2}$$

which is the modulus of the propagation vector **k**, c is the speed of light in vacuum, and n' and n'' are the real and imaginary parts of the refractive index, which determine the speed of light and the absorption in the medium, respectively.

The *polarization* of the wave is defined by the shape of the trajectory described by E in the xy plane. This shape depends on the ratio of the amplitudes $\tan\nu$ and the phase difference δ, defined as

$$\tan\nu = \frac{E_{0y}}{E_{0x}}; \delta = \delta_y - \delta_x. \tag{6.3}$$

This trajectory is in general elliptical and is represented in Fig. 6.1. Besides the parameters defined above, the ellipse can also be described by the orientation (azimuth) α of its major axis and its ellipticity ϵ, which is positive (negative) for left- (right-) handedness. The ellipticity ϵ varies between the two limits of zero (linearly polarized light) and $\pm 45°$ (circularly polarized light), representing the two limits of generally elliptical polarization. R. Clark

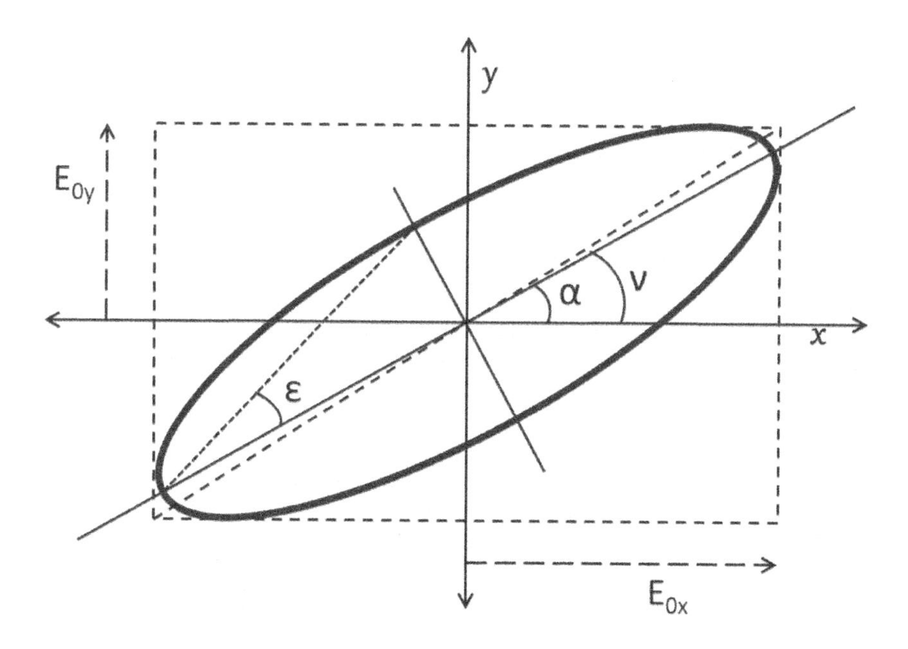

FIGURE 6.1: The polarization ellipse of a wave propagating in the z direction. Here E_{0x} and E_{0y} are the amplitudes of the x and y field oscillations; their ratio is given by $\tan\nu$. The parameter α is the azimuth of the major axis of the ellipse and ϵ is its ellipticity. ϵ is positive or negative for left- or right-handed polarization states, respectively.

Jones represented (between 1941 and 1947) the polarization state of a quasi-monochromatic transverse plane wave by a two-dimensional column matrix or a vector whose elements are complex amplitudes of the field vector along the two orthogonal directions, known as the Jones vector [33]. Accordingly, the Jones vector is defined as

$$\mathbf{E} = \begin{pmatrix} E_x \\ E_y \end{pmatrix} = \begin{pmatrix} E_{0x} \exp(-i\delta_x) \\ E_{0y} \exp(-i\delta_y) \end{pmatrix}. \tag{6.4}$$

Depending on the relative amplitudes and phases of the two orthogonal components of the electric field in Eq. (6.4), the Jones vectors corresponding to the different pure polarization states are listed in Table 6.1 (with H, V, P and M, for linear polarizations along the horizontal, vertical, +45°and −45° directions, respectively and L and R for left and right circular polarizations).

The intensity of a fully polarized wave characterized by the Jones vector is given by

$$I = I_x + I_y = \frac{1}{2}(E_{0x}^2 + E_{0y}^2) = \frac{1}{2}(E \cdot E^*), \tag{6.5}$$

where E^* is the conjugate of E.

Experimentally, one can determine the azimuth α of a linearly polarized light beam propagating along the z direction by observing its *extinction* through a linear analyzer set perpendicular to α. This type of characterization may also be extended to elliptically polarized beams as illustrated in Fig. 6.2.

To determine the ellipticity ϵ, a quarter-wave plate (QWP) is inserted in the beam path with its slow axis oriented at the azimuth α. Due to the 90° phase shift introduced by the QWP, the initial elliptical polarization state is transformed into a linear one, oriented at $\alpha + \epsilon$ from the x reference axis. Then, a linear analyzer with its pass axis oriented at ϵ from the fast axis of the QWP will lead to complete extinction of the beam. In practice, the extinction is achieved by a trial-and-error procedure, and the azimuth α and the ellipticity ϵ are eventually determined from the angular settings of the quarter-wave plate and the analyzer when maximum extinction is obtained.

Note that this vectorial description of polarization state enables the matrix treatment for describing the polarizing transfer of light in its interaction with any medium. An optical element, like a retardation plate or a partial polarizer, is therefore represented by a 2×2 matrix, whose four elements are generally complex. This can be represented as [33]

$$\mathbf{E}' = J\mathbf{E},$$
$$\begin{pmatrix} E'_x \\ E'_y \end{pmatrix} = \begin{pmatrix} J_{11} & J_{12} \\ J_{21} & J_{22} \end{pmatrix} \begin{pmatrix} E_x \\ E_y \end{pmatrix}, \tag{6.6}$$

where J is a 2×2 complex matrix, known as the Jones matrix of the interacting medium, and $\mathbf{E}$ and $\mathbf{E}'$ are the input and the output Jones vectors of light, respectively. Applying the associative properties of matrices, the matrix operator equivalent to a combination of several optical elements can then be

TABLE 6.1: Usual polarization states: Jones vectors, azimuths, ellipticities and shapes of the ellipses.

State	**H**	**V**	**P**	**M**	**L**	**R**	**Elliptical**
E	$\begin{pmatrix} 1 \\ 0 \end{pmatrix}$	$\begin{pmatrix} 0 \\ 1 \end{pmatrix}$	$\frac{1}{\sqrt{2}}\begin{pmatrix} 1 \\ 1 \end{pmatrix}$	$\frac{1}{\sqrt{2}}\begin{pmatrix} 1 \\ -1 \end{pmatrix}$	$\frac{1}{\sqrt{2}}\begin{pmatrix} 1 \\ i \end{pmatrix}$	$\frac{1}{\sqrt{2}}\begin{pmatrix} 1 \\ -i \end{pmatrix}$	$\begin{pmatrix} \cos\alpha\cos\epsilon - i\sin\alpha\sin\epsilon \\ \sin\alpha\cos\epsilon + i\cos\alpha\sin\epsilon \end{pmatrix}$
α	0	90°	+45°	−45°	Undefined	Undefined	α
ϵ	0	0	0	0	+45°	−45°	ϵ
Shape of the ellipse	⟶	↑	↗	↖	↺	↻	↓

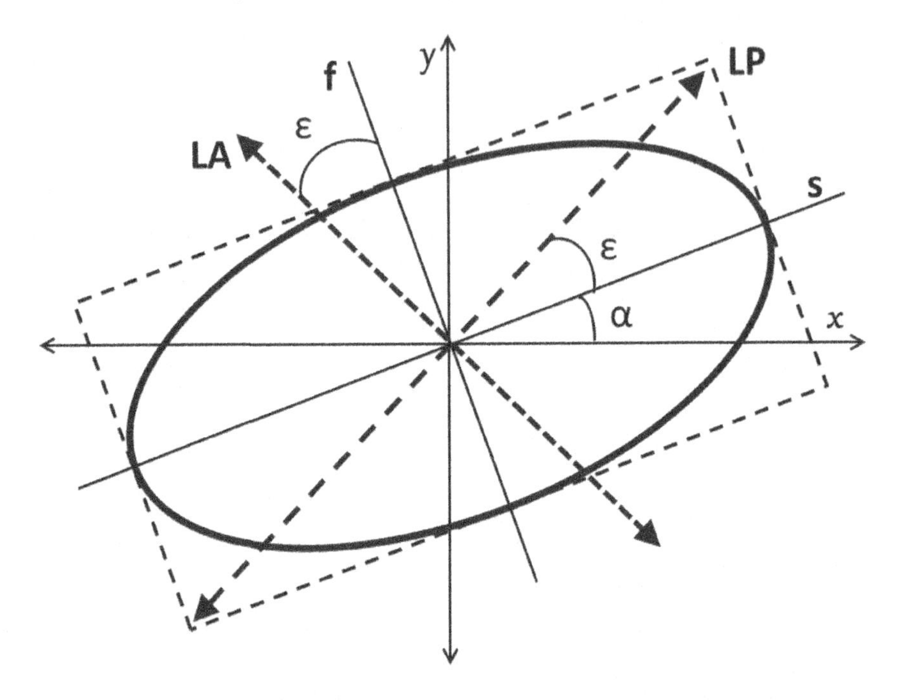

FIGURE 6.2: Extinction method for the analysis of arbitrary elliptical polarizations. The input elliptical polarization is transformed into a linear polarization state (**LP**) by inserting a quarter-wave plate with its slow axis **s** oriented at azimuth α. Complete extinction is then observed by setting a linear analyzer **LA** at perpendicular orientation to **LP**. The ellipticity ϵ is then measured as the angle between the analyzer axis for extinction and the fast axis **f** of the quarter-wave plate.

easily determined. It is the result of the multiplication of the matrices of each optical element, in the same order as that of light passing through. Hence, the Jones vector of an optical wave that emerges from a system of n optical systems can be written as

$$J = J_n J_{n-1} \dots J_2 J_1. \tag{6.7}$$

We shall address the Jones matrices corresponding to various polarization transforming interactions of medium in subsequent sections.

While theoretically interesting, the Jones formalism is limited in that it can only describe pure polarization states (completely polarized waves), and it is thus ill-suited for applications in which it is necessary to consider partial polarization or depolarizing interactions (polarization loss). Yet, quasi-monochromatic radiation is not necessarily completely polarized and many of the naturally occurring optical materials tend to be depolarizing. Such general cases can be better addressed by the coherency matrix and the Stokes-Mueller formalisms, as we describe next.

6.1.2 Partially polarized states

The previous subsection dealt with completely polarized waves. In such an idealized situation, the transverse components of the optical field (E_x and E_y) describe a perfect polarization ellipse (or some special form of an ellipse, such as a circle or a straight line, depending upon the relative amplitudes and phases) in the xy (transverse) plane. Note that the time scale at which the light vector traces out an instantaneous ellipse is of the order of 10^{-15} seconds. This period of time is clearly too short to allow us to follow the tracing of the ellipse. This fact, therefore, immediately prevents us from ever observing the polarization ellipse. More important, such a description is only applicable for light that is completely polarized, waves for which transverse components of the field amplitudes E_{ox}, E_{oy} and the associated phases δ_x and δ_y can be considered as constant during the measurement time. Yet, in nature, light is very often unpolarized or partially polarized. Thus, the polarization ellipse is an idealization of the true behavior of light; it is only correct at any given instant of time. These limitations force us to consider an alternative description of polarized light in which only observed average values or measured quantities ('intensities' rather than instantaneous field) enter.

Before we invoke mathematical formulation of partial polarization states via the measurable intensities (time average of the square of the amplitude), it might be useful to gain some qualitative idea on the phenomenon of partial polarization (or depolarization) from practical extinction measurements. For example, if we try the extinction method to characterize natural light directly coming from a source, such as the sun or a light bulb, the detected intensity can be independent of the settings of the quarter-wave plate and the analyzer. We can thus conclude that the light coming from the sun or the light bulb is *totally depolarized.* In other cases—for example, the light coming from a bulb but reflected from a floor en route to the observer—the intensity detected through the quarter-wave plate and the analyzer may vary between I_{min} and I_{max}. This provides an experimental definition of the *degree of polarization* (DOP) of the light beam, :

$$DOP = \frac{I_{max} - I_{min}}{I_{max} + I_{min}}. \tag{6.8}$$

For totally polarized states, I_{min} vanishes leading to $DOP = 1$. At the other extreme, for totally unpolarized light, $I_{min} = I_{max}$ and $DOP = 0$. For partially polarized states, on the other hand, the DOP may take any intermediate values between zero and one. For such partially polarized states, the motion of the electric field in the xy plane is no longer a perfect ellipse, but rather a somewhat *disordered* one. In case of a totally random motion of the electric vector $\mathbf{E}$, in the extinction procedure the analyzer would detect the same

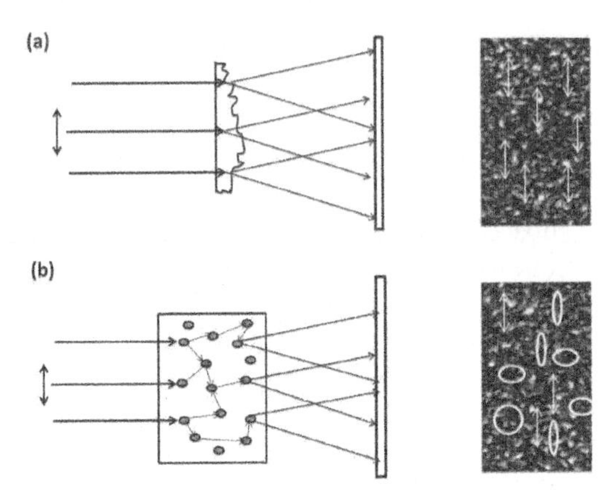

FIGURE 6.3: Scattering of a linearly polarized coherent light beam by static samples. Top: single scattering from an optically thin sample. The state of polarization of the speckle spots remains the same as that of the incident beam. Bottom: multiple scattering by an optically thick sample. The state of polarization varies considerably from speckle to speckle.

constant intensity. What is implicitly assumed in this description is that the light polarization may be defined at any instant, but may vary over time scales much shorter than the integration time of the detector. As a result, this detector takes the *temporal averages of the intensities*, which is sequentially generated by *different totally polarized states*. We note here that that the averaging of intensities (i.e., the *incoherent* sum) of polarized contributions is not necessarily temporal (it may be *spatial* as well, as illustrated below).

Consider the scattering experiments shown in Fig. 6.3. In one case (top panel) the object is optically thin and the laser undergoes single scattering by the rough surface. In the other case, the object is optically thick leading to strong multiple scattering effect. In both cases, the incident laser beam is spatially coherent, and the scattering objects are static (we ignore for the moment any possible thermal/Brownian motions). It is well known that in these conditions we can observe a speckle pattern in the screen due to the interferences (at each point of the screen) of many scattered waves having random (but static) amplitudes and relative phases.

The major difference between single and multiple scattering regimes is that for the former, the polarization of all scattered waves is the same as that of the incident wave, while the polarization states become random in the case of multiple scattering. Consequently, as outlined in Fig. 6.3, all the speckles feature the same polarization as the incident laser for single scattering, while in the other case, each speckle is still fully polarized, but this polarization varies randomly from one speckle to the next.

To summarize, *true* depolarization requires that the detected signal is the sum of intensities due to various polarized contributions with different state of polarization. The summing may take place temporally, spatially or even spectrally, and it depends not only on the sample itself but also on the characteristics of the illumination beam and of the detection system.

In 1852 Sir George Gabriel Stokes discovered that the polarization behavior could be represented in terms of observables [34]. He found that any state of polarized light could be completely described by four measurable quantities now known as the Stokes polarization parameters. As we saw earlier, the amplitude of the optical field cannot be observed; rather, the quantity that can be observed is the intensity, which is derived by taking a time average of the square of the amplitude. This suggests that if we take a time average of the unobserved polarization ellipse, we will be led to the observables of the polarization ellipse. As we shall show shortly, these observables of the polarization ellipse (measured as four sets of intensity values) are exactly the Stokes polarization parameters. Importantly, these Stokes parameters can encompass any polarization state of light, whether it is natural, totally or partially polarized (and can thus deal with both polarizing and depolarizing optical interactions). Before we address that, we shall introduce the concept of the *coherency matrix*, which deals with the time-averaged description of the transverse field components (amplitudes and phases). The definition of degree of polarization (DOP) will be introduced via this so-called coherency matrix formalism and it will be shown that the four measurable Stokes polarization parameters actually follow from combinations of the various elements of the coherency matrix.

6.1.3 Concept of 2×2 coherency matrix

The coherency matrix (also called the matrix of polarization in the literature) includes partial polarization effects by taking the *temporal average* of the direct product of the Jones vector by its Hermitian conjugate. In this way, the 2×2 coherency matrix ϕ is defined as [30, 35, 36, 37]

$$\phi = \left\langle \mathbf{E} \otimes \mathbf{E}^{\dagger} \right\rangle = \begin{bmatrix} \langle E_x E_x^* \rangle & \langle E_x E_y^* \rangle \\ \langle E_y E_x^* \rangle & \langle E_y E_y^* \rangle \end{bmatrix} = \begin{bmatrix} \phi_{xx} & \phi_{xy} \\ \phi_{yx} & \phi_{yy} \end{bmatrix}, \tag{6.9}$$

where $\langle \cdots \rangle$ denotes temporal (ensemble) average, $\otimes$ denotes the tensorial or Kronecker product, $\mathbf{E}^*$ is the conjugate of $\mathbf{E}$ and $\mathbf{E}^{\dagger}$ is the transpose conjugate of the Jones vector $\mathbf{E}$. The two defining properties of the coherency matrix are its Hermiticity ($\phi = \phi^{\dagger}$, by its definition) and non-negativity ($\phi \geq 0$): every 2×2 matrix obeying these two conditions is a valid coherency matrix and represents some physically realizable polarization state. The non-negativity condition for this 2×2 matrix can also be written as

$$tr\phi > 0 \text{ and } det\phi \geq 0. \tag{6.10}$$

It is pertinent to note that the trace of the coherency matrix ($tr\phi$) represents an experimentally measurable quantity, the total intensity of light that corresponds to the addition of the two orthogonal component intensities. As it will be discussed shortly (in context with the Stokes parameters), this corresponds to the sum of intensities measured using two orthogonal orientations of a polarizer. Thus, the first non-negativity condition of the coherency matrix ($tr\phi > 0$) directly follows from the non-negativity of total intensity. The second condition ($det\phi \geq 0$), on the other hand, follows from the limiting condition of degree of polarization ($0 \leq DOP \leq 1$). This can be understood by noting that the off-diagonal elements of ϕ are defined by taking the time average over the product of a field component with the conjugate of its transverse component. The quantity $det\phi$ therefore represents the fluctuations in the phases of the field components. As we can easily see, for a perfectly coherent source (where the phases of the transverse field components and their difference can be considered as a constant over a finite measurement time), the determinant of the coherency matrix should vanish ($det\phi = 0$). This corresponds to the fully polarized light (the ideal situation that we discussed in context with the Jones formalism and polarization ellipse). For partially coherent (or incoherent) sources, on the other hand, $det\phi > 0$, representing partially (mixed) polarized states or even completely unpolarized states. In fact, the quantity $det\phi$ (its square root, rather, as we shall show later) is a quantitative measure of the unpolarized intensity component of any partially polarized light (the natural intensity component that is independent of the polarizer/wave plate orientation in the experiment described in the beginning to define the partial polarization states). This definition of the state of polarization via the coherency matrix thus enables us to relate the DOP of light to the coherence characteristics of the source. It follows that the coherence property of the source itself limits the maximum achievable polarization ($DOP = 1$ can only be produced by an idealized perfectly coherent source). The definition of DOP can in fact be invoked from the coherency matrix using the ratio of determinant of ϕ (representing the completely unpolarized component of intensity) and the trace of ϕ (representing the total intensity) as

$$DOP = \frac{I_{pol}}{I_{tot}} = \sqrt{1 - \frac{4det(\phi)}{[Tr(\phi)]^2}}. \tag{6.11}$$

Here, I_{pol} is the polarized fraction of the intensity and I_{tot} is the total intensity. As we can observe now, the empirical definition of DOP (Eq. 6.8), which was introduced rather naively in context to the experiment discussed in the previous section, is consistent with the definition of DOP from the coherency matrix. It is clear that fully polarized light corresponds to $det\phi = 0$ ($DOP = 1$) and partially polarized or mixed polarization states correspond to $det\phi > 0$ ($DOP < 1$). The physical significance of these and other relevant issues dealing with DOP will become more apparent when we discuss (in the following section) the relationship between the coherency matrix elements and the experimentally measurable Stokes polarization parameters. In passing, we note that for polarization preserving (nondepolarizing) interactions, the transformation of $\phi(\phi \rightarrow \phi')$ (the changes in polarization state of light represented by coherency matrix transformation) can be represented by the action of the Jones matrix J as $\phi' = J\phi J^{\dagger}$, implying Jones systems map pure states ($det\phi = 0$) into pure states. Depolarizing transformation involving mixed (partial) polarization states ($det\phi > 0$), on the other hand, is handled by Stokes-Mueller formalism, as discussed subsequently in Section 6.2.3.

6.1.4 Stokes parameters: Intensity-based representation of polarization states

Following the presentation above, polarized states are not characterized in terms of well-determined field amplitudes, but rather by intensities (the time average of the square of the field amplitudes). These measurable intensities are grouped in a 4×1 vector (four row, single-column array) known as the Stokes vector **S**, which is sufficient to characterize any polarization state of light (pure, partial or unpolarized). These are defined as [30, 34, 35, 36]

$$\mathbf{S} = \begin{bmatrix} I \\ Q \\ U \\ V \end{bmatrix} = \begin{bmatrix} \langle E_x E_x^* \rangle + \langle E_y E_y^* \rangle \\ \langle E_x E_x^* \rangle - \langle E_y E_y^* \rangle \\ \langle E_x E_y^* \rangle + \langle E_y E_x^* \rangle \\ i\left(\langle E_y E_x^* \rangle - \langle E_x E_y^* \rangle\right) \end{bmatrix} = \begin{bmatrix} \langle E_{0x}^2 + E_{0y}^2 \rangle \\ \langle E_{0x}^2 - E_{0y}^2 \rangle \\ \langle 2E_{0x}E_{0y}cos\delta \rangle \\ \langle 2E_{0x}E_{0y}sin\delta \rangle \end{bmatrix}, \tag{6.12}$$

where once again, $\langle \cdots \rangle$ denotes temporal average and the electric field components (E_{0x} and E_{0y}) and the corresponding phase difference $\delta(= \delta_y - \delta_x)$ are also temporally averaged over the measurement time. As apparent, these four Stokes parameters are real experimental quantities (intensities) typically measured with conventional square-law photo-detectors, usually in energy-like dimensions. I is the total detected light intensity that corresponds to the addition of the two orthogonal component intensities; Q is the difference in intensity between horizontal and vertical polarization states; U is the difference between the intensities of linear $+45°$ and $+45°(135°)$ polarization states; and V is the difference between intensities of right circular and left circular polarization states (note that if a difference was replaced by a sum in any of these pairs, total intensity I would result). Thus these parameters can be directly determined by the following six intensity measurements (I) performed with ideal polarizers: I_H, horizontal linear polarizer ($0°$); I_V, vertical linear polarizer ($90°$); I_P, $45°$ linear polarizer; I_M, $135°$ ($-45°$) linear polarizer; I_R, right circular polarizer, and I_L, left circular polarizer.

$$\mathbf{S} = \begin{bmatrix} I \\ Q \\ U \\ V \end{bmatrix} = \begin{bmatrix} I_H + I_V \\ I_H - I_V \\ I_P + I_M \\ I_R - I_L \end{bmatrix}. \tag{6.13}$$

TABLE 6.2: Normalized Stokes vectors for usual totally polarized states (of Table 6.1)

State	H	V	P	M	R	L	Elliptical
S	$\begin{bmatrix}1\\1\\0\\0\end{bmatrix}$	$\begin{bmatrix}1\\-1\\0\\0\end{bmatrix}$	$\begin{bmatrix}1\\0\\1\\0\end{bmatrix}$	$\begin{bmatrix}1\\0\\-1\\0\end{bmatrix}$	$\begin{bmatrix}1\\0\\0\\1\end{bmatrix}$	$\begin{bmatrix}1\\0\\0\\-1\end{bmatrix}$	$\begin{bmatrix}1\\ \cos 2\alpha \cos 2\epsilon\\ \sin 2\alpha \cos 2\epsilon\\ -\sin 2\epsilon\end{bmatrix}$

Also note that **S** is not a vector in the geometric space; rather, this array of intensity values represent a directional vector in the polarization state space (the Poincaré sphere, described subsequently). For totally polarized states defined by Jones vectors of the form given by Eq. (6.4), the corresponding Stokes vectors are

$$\mathbf{S} = \begin{pmatrix} E_{0x}^2 + E_{0y}^2 \\ E_{0x}^2 - E_{0y}^2 \\ 2E_{0x}E_{0y}\cos\delta \\ 2E_{0x}E_{0y}\sin\delta \end{pmatrix}. \tag{6.14}$$

Usually, Stokes vectors are represented in intensity normalized form (normalized by the first element I). The normalized Stokes vectors for fully polarized states are listed in Table 6.2. At the other extreme, for *totally unpolarized states*, $Q = U = V = 0$, which corresponds to the fact that no matter how the analyzer is oriented, for such states the transmitted intensity is always the same, equal to one-half of the total intensity.

Using this formalism, the following polarization parameters of any light beam are defined:

- net degree of polarization

$$DOP = \frac{\sqrt{(Q^2 + U^2 + V^2)}}{I}, \tag{6.15}$$

- degree of linear polarization

$$DOP = \frac{\sqrt{(Q^2 + U^2)}}{I}, \tag{6.16}$$

- degree of circular polarization

$$DOP = \frac{V}{I}. \tag{6.17}$$

Note that the degree of polarization of light should not exceed unity. This therefore imposes the following restriction on the Stokes parameters,

$$I \geq \sqrt{Q^2 + U^2 + V^2}, \tag{6.18}$$

where the equality and the inequality signs correspond to completely and partially polarized states, respectively. It is worth noting that this restriction of DOP actually originates from the non-negativity condition of the coherency matrix (Eq. (6.10)), which implies the physical realizability of any polarization state. Moreover, the definition of DOP (Eq. (6.15)) is also commensurate with the definition based on the coherency matrix elements (Eq. (6.11)). In order to understand this, we relate the Stokes vector elements with the elements of the coherency matrix (Eq. (6.9)) as

$$\mathbf{S} = \begin{bmatrix} I \\ Q \\ U \\ V \end{bmatrix} = \begin{bmatrix} \langle E_x E_x^* \rangle + \langle E_y E_y^* \rangle \\ \langle E_x E_x^* \rangle - \langle E_y E_y^* \rangle \\ \langle E_x E_y^* \rangle + \langle E_y E_x^* \rangle \\ i \left(\langle E_y E_x^* \rangle - \langle E_x E_y^* \rangle \right) \end{bmatrix} = \begin{bmatrix} \phi_{xx} + \phi_{yy} \\ \phi_{xx} - \phi_{yy} \\ \phi_{xy} + \phi_{yx} \\ i(\phi_{yx} - \phi_{xy}) \end{bmatrix}. \tag{6.19}$$

In the literature, the coherency matrix is also sometimes written as a 4×1 vector (four row, single-column array, analogous to the Stokes vector) and is denoted as the coherency vector $\mathbf{L}$. Thus, $\mathbf{S}$ and $\mathbf{L}$ are related by the 4×4 matrix A as

$$\mathbf{S} = A \begin{bmatrix} \phi_{xx} \\ \phi_{xy} \\ \phi_{yx} \\ \phi_{yy} \end{bmatrix} = A\mathbf{L}, \quad A = \begin{pmatrix} 1 & 0 & 0 & 1 \\ 1 & 0 & 0 & -1 \\ 0 & 1 & 1 & 0 \\ 0 & -i & i & 0 \end{pmatrix}. \tag{6.20}$$

We are now in a position to inspect the non-negativity condition of 2×2 coherency matrix and the corresponding definition of DOP via the coherency matrix elements. By performing simple algebraic manipulations using Eq. (6.20), we can see that the determinant of the coherency matrix can be written in terms of the Stokes parameters as

$$det\phi = \frac{1}{4} \left[I^2 - (Q^2 + U^2 + V^2) \right]. \tag{6.21}$$

Thus, the non-negativity condition of coherency matrix is equivalent to the condition that the DOP should not exceed unity (Eq. (6.18)). Moreover, the quantity $(Q^2 + U^2 + V^2)^{1/2}$ signifies polarized component of the detected intensity (as each of the quantities, Q, U and V are differences in intensities between orthogonal polarizations). Thus, either in Eq. (6.11) or in Eq. (6.15), the degree of polarization is defined as the ratio of the polarized component of the detected intensity (I_{pol}) to the total detected intensity (I_{tot}). It thus follows, as we noted earlier, the determinant of the coherency matrix (the quantity $det\phi$) is an absolute measure of the unpolarized intensity of any partially polarized light:

$$det\phi = \frac{1}{4} \left[I_{tot}^2 - I_{pol}^2 \right]. \tag{6.22}$$

6.1.5 The Poincaré sphere representation of Stokes polarization parameters

The Poincaré sphere is a very convenient geometrical representation of all possible polarization states. The intensity-normalized Stokes parameters (q, u and v) are used as coordinate axes to form the Poincaré sphere. The intensity-normalized form of Stokes vector is [30]

$$\mathbf{S}^T = I\left(1, \frac{Q}{I}, \frac{U}{I}, \frac{V}{I}\right) = I(1, q, u, v) = I(1, \mathbf{s}^T). \tag{6.23}$$

This is illustrated in Fig. 6.4. In this space, the DOP is nothing else but the distance of the representative point from origin. Thus the physical realizability condition given by Eq. (6.18) implies that all acceptable Stokes vectors are represented by points located within the unit radius sphere, the *Poincaré sphere.* Totally polarized states are found at the surface of the sphere (point A) while partially polarized states are inside (point B). The other spherical coordinates, the points 'latitude' and 'longitude,' are nothing else but twice the azimuth α and ellipticity ϵ, as shown by the last column of Table 6.2 for totally polarized states. It is clear that in this geometric representation, the equatorial circle of the sphere represents the set of linear polarization states (with zero ellipticity); the poles are the points of ellipticity ∓ 1 representing right (north pole) and left (south pole) circular polarization states, respectively; the north hemisphere and the south hemisphere correspond to right-handed and left-handed elliptical polarizations, respectively. This geometrical representation provides simple and intuitive descriptions of many aspects of the interactions between polarized light and samples and/or instruments. As we shall discuss subsequently, any type of polarization transformation introduced by interaction with a medium can be conveniently described by a characteristic trajectory in the Poincaré sphere.

6.1.6 Decomposition of mixed polarization states

The Stokes formalism enables us to express the incoherent superposition of two light waves. The Stokes vector $\mathbf{S}$ of a partially polarized wave can be decomposed into a completely polarized part and an unpolarized part; this type of decomposition is unique [30, 37].

$$\mathbf{S} = \begin{bmatrix} I \\ Q \\ U \\ V \end{bmatrix} = \begin{bmatrix} I - \sqrt{Q^2 + U^2 + V^2} \\ 0 \\ 0 \\ 0 \end{bmatrix} + \begin{bmatrix} \sqrt{Q^2 + U^2 + V^2} \\ Q \\ U \\ V \end{bmatrix} \tag{6.24}$$

(Partially polarized wave = Unpolarized wave + Completely polarized wave.) While the above decomposition of the Stokes vector in Eq. (6.24) is relatively easy to implement, analogous decomposition of the coherency matrix is much more important in conceptual and practical grounds, as we discuss here. The

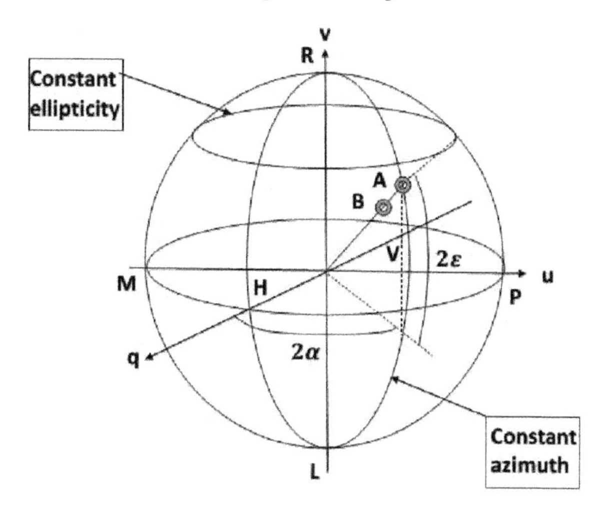

FIGURE 6.4: Geometrical representation of Stokes vectors within the Poincaré sphere. Any given polarization state is represented by a point whose Cartesian coordinates are the intensity-normalized coordinates (q, u, v). The radial coordinate is the *DOP*, and the 'longitude' and 'latitude' are, respectively, 2α and 2ϵ. Totally polarized states are found at the surface of the unit radius sphere, while partially polarized states are inside (e.g., points A and B, respectively). Linearly polarized states, among which are the H, V, P and M states, are on the 'equator' while the L and R circular states are found at the 'poles.'

coherency matrix of a partially polarized wave can indeed be decomposed into incoherent superposition of two independent completely polarized waves. Here, we briefly outline the steps of such a decomposition, which leads to the useful concept of polarization entropy related to the degree of polarization of a wave. Since the coherency matrix is Hermitian by construction, a unitary matrix can always be found permitting its diagonalization. The diagonalized coherency matrix can be written in the form

$$\phi = \begin{bmatrix} \lambda_1 & 0 \\ 0 & \lambda_2 \end{bmatrix}. \tag{6.25}$$

The two eigenvalues, λ_1 and λ_2, of the diagonalized coherency can be obtained from the solution of the characteristic equation as

$$\lambda_1 = \frac{1}{2}\mathrm{tr}\phi\left[1 + \left[1 - \frac{4\det(\phi)}{[Tr\phi]^2}\right]^{1/2}\right] \quad \lambda_2 = \frac{1}{2}\mathrm{tr}\phi\left[1 - \left[1 - \frac{4\det(\phi)}{[Tr\phi]^2}\right]^{1/2}\right]. \tag{6.26}$$

It is clear that the two constituent completely polarized waves of the coherency matrix are polarized in the direction corresponding to the eigenvectors associated to λ_1 and λ_2, respectively, and the resulting decomposition can be written as

$$\frac{1}{\lambda_1+\lambda_2}\phi = \frac{1}{\lambda_1+\lambda_2}\begin{bmatrix} \lambda_1 & 0 \\ 0 & 0 \end{bmatrix} + \frac{1}{\lambda_1+\lambda_2}\begin{bmatrix} 0 & 0 \\ 0 & \lambda_2 \end{bmatrix}. \tag{6.27}$$

Eq. (6.27) leads to the probabilistic interpretation of λ_1 and λ_2, which consequently leads to the concept of entropy in the study of partial polarization. Note that entropy ζ of a given system describes the degree of disorder of the system and is usually defined as

$$\zeta = \sum_{i=1}^{N} p_i \log p_i = \frac{\lambda_i}{\sum_{i=1}^{N} \lambda_j}. \tag{6.28}$$

Considering the condition $(\lambda_1+\lambda_2) = constant = C$, we can obtain conditions for minimum and maximum values for entropy ζ with Eq. (6.29):

$$\begin{aligned} \zeta_{min} =& 0, \text{ when } \lambda_j = C, \ \lambda_i = 0 \ (i \neq j; \ i,j = 1,2), \\ \zeta_{max} =& 1, \text{ when } \lambda_j = C/2, \text{ for any } i = 1, \ 2. \end{aligned} \tag{6.29}$$

The minimum and maximum values for entropy $\zeta_{min} = 0$ and $\zeta_{max} = 1$ correspond to completely polarized waves (having a single eigenvalue of the diagonalized coherency matrix) and completely unpolarized waves (having two equal eigenvalues), respectively. The concept of entropy is thus directly related to the degree of polarization of the wave. For any completely polarized input wave, any depolarizing interactions with the medium thus leads to entropy production, and accordingly, the depolarization property of the medium can be quantified by the entropy of the output polarization state (which can be obtained from the decomposition of the coherency matrix above).

Having used the Jones vector, coherency matrix and Stokes vector formalism to describe the state of polarization of light waves, we now turn to the more interesting problem of transformation (or even loss) of polarization of light waves in their interaction with any material medium. In the following section, we define the various medium polarimetry characteristics and their mathematical representation through Jones matrix and Mueller matrix formalisms.

6.2 Interaction of polarized light with material media

As we discussed in Section 6.1.1 (in the context of Jones vector representations of pure polarization states of light), a vectorial description of the polarization state enables the matrix treatment to describe the polarizing transfer of light in its interaction with any medium. This is true for either the Jones vector (representing completely polarized states) or the Stokes vector (representing both complete and partial polarization states). As it is clear now, the interactions that only transform a pure polarization state into another pure polarization state (polarization-preserving interactions keeping the degree of polarization unity) can be tackled using the Jones formalism [33]. In contrast, the Stokes-Mueller formalism can deal with both the polarization-preserving and depolarizing interactions (which lead to loss of polarization, leading to reduction in DOP) [34, 38]. In this section, we shall address both these formalisms. First, we shall introduce the fundamental medium polarimetric characteristics, and then these medium polarization properties will be represented via their characteristic transformation matrices, namely, the Jones matrix (2×2 field transformation matrix) and the Mueller matrix (4×4 intensity-based Stokes vector transformation matrix). On the way, we shall establish useful relationships between the Jones and Mueller matrices (for polarization-preserving interactions).

6.2.1 Basic medium polarimetry characteristics

The three basic medium polarization properties are *retardance*, *diattenuation* and *depolarization*. The first two effects represent polarization-preserving interaction and can accordingly be modeled using both Jones and Stokes-Mueller formalisms. The third one, on the other hand, leads to loss of polarization and thus cannot be handled using Jones formalism.

The two polarization effects *retardance* and *diattenuation* arise from differences in the refractive indices for different polarization states, and they are often described in terms of ordinary and extraordinary axes and indices. Differences in the real parts of refractive indices result in linear and circular birefringence (retardance), whereas differences in the imaginary parts can cause linear and circular dichroism (which manifests itself as diattenuation, described below) [30, 31, 32]. Mathematically, retardance and birefringence are related simply via $R = k.L.\Delta n$, where R is the retardance, k is the wave vector of the light, L is the pathlength in the medium and Δn is the difference in the real parts of the refractive index known as birefringence. *Linear retardance*, denoted δ, is therefore the relative phase shift between orthogonal linear polarization components (between vertical and horizontal, or between $+45°$ and $-45°$) upon propagation through any medium. The different types of wave plates (half-wave plate, quarter-wave plate, etc.) made of anisotropic materials are examples of perfect linear retarders. Usually, a linear retarder converts input linearly polarized light into elliptically polarized light by introducing phase difference between orthogonal linear polarization components. The output state of polarization depends upon the magnitude of linear retardance and orientation angle (θ) of the principal axis of the retarder with respect to the input linear polarization direction. Analogously, *circular retardance* (δ_C) arises from phase differences between right circularly polarized

(RCP) and left circularly polarized (LCP) states. Such effects are usually introduced by asymmetric chiral structures, and they are manifested as rotation of input linear polarization ($\delta_C = 2 \times$ optical rotation ψ).

The *diattenuation (D)* of an optical element is a measure of the differential attenuation of orthogonal polarization states for both linear and circular polarization. This is analogous to dichroism, which is the differential absorption of two orthogonal polarization states (linear or circular); however, diattenuation is more general, since the differential attenuation need not be caused by absorption alone, rather, it can be the result of various other effects (e.g., scattering, reflection, refraction, etc.). *Linear diattenuation* is defined as differential attenuation of two orthogonal linear polarization states and *circular diattenuation* is defined as differential attenuation of RCP and LCP states. Like linear retardance, polarization transformation by a linear diattenuator also depends upon the magnitude of diattenuation and the orientation angle (θ) of the principal axis of the diattenuator. The simplest form of a diattenuator is the ideal polarizer that transforms incident unpolarized light to completely polarized light ($D = 1$ for ideal polarizer), although often with a significant reduction in the overall intensity.

If an incident state is completely polarized and the exiting state after interaction with the sample has a degree of polarization less than unity, then the sample possesses the *depolarization* property. Depolarization is usually encountered due to multiple scattering of photons (although randomly oriented uniaxial birefringent domains can also depolarize light); incoherent addition of amplitudes and phases of the scattered field results in scrambling of the output polarization state.

6.2.2 Relationship between Jones and Mueller matrices

In Section 1.1, we defined the 2×2 Jones (Eq. (6.6)) to represent the polarization transfer function of a medium in its interaction with completely polarized light. The Jones matrix J is generally complex and contains eight independent parameters (real and imaginary parts of each of the four matrix elements), or seven parameters if the absolute phase is excluded. The polarizing interactions of any medium are contained in the elements of this matrix J; the medium polarization characteristics associated with alterations of relative amplitudes and phases (of orthogonal polarization states) are encoded in the real and imaginary parts of the elements, respectively. As we noted, matrix algebra enables us to compute the Jones matrix of an optical system formed by a series of elements through sequential multiplication of the individual matrices of these elements. Moreover, rotation (by an angle α) of any optical element can also be conveniently modeled by the rotational transformation of Jones matrices ($J \to J'$) via the usual coordinate rotation matrix $R(\alpha)$:

$$R(\alpha) = \begin{bmatrix} \cos(\alpha) & \sin(\alpha) \\ -\sin(\alpha) & \cos(\alpha) \end{bmatrix}, \qquad J' = R^{-1}(\alpha) J R(\alpha). \tag{6.30}$$

Analogous to the Jones matrix, a 4×4 matrix M known as the Mueller matrix (developed by Hans Mueller in the 1940s) describes the transformation of the Stokes vector (polarization state) in its interaction with a medium [36, 37, 39]:

$$\mathbf{S_o} = M\mathbf{S_i}, \tag{6.31}$$

$$\begin{bmatrix} I_o \\ Q_o \\ U_o \\ V_o \end{bmatrix} = \begin{bmatrix} m_{11} & m_{12} & m_{13} & m_{14} \\ m_{21} & m_{22} & m_{23} & m_{24} \\ m_{31} & m_{32} & m_{33} & m_{34} \\ m_{41} & m_{42} & m_{43} & m_{44} \end{bmatrix} \begin{bmatrix} I_i \\ Q_i \\ U_i \\ V_i \end{bmatrix},$$

with $\mathbf{S_i}$ and $\mathbf{S_o}$ being the Stokes vectors of the input and the output light, respectively. The 4×4 real Mueller matrix M has at most sixteen independent parameters (or fifteen if the absolute intensity is excluded), including depolarization information. All the medium polarization properties are encoded in the various elements of the Mueller matrix, which can thus be thought of as the complete optical polarization fingerprint of a sample. Similar to the Jones formalism, matrix properties allow us to determine the resultant Mueller matrix equivalent to a system formed by a series of optical elements through sequential multiplication of the individual Mueller matrices of the elements.

The fundamental requirement real Mueller matrices must meet is that they map physical incident Stokes vectors into physically realizable resultant Stokes vectors (satisfying Eq. (6.18)). Similarly, a Mueller matrix cannot output a state with negative flux. In fact, conditions for physical realizability of Mueller matrices have been studied extensively in the literature, and many necessary conditions have been derived [39, 40, 41, 42, 43]; this is outside the scope of this book. We note below the other important necessary condition for physical realizability of a Mueller matrix (Eq. (6.32)), and we refer the reader to references [30, 39, 40, 41, 42, 43] for a more detailed account of the necessary and sufficient conditions that any 4×4 real matrix should satisfy to qualify as a Mueller matrix of any physical system.

$$tr(MM^T) = \sum_{i,j=1}^{4} m_{ij} \leq 4m_{11}^2, \tag{6.32}$$

where M^T is the transpose of matrix M and the indices $i, j = 1, 2, 3, 4$ denote its rows and columns, respectively. Here the equality and the inequality signs correspond to nondepolarizing and depolarizing systems, respectively.

Relationships between the Jones formalism and the Stokes-Mueller formalism are worth a brief mention here. For the special case of a nondepolarizing linear optical system (a deterministic system, satisfying the equality in Eqs. (6.10) and (6.32)), a one-to-one correspondence between the real 4×4 Mueller matrix M and the complex 2×2 Jones matrix J can be derived via the coherency matrix formalism. Such a relationship can be obtained by using the following set of equations describing the transformation of the input Jones vector ($\mathbf{E_i}$), coherency vector ($\mathbf{L_i}$, defined in Eq. (6.20)) and Stokes vector ($\mathbf{S_i}$).

$$\mathbf{E_0} = J\mathbf{E_i}, \qquad \mathbf{L_0} = W\mathbf{L_i}, \qquad \mathbf{S_0} = M\mathbf{S_i}. \tag{6.33}$$

Here, $\mathbf{E_o}$, $\mathbf{L_o}$ and $\mathbf{S_o}$ are the output Jones, coherency and Stokes vectors, respectively, after medium interaction. The Jones and Mueller matrices J and M have been defined earlier. The matrix W is a 4×4 matrix that describes the transformation of the coherency vector in its interaction with the medium and is known as the Wolf matrix. Using Eqs. (6.9) and (6.20) and by performing simple algebraic manipulations, we can show that the Wolf matrix W and the Mueller matrix M are related to the Jones matrix J as

$$W = J \otimes J^*, \quad M = A \cdot (J \otimes J^*) \cdot A^{-1}. \tag{6.34}$$

Here, A is the 4×4 matrix defined in Eq. (6.20), relating the Stokes vector and the coherency vector.

Thus every Jones matrix (that can only describe a special case of a nondepolarizing optical system) can be transformed into an equivalent Mueller matrix (and a Wolf matrix); however, the converse is not necessarily true. The resulting nondepolarizing Mueller matrix contains seven independent parameters and is accordingly termed a Mueller-Jones matrix. The examples of such Mueller-Jones matrices are the matrices for retardance (both linear and circular) and diattenuation (linear and circular) effects. We show below an interesting example of transforming the Jones matrix to the Mueller-Jones matrix. Consider the rotational transformation of Jones matrices ($J \rightarrow J'$) via the usual coordinate rotation matrix $R(\alpha)$ (Eq. (6.30)). Apparently, $R(\alpha)$ represents coordinate rotation of the electric field vector. This warrants that analogous rotational transformation should also exist for Mueller-Jones matrices. Employing Eq. (6.34), we can determine the analogous rotational transformation of the Mueller-Jones matrix ($M \rightarrow M'$) as

$$M' = T^{-1}(\alpha) M T(\alpha), \qquad T(\alpha) = \begin{pmatrix} 1 & 0 & 0 & 0 \\ 0 & \cos 2\alpha & \sin 2\alpha & 0 \\ 0 & -\sin 2\alpha & \cos 2\alpha & 0 \\ 0 & 0 & 0 & 1 \end{pmatrix}, \tag{6.35}$$

where the rotation matrix $T(\alpha)$ implies rotation of the Stokes vector in the polarization state space (i.e., in the Poincaré sphere; see Fig. 6.4) rather than in the coordinate space. This also implies that a rotation of the field vector by an angle α leads to a rotation of 2α of the Stokes vector (around the v-axis describing circular polarization) in the Poincaré sphere.

We conclude this section by once again noting, while both the Jones and the Stokes-Mueller formalisms describe polarization change using matrix/vector equations, the latter provides a framework with which partial polarization states can be handled and depolarizing materials can be described. Since in nature, light is often partially polarized (or unpolarized) and in most practical situations, loss of polarization is unavoidable, the Stokes-Mueller formalism has been used in most practical polarimetry applications. In contrast, the use of the Jones formalism has been limited as a complementary theoretical approach to the Mueller matrix calculus, or to studies in clear media, specular reflections and thin films where polarization loss is not an issue.

6.2.3 Jones matrices for nondepolarizing interactions: Examples and parametric representation

We now provide explicit expressions for the Jones matrices corresponding to the two polarization preserving effects, *retardance (linear and circular)* and *diattenuation (linear and circular)*, and we briefly discuss the resulting effect on the state of polarization introduced by these transformations.

Retardance (birefringence): Linear retardance originates from the difference in the real part of the refractive index between two orthogonal linear polarization states and accordingly leads to a difference in phase between these states while propagating through an 'anisotropic' medium exhibiting this effect. The Jones matrix for this effect can be written as [28, 30, 33];

$$J_{LR} = \begin{bmatrix} e^{i\phi_x} & 0 \\ 0 & e^{i\phi_y} \end{bmatrix}. \tag{6.36}$$

Here, ϕ_x and ϕ_y are the respective phases of the two orthogonal linear polarization states (x- and y-polarized, respectively; corresponding Jones vectors are noted as H and V states in Table 6.1). The resulting magnitude of linear retardance is

$$\delta = \frac{2\pi}{\lambda}(n_y - n_x)L,$$

where n_x and n_y are the real part of the refractive indices for x- and y-polarized light, respectively; L is the pathlength. Note that this diagonal form of the Jones matrix (J_{LR} in Eq. (6.36)) is obtained for an anisotropic medium whose principal axis is oriented along the laboratory x/y direction. In general, J_{LR} may have off-diagonal elements based on the orientation angle (θ) of the principal axis with respect to the laboratory x-/y-axes. As noted in Eq. (6.30), the general form of the Jones matrix for the arbitrary orientation angle θ can be obtained using rotational transformation as

$$\begin{aligned} J_{LR}(\delta,\theta) &= R^{-1}(\theta)J_{LR}R(\theta) \qquad (6.37)\\ &= \begin{pmatrix} e^{i\phi_x}\cos^2\theta + e^{i\phi_y}\sin^2\theta & (e^{i\phi_x} - e^{i\phi_y})\cos\theta\sin\theta \\ (e^{i\phi_x} - e^{i\phi_y})\cos\theta\sin\theta & e^{i\phi_x}\sin^2\theta + e^{i\phi_y}\cos^2\theta \end{pmatrix}, \end{aligned}$$

where $R(\theta)$ is the rotation matrix of Eq. (6.30).

For example, the Jones matrix of a quarter-waveplate ($\delta = \pi/2$) with its principal axis aligned along the laboratory x-axis ($\theta = 0°$) is

$$\begin{pmatrix} 1 & 0 \\ 0 & i \end{pmatrix}.$$

The state of polarization of light emerging from such a medium exhibiting linear birefringence obviously depends upon the input polarization state, the magnitude of retardance δ and orientation angle of the principal axis θ. For example, if input $+45°$ linearly polarized light characterized by Jones vector $\frac{1}{\sqrt{2}}\begin{bmatrix} 1 & 1 \end{bmatrix}^T$) is incident on the above quarter-waveplate, the output polarization state will be left circularly polarized (LCP) with Jones vector $\frac{1}{\sqrt{2}}\begin{bmatrix} 1 & i \end{bmatrix}^T$).

Analogously, circular retardance (δ_C) originates from the difference in the real part of the refractive index ($n_L - n_R$) between two orthogonal circular polarization states (LCP/RCP) and is manifested as the rotation of the plane of polarization (optical rotation $\psi; \delta_C = 2\psi$):

$$\delta_C = \frac{2\pi}{\lambda}(n_L - n_R)L.$$

The Jones matrix corresponding to this effect is a pure rotation matrix:

$$J_{CR}(\psi) = \begin{bmatrix} \cos(\psi) & \sin(\psi) \\ -\sin\psi & \cos(\psi) \end{bmatrix}. \tag{6.38}$$

Diattenuation (dichroism): As previously mentioned diattenuation arises due to differential attenuation of orthogonal polarization states (for both linear and circular) and originates from the differences in the imaginary part of the refractive index for orthogonal polarization states. The Jones matrix for linear diattenuation effect can be written as [28, 30, 33]

$$J_{LD} = \frac{1}{\sqrt{a^2 + b^2}} \begin{bmatrix} a & 0 \\ 0 & b \end{bmatrix}. \tag{6.39}$$

Here, a and b are real numbers because they are related to the amplitudes of the two orthogonal linear polarization states (x- and y-polarized, respectively). The magnitude of linear diattenuation ($-1 \leq D \leq +1$) can be written as

$$D = \frac{a^2 - b^2}{a^2 + b^2}.$$

Note that like the linear retarder, the Jones matrix of a linear diattenuator also depends upon the magnitude of diattenuation D and the orientation angle (θ) of the principal axis of the diattenuator, and the general form of the diattenuator oriented at an angle θ can be obtained as

$$J_{LD}(d, \theta) = R^{-1}(\theta) J_{LD} R(\theta). \tag{6.40}$$

An example of a perfect diattenuator matrix is that of a linear polarizer (magnitude of diattenuation $D = \pm 1$):

$$J_{LD}(D = \pm 1, \theta) = \begin{pmatrix} \cos^2\theta & \sin\theta\cos\theta \\ \sin\theta\cos\theta & \sin^2\theta \end{pmatrix} \Rightarrow \begin{pmatrix} 1 & 0 \\ 0 & 0 \end{pmatrix}. \quad (6.41)$$

Apparently, light emerging from a perfect linear diattenuator is always linearly polarized along the direction of the principal axis of the diattenuator, irrespective of the polarization state of the incident light.

6.2.4 Standard Mueller matrices for basic interactions (diattenuation, retardance, depolarization): Examples and parametric representation

Having defined the Jones matrices for the various polarization-preserving interactions, we now turn to the corresponding representation using Mueller matrices. We must note that although both the Jones and the Stokes-Mueller approaches rely on linear algebra and matrix formalisms, they differ in many aspects. Specifically, the Stokes-Mueller formalism has certain advantages. First of all, it can encompass any polarization state of light, whether it is natural, totally or partially polarized (can thus deal with both polarizing and depolarizing optical systems). Second, the Stokes vectors and Mueller matrices can be measured with relative ease using intensity-measuring conventional (square-law detector) instruments, including most polarimeters, radiometers and spectrometers.

We also note that in the conventional Mueller matrix representation of the retardance and diattenuation effects, the optical elements exhibiting these two effects are often referred to as the *homogeneous retarder* and the *homogenous diattenuator*. In this convention, polarimetric elements are called homogeneous if they exhibit two fully polarized orthogonal eigenstates, i.e., two polarization states that are transmitted without alteration and that do not interfere with each other. In practice, such light states are linearly polarized along two perpendicular directions, or circularly polarized and rotating in opposite senses. The normalized Stokes vectors $\mathbf{S_1}$ and $\mathbf{S_2}$ of such orthogonal states are of the form

$$\mathbf{S_1}^T = (1, \mathbf{s}^T), \mathbf{S_2}^T = (1, -\mathbf{s}^T), \quad (6.42)$$

with $\parallel \mathbf{s} \parallel = 1$ as these states are fully polarized. Orthogonal states are thus found on the surface of the Poincaré sphere at diametrically opposed positions. For any homogeneous polarimetric element, there are thus two (and only two) such states that are left invariant on the Poincaré sphere.

Homogeneous retarders: The elements exhibiting the retardance effects are characterized by two orthogonal eigenpolarization states, each of which is transmitted without modification. The corresponding orthogonal Stokes eigenvectors are of the form given by Eq. (6.42). Homogenous retarders transmit both eigenstates with the same intensity coefficients, but different phases. This phase difference is the scalar retardation δ, as we defined earlier in context with Jones matrix representation. A pure retarder can be described geometrically as rotation in the space of Stokes vectors. Mathematically, the Mueller matrix M_R of the retarder can be written as [44, 45, 46]

$$M_R = \begin{pmatrix} 1 & 0^T \\ 0 & m_R \end{pmatrix}, \tag{6.43}$$

where 0 represents the null vector and the 3×3 submatrix, m_R, is a rotation matrix in the Poincaré (q,u,v) space. The action of a retarder on an arbitrary incident Stokes vector $\mathbf{S}$ is a rotation of its representative point on the Poincaré sphere, described by m_R. Moreover, the axis of this rotation is defined by the two diametrically opposed points representing the two eigenpolarizations, and the rotation angle is the retardation δ.

For linear retarders with eigenstates linearly polarized along θ and $\theta + 90°$ azimuths, the form of the Mueller matrix $(M_{LR}(\tau, \delta, \theta))$ can be obtained by applying the transformation of Eq. (6.34) on the Jones matrix of a retarder (Eq. 6.37) [44, 45, 46]

$$\tau \begin{pmatrix} 1 & 0 & 0 & 0 \\ 0 & \cos^2 2\theta + \sin^2 2\theta \cos\delta & \cos 2\theta \sin 2\theta (1-\cos\delta) & -\sin 2\theta \sin\delta \\ 0 & \cos 2\theta \sin 2\theta (1-\cos\delta) & \sin^2 2\theta + \cos^2 2\theta \cos\delta & \cos 2\theta \sin\delta \\ 0 & \sin 2\theta \sin\delta & -\cos 2\theta \sin\delta & \cos\delta \end{pmatrix}, \tag{6.44}$$

where τ is the intensity transmission for incident unpolarized light, and can be taken to be unity if the optical material is nonabsorbing (lossless).

A straightforward calculation indeed shows that two fully polarized orthogonal eigenstates (linearly polarized states with azimuths θ and $\theta + 90°$) are transmitted unchanged:

$$M_{LR}(\tau, \delta, \theta) \begin{bmatrix} 1 \\ \pm\cos 2\theta \\ \pm\sin 2\theta \\ 0 \end{bmatrix} = \tau \begin{bmatrix} 1 \\ \pm\cos 2\theta \\ \pm\sin 2\theta \\ 0 \end{bmatrix}. \tag{6.45}$$

An example of a Mueller matrix of a homogeneous linear retarder is that of a quarter-wave plate (with $\theta = 0°$)

$$\begin{bmatrix} 1 & 0 & 0 & 0 \\ 0 & 1 & 0 & 0 \\ 0 & 0 & 0 & 1 \\ 0 & 0 & -1 & 0 \end{bmatrix}.$$

As noted before, the Mueller matrix of the quarter wave plate above for its principal axis oriented at any arbitrary angle θ can easily be determined using the rotational transformation of Eq. (6.35) (yielding Eq. (6.44) with $\delta = \pi/2$).

We now consider *circular retarders*, i.e., elements for which the eigenpolarizations are the opposite circular polarization states. The Mueller matrices of such elements are of the form [44, 45, 46]

$$M_{CR}(\psi) = \tau \begin{pmatrix} 1 & 0 & 0 & 0 \\ 0 & \cos 2\psi & \sin 2\psi & 0 \\ 0 & -\sin 2\psi & \cos 2\psi & 0 \\ 0 & 0 & 0 & 1 \end{pmatrix}. \tag{6.46}$$

When a linearly polarized wave interacts with a circular retarder, its polarization remains linear, but it is rotated by an angle ψ (known as *optical rotation*). The effect may also be interpreted as a rotation of the incident linearly polarized Stokes vector $\mathbf{S}$ in the Poincaré sphere by an amount equal to the circular retardance ($\delta_C = 2\psi$).

Finally, we point out that the scalar retardation of any homogeneous retarder (linear or circular) can be determined from the general Mueller matrix M_R of a retarder as

$$\delta, \psi = \cos^{-1}\left(\frac{Tr(M_R)}{2} - 1\right). \tag{6.47}$$

The formula above is valid for media exhibiting both linear retardance δ and circular retardance (optical rotation ψ). We can readily verify this from the Mueller matrix of the combined effect by multiplying individual matrices for the linear and circular retarder (in either order). The total retardance in such case can be obtained employing Eq. (6.47) on the Mueller matrix representing the combined effects.

Homogeneous diattenuators: For a diattenuating system, the output intensity depends on the input polarization state. If we consider an intensity normalized input Stokes vector $\mathbf{S}$ such that

$$\mathbf{S_{in}}^T = (1, \mathbf{s}^T), \qquad \text{with} \qquad \| \mathbf{s} \| = DOP \leq 1, \tag{6.48}$$

which corresponds to arbitrary polarizations at constant intensity (normalized to unity), then the output intensity (i.e., the first component of S_{out}) is simply given by

$$I_{out} = m_{11}(1 + \mathbf{D} \cdot \mathbf{s}). \tag{6.49}$$

This output intensity reaches its maximum (minimum) value I_{max} (I_{min}) when the scalar product $\mathbf{D} \cdot \mathbf{s}$ is maximum (minimum) under the constraint $\| \mathbf{s} \| = DOP \leq 1$, i.e., when $s = \pm \frac{D}{\|D\|}$. We thus obtain:

$$\mathbf{S_{max}}^T = \left(1, \frac{\mathbf{D}^T}{\| \mathbf{D} \|}\right) \text{ and } I_{max} = m_{11}(1 + \| \mathbf{D} \|),$$

$$\mathbf{S_{min}}^T = \left(1, -\frac{\mathbf{D}^T}{\| \mathbf{D} \|}\right) \text{ and } I_{min} = m_{11}(1 - \| \mathbf{D} \|), \tag{6.50}$$

from which we immediately obtain the *scalar diattenuation* D:

$$D = \frac{I_{max} - I_{min}}{I_{max} + I_{min}} = \parallel \mathbf{D} \parallel . \tag{6.51}$$

The diattenuator vector $\mathbf{D}$ thus defines both the scalar diattenuation D and the polarization states transmitted with the highest (or the lowest) intensity. Note that the two polarization states giving these extremal intensity transmission values are totally polarized, and they are located at diametrically opposite positions on the Poincaré sphere.

The elements of the Mueller matrix of a diattenuator are uniquely determined by their diattenuation vector $\mathbf{D}$. Their (totally polarized) eigenpolarization states corresponding, respectively, to maximum and minimum transmissions are given by Eqs. (6.50). The corresponding Mueller matrix is then given by

$$M_D = \tau \begin{pmatrix} 1 & \mathbf{D}^T \\ \mathbf{D} & m_d \end{pmatrix}, \quad \text{where} \quad m_D = \sqrt{1-D^2} I_3 + (1 - \sqrt{1-D^2})\mathbf{D}\mathbf{D}^T . \tag{6.52}$$

Once again, τ is the intensity transmission for incident unpolarized light. Diattenuation may occur due to reflection and/or refraction at an interface, or to propagation in anisotropic or chiral materials. Anisotropy may introduce *linear dichroism.* For any propagation direction, the imaginary part of the wave vector (corresponding to the imaginary part of the refractive index n'') may take two different values, n''_L and n''_H; the former, corresponding to the lowest absorption, is valid for a wave linearly polarized at azimuth θ and the latter for the orthogonal polarization, at $\theta + 90°$. The linear (scalar) dichroism is then defined as

$$\Delta n'' = n''_H - n''_L > 0. \tag{6.53}$$

For a parallel slab of thickness L, the intensity transmissions for the two eigenpolarizations are, respectively,

$$T_{max} = \exp(-2n''_L L), \quad T_{min} = \exp(-2n''_H L), \tag{6.54}$$

yielding a scalar diattenuation

$$D = \frac{T_{max} - T_{min}}{T_{max} + T_{min}} = \sinh(\Delta n'' L), \tag{6.55}$$

where $\Delta n'' L$ is the dichroism integrated over the slab thickness L.

Chiral media (e.g., a biological fluid such as glucose) can also exhibit dichroism, but usually it is *circular dichroism.* The formulas above are still valid, but in this case the eigenpolarizations for which the absorption coefficients are well defined are left and right circular ones.

The Mueller matrix for a linear diattenuator ($M_{LD}(\tau, D, \theta)$) can be written in the following symmetric form [44, 45, 46];

$$\frac{\tau}{2}\begin{bmatrix} 1 & D\cos 2\theta & D\sin 2\theta & 0 \\ D\cos 2\theta & \cos^2 2\theta + \sqrt{1-D^2}\sin^2 2\theta & (1-\sqrt{1-D^2})\cos 2\theta \sin 2\theta & 0 \\ D\sin 2\theta & (1-\sqrt{1-D^2})\cos 2\theta \sin 2\theta & \sin^2 2\theta + \sqrt{1-D^2}\cos^2 2\theta & 0 \\ 0 & 0 & 0 & \sqrt{1-D^2} \end{bmatrix}, \tag{6.56}$$

implying that the maximum and minimum intensity transmittances are obtained for linearly polarized states with azimuths θ and $\theta + 90°$. It can be easily checked that these eigenpolarization states are unchanged by $M_{LD}(\tau, D, \theta)$ but are transmitted with intensity factors $\frac{\tau}{2}(1 \pm D)$.

Similarly, for circular diattenuators, the general form of the Mueller matrix is

$$M_{CD}(\tau, D) = \frac{\tau}{2}\begin{pmatrix} 1 & 0 & 0 & D \\ 0 & \sqrt{1-D^2} & 0 & 0 \\ 0 & 0 & \sqrt{1-D^2} & 0 \\ D & 0 & 0 & 1 \end{pmatrix}, \tag{6.57}$$

and of course in this case there is no need to define any partial azimuth θ.
Depolarizers: A depolarizer is an object that reduces the degree of polarization of the incoming light. The simplest depolarizers are those for which the Mueller matrix M_Δ is diagonal [44, 45, 46];

$$M_\Delta = \tau\begin{pmatrix} 1 & 0 & 0 & 0 \\ 0 & a & 0 & 0 \\ 0 & 0 & b & 0 \\ 0 & 0 & 0 & c \end{pmatrix}, \tag{6.58}$$

with absolute values of a, b and c smaller than unity. If so, any incident Stokes vector S_i of the form

$$\mathbf{S_i}^T = I(1, q, u, v)$$

is transformed into

$$\mathbf{S_{out}}^T = \tau I(1, aq, bu, cv),$$

which gives the output degree of polarization

$$DOP_{out} = \sqrt{a^2q^2 + b^2u^2 + c^2v^2} \leq \sqrt{q^2 + u^2 + v^2} = DOP_{in}. \tag{6.59}$$

In the geometrical representation, the action of a depolarizer defined in Eq. (6.58) is to pull the representative point of the incoming Stokes vector toward the origin. As a result, the Poincaré sphere is transformed into an ellipsoid limited by the segments $[-a, a], [-b, b]$ and $[-c, c]$ along the q, u,v-axes.

As discussed previously, depolarization occurs due to incoherent addition of intensities of polarized states with different polarizations. Depolarization may occur due to multiple scattering in the first place, together with spatially varying, randomly oriented birefringent domains. However, these effects alone are not sufficient to cause real depolarization but would give rise to a speckle pattern with $DOP = 1$ everywhere but with different polarizations from one point to another. True depolarization occurs if this speckle pattern is blurred by the motion of the scattering sample, the lack of spatial coherence of the illumination beam, the sample motion, and the like.

In the most general case, the Mueller matrix of a depolarizer, M_Δ, is given in compact notation as

$$M_\Delta = \begin{pmatrix} 1 & \mathbf{0}^T \\ \mathbf{0} & m_\Delta \end{pmatrix}, \tag{6.60}$$

where m_Δ is a 3×3 real symmetric matrix. This matrix can be diagonalized to recover the form given by Eq. (6.58) where the eigenvalues a, b, c are real numbers varying between -1 and 1. Thus the Mueller matrix of the most general depolarizer depends on six parameters (as can be seen from the very definition of the m_Δ matrix as a 3×3 symmetric matrix, or by the fact that the diagonalization process involves not only the three eigenvalues but also the basis formed by the eigenvectors of m_Δ). General depolarizers are thus rather complex mathematical objects, this complexity being related to situations like multiple scattering in anisotropic media. Here we will not discuss the properties of general depolarizers any further, and in the following section we will only consider depolarizers of the form given by Eq. (6.58). We thus define the depolarizing power of such samples as

$$\Delta = 1 - \frac{1}{3}(\mid a \mid + \mid b \mid + \mid c \mid), \tag{6.61}$$

which can be further separated in depolarizing powers for linear and circular polarizations as

$$\Delta_L = 1 - \frac{1}{2}(\mid a \mid + \mid b \mid) \text{ and } \Delta_C = 1 - \mid c \mid . \tag{6.62}$$

Another widely used approach of defining depolarization property of a sample using Mueller matrix M uses the following definition of the depolarization index P_d:

$$P_d = \sqrt{\frac{\sum_{i=1}^{4}\sum_{j=1}^{4} M_{ij}^2 - M_{11}^2}{3M_{11}^2}} = \sqrt{\frac{Tr(M^T M) - M_{11}^2}{3M_{11}^2}}. \tag{6.63}$$

As a final remark, the depolarization powers Δ defined in Eq. (6.61) vary from zero, for nondepolarizing samples, to one for totally depolarizing ones, while the opposite holds for the general depolarization index P_d defined in Eq. (6.63).

To summarize, thus far in this chapter we have discussed mathematical formalisms to deal with the state of polarization of light and interaction of polarized light with material media. Specifically, the two most widely used formalisms, namely, the Jones calculus and the Stokes-Mueller calculus, have been discussed. As previously noted, the former is a field-based model and is limited to describing pure polarization states (a completely polarized wave) and polarization-preserving (nondepolarizing) interactions only. The latter, on the other hand, is an intensity-based model and is more encompassing in the sense that it provides a framework with which partial polarization states can be handled and depolarizing interactions can also be modeled. Caution must, however, be exercised while implementing these formalisms either for practical purposes of polarimetric measurements or for conceptual reasons for polarimetric modeling. It may be worthwhile to spend a few words on the validity regime of such algebra. Note that both the Jones vector (in the field-based representation) and the Stokes vector (in the intensity-based representation) deal with a two-dimensional electromagnetic field, and are applicable when the light wave is completely transverse in nature (plane electromagnetic waves or more generally to uniformly polarized elementary beams). Note that even for paraxial beam-like fields, the spatial mode (distribution of field) and polarization are not always separable (unlike plane waves or elementary beams) and accordingly, we need different algebra to describe such inhomogeneous polarization. This so-called *classical entanglement* between polarization and spatial mode is handled by defining the beam coherency polarization matrix (a variant of the 2×2 coherency matrix incorporating simultaneously both the field polarization and its spatial distribution) [47]. In other general cases involving three-dimensional fields (as encountered in tight focusing, scattering and the near field), the two-dimensional polarimetry formalisms have been extended via the definition of the 3×3 coherency matrix and the generalized nine-element Stokes vector [48]. Moreover, there is other emerging 'un-conventional' polarization algebra involving vector beams, geometric phases (Pancharatnam-Berry phase) arising from spin-orbit interactions of light, radial and azimuthal polarization of light beams and so forth [49, 50]. Some of these issues related to advanced topics in polarization optics are discussed in Chapter 11. For now, we restrict our discussion to the conventional polarization algebra. In the following, we briefly introduce the concepts of polarimetric measurements (based on conventional polarization algebra), and we touch upon representative applications of experimental polarimetry.

6.3 Experimental polarimetry and representative applications

Polarimeters can be regarded as optical instruments used for the determination of polarization characteristics of light and the sample. Based on this definition, experimental polarimetry systems can be broadly classified into two categories:

1. The light-measuring polarimeters, and
2. The sample-measuring polarimeters.

The light-measuring polarimeters (Stokes polarimeters) determine the polarization state of a light beam by measuring the four Stokes parameters (Stokes vector $\begin{bmatrix} I & Q & U & V \end{bmatrix}^T$). In contrast, the sample measuring polarimeters (Mueller matrix polarimeters) aim to determine the complete 4×4 Mueller matrix of the sample. A variety of experimental schemes have been developed to maximize measurement sensitivity and to measure the Stokes vector of a beam upon interacting with the sample in question, and/or the Mueller matrix of the sample itself. Here, in this section, we briefly discuss some of the common strategies employed in these polarimeters. For a detailed account of these, we refer the reader to the relevant literature available [39, 44, 51, 52, 53, 54, 55].

6.3.1 Stokes vector (light-measuring) polarimeters

As discussed in Section 6.1.4, the Stokes parameters corresponding to a beam of light can be determined by performing six intensity measurements (I) through linear and circular polarizers ($I_H, I_V, I_P, I_M, I_R, I_L$). According to Eq. (6.13), these intensity measurements are I_H, horizontal linear polarizer (0°); I_V, vertical linear polarizer (90°); I_P, 45° linear polarizer; I_M, 135° (−45°) linear polarizer; I_R, right circular polarizer; and I_L, left circular polarizer. A schematic of the experimental setup for this classical method is shown in Fig. 6.5. For Stokes parameters measurement, we are only concerned about the polarization state analyzer (PSA) part of the setup shown in Fig. 6.5. The polarization state generator (PSG) is used to generate one particular incident polarization state (either linear or circular). As shown in the figure, the PSA usually comprises a linear polarizer (P_2) and a wave plate (WP_2) for performing the required six intensity measurements. In this case, the wave plate is a quarter-wave retarder. Note that the four intensity measurements involving linear polarization states (I_H, I_V, I_P, I_M) are performed by removing the quarter wave plate, whereas the quarter-wave plate is inserted for the two remaining intensity measurements involving circular polarization states (I_R, I_L). By exploiting the property that $I_H + I_V = I_P + I_M = I_R + I_L$, it is possible to determine the Stokes parameters of a beam with only four intensity measurements. Briefly, in this approach, a circular polarizer (the PSA here) is designed consisting of a linear polarizer whose transmission axis is set at +45° with respect to the horizontal direction, followed by a quarter-wave plate with its fast axis parallel to the horizontal direction. Three sets of intensity measurements (denoted by $I_{cir}(\alpha)$, where α is the angle of the combined polarizer's fast axis above the horizontal) are performed by varying the angle (α) of the circular polarizer with respect to the horizontal axis at

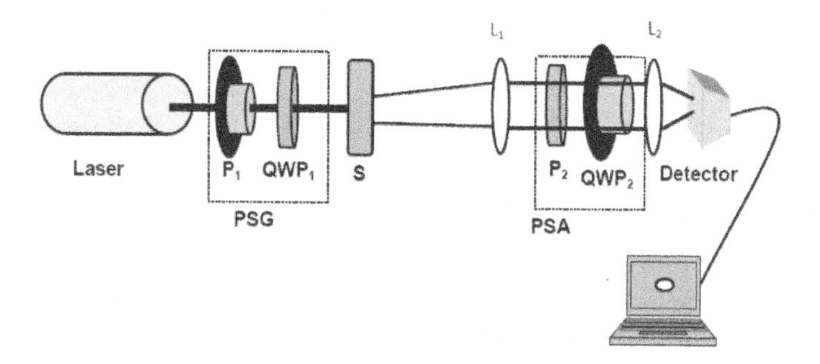

FIGURE 6.5: A schematic of the experimental Stokes polarimeter setup; P_1, P_2 linear polarizers; QWP_1, QWP_2, removable quarter-wave plates; and L_1. L_2 lenses, respectively.

$\alpha = 0°, 45°$and $90°$. The combined polarizer is then flipped to the other side and the final intensity measurement $[I_L(\alpha)]$ is made by setting α to $0°$. The Stokes parameters can be inferred from these intensity measurements by

$$\mathbf{S} = \begin{bmatrix} I \\ Q \\ U \\ V \end{bmatrix} = \begin{bmatrix} I_{cir}(0°) + I_{cir}(90°) \\ I - 2I_{cir}(45°) \\ I_{cir}(0°) - I_{cir}(90°) \\ -I + 2I_{lin}(0°) \end{bmatrix}. \quad (6.64)$$

Although this method has been widely employed to measure Stokes parameters of light transmitted (or reflected) from nonscattering media, for polarimetric measurements in strongly depolarizing scattering media, a more sensitive detection schemes is desirable. This follows because multiple scattering in a turbid medium leads to depolarization of light, creating a large depolarized source of noise that hinders the detection of the small remaining information-carrying polarization signal. One possible method for improving the sensitivity of the measurement procedure is the use of polarization modulation with synchronous detection. Various experimental strategies based on polarization modulation and synchronous detection scheme have therefore been developed. Generally in this approach, the polarization state analyzer contains a polarization modulator, a rapidly changing (with time) polarization element. The output of PSA is thus a rapidly fluctuating intensity (oscillating at a frequency of $\omega_p = 2\pi f_p$, set by the modulator) on which the polarization information is coded. The polarization information is then extracted by synchronously detecting the time-varying signal at the fundamental modulation frequency and its different harmonics. Various types of resonant devices like the electro-optical modulator, the magneto-optical modulator and the photoelastic modulator (PEM) have been employed for polarization modulation. Among these, the PEMs have been the most widely used. The synchronous detection of the modulated signal can be conveniently done by using a lock-in amplifier.

6.3.2 Mueller matrix (sample-measuring) polarimeter

As noted previously, the sample-measuring polarimeters measure the complete 4×4 Mueller matrix of the sample. For Mueller matrix measurements also, both dc measurements (involving sequential measurement) and modulation-based measurement procedures have been employed. The former approach involves sequential measurements with different combinations of source polarizers and detection analyzers. Because a general 4×4 Mueller matrix has sixteen independent elements, at least sixteen measurements are required for the construction of a Mueller matrix. The process for constructing of Mueller matrix from sixteen such combinations of intensity measurements are listed in Table 6.3. In Table 6.3, the first and the second letters (H, V, P, R) in the intensity measurements correspond to incident and detection polarization states, respectively. Note that methods involving a greater number of polarization measurements such as thirty-six polarization measurements and forty-nine polarization measurements have also been explored for the construction of a Mueller matrix. As the number of the measurements increases, the accuracy also increases because in the methods with lesser measurements, error in the measurements of one element of the Mueller matrix propagates to cause further errors in other elements (which are indirectly obtained). In general, with some additional measurements and analysis, the Stokes polarimeter shown in Fig. 6.5 can be used to measure the Mueller matrix of a sample by sequentially cycling the input polarization between four states using the PSG unit (e.g., linear polarization at $0°, 45°, 90°$ and right circular polarization) and by measuring the output Stokes vector for each respective input states using the PSA unit. The elements of the resulting four measured Stokes vectors (sixteen values) can be algebraically manipulated to solve for the sample Mueller matrix

$$M(i,j) = \begin{bmatrix} \frac{1}{2}(I_H + I_V) & \frac{1}{2}(I_H - I_V) & I_P - M(1,1) & I_R - M(1,1) \\ \frac{1}{2}(Q_H + Q_V) & \frac{1}{2}(Q_H - Q_V) & Q_P - M(2,1) & Q_R - M(2,1) \\ \frac{1}{2}(U_H + U_V) & \frac{1}{2}(U_H - U_V) & U_P - M(3,1) & U_R - M(3,1) \\ \frac{1}{2}(V_H + V_V) & \frac{1}{2}(V_H - V_V) & V_P - M(4,1) & V_R - M(4,1) \end{bmatrix}. \tag{6.65}$$

Here, the four input states are denoted with the subscripts $H(0°)$, $P(45°), V(90°)$ and R (right circularly polarized; left circular incident can be used as well, resulting only in a sign change). The indices $i, j = 1, 2, 3, 4$ denote rows and columns, respectively.

Various polarization modulation schemes have also been employed for simultaneous determination of all the sixteen Mueller matrix elements. As for the case of Stokes polarimeters, here also various optical elements like liquid crystal variable retarders, photoelastic modulators (PEM), etc., have been used for modulating either the polarization state of light that is incident on

TABLE 6.3: Construction of the Mueller matrix from sixteen combinations of intensity measurements. The first and second letters in the intensity measurements correspond to incident and detection polarization states, respectively. The different polarization states are H = Horizontal, V = Vertical, P = +45° and R = Right circular polarization.

$M_{11} = HH + HV + VH + VV$	$M_{12} = HH + HV - VH - VV$	$M_{13} = 2PH + 2PV - M_{11}$	$M_{14} = 2RH + 2RV - M_{11}$
$M_{21} = HH - HV + VH - VV$	$M_{22} = HH - HV - VH + VV$	$M_{23} = 2PH - 2PV - M_{21}$	$M_{24} = 2RH + 2RV - M_{21}$
$M_{31} = 2HP + 2VP - M_{11}$	$M_{32} = 2HP - 2VP - M_{21}$	$M_{33} = 4PP - 2PH - 2PV - M_{31}$	$M_{34} = 4RP - 2RH - 2RV - M_{31}$
$M_{41} = 2HR + 2VR - M_{11}$	$M_{42} = 2HR - 2VR - M_{21}$	$M_{43} = 4PR - 2PH - 2PV - M_{41}$	$M_{44} = 4RR - 2RH - 2RV - M_{41}$

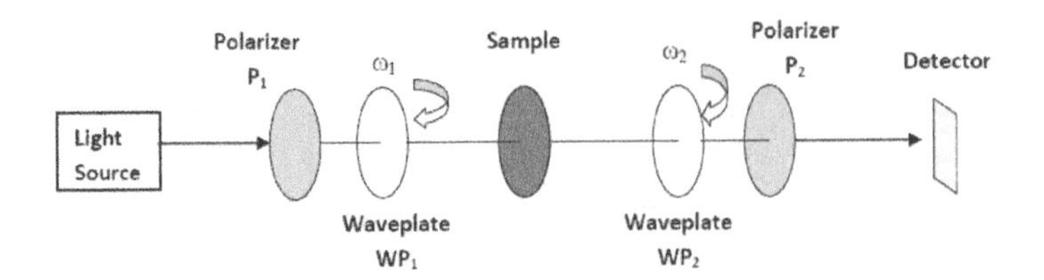

FIGURE 6.6: A schematic of the dual rotating retarder Mueller matrix polarimeter: P_1, P_2 linear polarizers; WP_1, WP_2, rotating wave plates with linear retardations δ_1 and δ_2 and rotation speeds ω_1 and ω_2, respectively.

the sample (by keeping the polarization modulator between the source and the sample) or the sample-emerging light (by placing the polarization modulator between the sample and the detector) or both.

Among the various modulation-based Mueller matrix polarimeters, the dual rotating retarder polarimeter has been the most widely used. A schematic of the setup is shown in Fig. 6.6. In this approach, the polarization of the sample-incident light is modulated by passing a beam through a fixed linear polarizer, followed by a wave plate (with retardation δ_1) rotating at an angular velocity of ω_1. This beam is then incident on the sample, and the resulting light is directed through the analyzing optics, which consist of another rotating wave plate (with retardation δ_2, rotating synchronously at angular velocity ω_2) and a linear polarizer, which is held fixed. The rotation of the two wave plates results in a periodic variation in the measured intensity, which can be analyzed by multiplying the Mueller matrices corresponding to the elements in the optical path (i.e., the polarizers, wave plates and the sample). In the most common configurations, the axes of the polarizers are set parallel, the wave plates are both chosen to have a quarter-wave retardance ($\delta_1 = \delta_2 = \frac{\pi}{2}$) and their angular velocities are set at a $5:1$ ratio ($5\omega_1 = \omega_2$). It has been shown that this ratio of angular velocities allows for the recovery of all sixteen Mueller matrix elements from the amplitudes and phases of the twelve frequencies in the detected intensity signal. In order to compute these elements, the detected signal is Fourier analyzed, and the elements of the Mueller matrix can be inferred from the resulting coefficients. A more generalized version of this measurement scheme could use arbitrary values of retardation and polarizer orientations, as well as a different ratio of angular velocities, in order to determine prioritized Mueller matrix elements with greater precision and/or higher signal-to-noise ratio (SNR).

While the modulation-based approaches yield the desired high sensitivity (capable of detecting very weak polarization retaining signal transmitted/backscattered from depolarizing turbid medium), these are poorly suited for applications involving large-area imaging. Yet, in many applications spatial maps of the polarization parameters (depolarization, diattenuation and retardance) are extremely useful. Therefore, several nonmodulation-based approaches have also been developed for such imaging applications. One such method involves liquid crystal variable retarders, which enables the measurement of Mueller matrices to have high sensitivity and precision. Such a polarimeter is comprised of a polarization state generator (PSG) unit, a polarization state analyzer (PSA) unit and an imaging camera for spatially resolved signal detection. The PSG, which polarizes the light incident on the sample, is composed of a linear polarizer (P_1) and two liquid crystal variable retarders (LC_1 and LC_2), with adjustable retardances of δ_1, δ_2, respectively, whose birefringent axes are aligned at angles θ_1, θ_2, respectively, with the axis of the linear polarizer. The liquid crystal variable retarders are transmissive optical elements whose retardance (birefringence) levels can be electronically controlled by applying appropriate voltages across a liquid crystal cell (uniaxial birefringent layers formed using anisotropic liquid crystal molecules). A schematic of the liquid crystal variable retarder imaging polarimeter is shown in Fig. 6.7.

In a widely used configuration under this scheme, the angles θ_1 and θ_2 are chosen to be 45° and 0°, respectively. The Stokes vector generated from this arrangement can once again be determined by sequential multiplication of Mueller matrices of the polarizer and the LC retarders;

$$\mathbf{S_{in}} = \begin{bmatrix} 1 & \cos\delta_1 & \sin\delta_1 \sin\delta_2 & \sin\delta_1 \cos\delta_2 \end{bmatrix}^T. \tag{6.66}$$

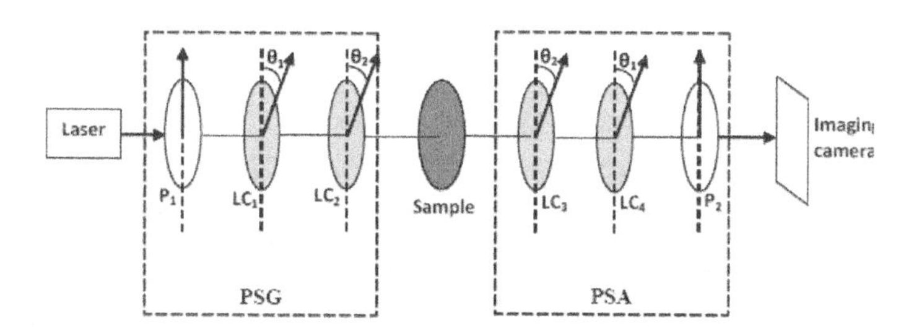

FIGURE 6.7: A liquid crystal variable retarder imaging polarimeter, where P_1 and P_2 are linear polarizers and $LC_1 - LC_4$ are liquid crystal variable retarders. Together, P_1, LC_1 (having a retardance of δ_1 and an orientation angle θ_1) and LC_2 (having a retardance of δ_2 and an orientation angle θ_2) comprise the polarization state generator (PSG). Likewise, LC_3 (having a retardance of δ_2 and an orientation angle θ_2), LC_4 (having a retardance of δ_1 and an orientation angle θ_1) and P_2 collectively form the polarization state analyzer (PSA). While the schematic above depicts transmission measurements, other detection geometries are possible using this measurement scheme.

Clearly, any possible polarization state on the Poincaré sphere can be generated with proper choices of δ_1 and δ_2. The system's PSA consists of a similar arrangement of liquid crystal variable retarders (LC_3 and LC_4) and a linear polarizer but positioned in the reverse order with respect to the incoming light. It is followed by a detector, which, for imaging applications, is a CCD camera.

The PSG unit can be used to generate four unique Stokes vectors, which can be grouped into a 4×4 generator matrix W, where the ith column of W corresponds to the polarization state. Similarly, after sample interactions, the PSA results can be described by a 4×4 analyzer matrix A. The Stokes vectors of the light to be analyzed are projected onto the four basis states, given by the rows of A. The sixteen intensity measurements required for the construction of a full Mueller matrix are grouped into the measurement matrix M_i, which can be related to PSA/PSG matrices and W and A, as well as the sample Mueller matrix M by

$$M_i = AMW. \tag{6.67}$$

The sample Mueller matrix can be represented as a sixteen-element column vector M^{vec}, which can be related to the corresponding 16×1 intensity measurement vector M_i^{vec} as

$$M_i^{vec} = QM^{vec}, \tag{6.68}$$

where Q is the 16×16 matrix given by the Krönecker product of A with the transpose of W:

$$Q = A \otimes W^T. \tag{6.69}$$

Once the exact forms of the system W and A matrices are known, all the sixteen elements of the sample Mueller matrix (M^{vec}, written in column vector form) can be determined from the sixteen measurements (M_i^{vec}) using Eq. (6.68). The 4×4 sample Mueller matrix can then be obtained by rearranging the elements of M^{vec}.

Based on this approach, several measurements schemes are possible. In fact, the choice of the values for retardance δ_1 and δ_2, the orientation angles of the retarders with respect to the polarizers (analyzers), can be optimized to minimize the noise in the resulting Mueller matrix M. Further, this approach also allows for calibration and determination of the exact forms of the system PSG and PSA matrices.

In this section, we briefly reviewed the basic principles and schemes of polarimetric instrumentation, and we outlined the most widely used practical implementations. These may take a variety of forms, depending on the selected optical systems (imaging or nonimaging) and the way the polarization is encoded and detected. There still remain many outstanding challenges in experimental polarimetry. These include development of novel schemes, system optimization, error calibration/analysis for achieving desirable sensitivity and accuracy for performing robust polarimetric measurements in applications involving high numerical aperture polarimetry (quantitative polarization microscopy), spectroscopic and imaging polarimetry, turbid medium polarimetry and so forth. Nevertheless, the availability of so many possibilities (the way the polarization is encoded and detected) is very valuable in practice, as it allows us to 'tailor' the polarimetric system for the specific needs of the envisioned application. Therefore, a considerable amount of current research is directed toward developing/optimizing polarization schemes for specific applications. Interpretation of polarimetric data and development of appropriate inverse analysis methods for the characterization of complex systems exhibiting simultaneous several polarization effects is another area where considerable research is still ongoing. These are, however, beyond the scope of this book. Finally, we conclude this section by noting that polarimetric instruments (either to measure the Stokes vector of a beam upon interacting with the sample in question, and/or the Mueller matrix of the sample itself) have long being pursued for numerous practical applications in various branches of science and technology. Remote sensing in meteorology and astronomy, characterization of thin films, quantification of protein properties in solutions, testing purity of pharmaceutical drugs, optical stress analysis of structures, crystallography of biochemical complexes and biological tissue characterization/diagnosis are just a few examples of their diverse uses.

Chapter 7

Interference and diffraction

Optical interference corresponds to the interaction of two or more electromagnetic waves yielding a resultant irradiance that is different from the sum of the component irradiances. We will divide interferometric devices into two broad categories, namely, (i) wavefront splitting and (ii) amplitude splitting. In the first case, portions of the primary wavefront are used either directly as sources of secondary radiation or by means of other optical elements used to produce virtual sources of secondary radiation. The radiation from the secondary sources are brought together to interfere. In the case of amplitude splitting, the primary wave is divided into two segments, which travel different paths and interfere.

7.1 A general approach to interference

Since electromagnetic waves are essentially represented by vector fields, we need to generalize the superposition principle in the context of scalar waves to vector waves. In this section we consider the superposition of two vector waves in order to bring out the essential prerequisites and effects of interference. In accordance with the superposition principle, the field $\mathbf{E}$ at a point in space arising out of the interaction of two waves $\mathbf{E}_1$ and $\mathbf{E}_2$ is given by

$$\mathbf{E} = \mathbf{E}_1 + \mathbf{E}_2. \tag{7.1}$$

The optical disturbance varies at a very rapid rate, rendering the amplitude an impractical quantity to detect. On the contrary, the irradiance I can be measured by a variety of sensors or detectors. The study of interference is therefore in terms of the experimental detection of the variation of irradiance. For the sake of simplicity, we consider two point sources S_1 and S_2 far away from the point of observation P, such that the wavefronts can be considered to be planar. Considering only linearly polarized light, we can write the following expressions for the two fields:

$$\mathbf{E}_1(\mathbf{r}, t) = \mathbf{E}_{01} \cos(\mathbf{k}_1 \cdot \mathbf{r} - \omega t + \phi_1), \tag{7.2}$$

$$\mathbf{E}_2(\mathbf{r}, t) = \mathbf{E}_{02} \cos(\mathbf{k}_2 \cdot \mathbf{r} - \omega t + \phi_2), \tag{7.3}$$

where ϕ_1 and ϕ_2 are the constant phases. The irradiance at P is given by

$$I = \varepsilon v \langle \mathbf{E}^2 \rangle, \tag{7.4}$$

where the angular brackets imply time averaging. Henceforth we will drop the coefficient in Eq. (7.4) and write irradiance as

$$I = \langle \mathbf{E}^2 \rangle. \tag{7.5}$$

The square of the resultant field can be written as

$$\mathbf{E}^2 = \mathbf{E}_1^2 + \mathbf{E}_2^2 + 2\mathbf{E}_1 \cdot \mathbf{E}_2. \tag{7.6}$$

Thus taking the average we have

$$I = I_1 + I_2 + I_{12}, \tag{7.7}$$

where

$$I = \langle \mathbf{E}^2 \rangle, \ I_{1,2} = \langle \mathbf{E}_{1,2}^2 \rangle \text{ and } I_{12} = 2\langle \mathbf{E}_1 \cdot \mathbf{E}_2 \rangle. \tag{7.8}$$

The last term is known as the interference term. In the absence of the interference term (or in the absence of interference), the resultant intensity is just the sum of the component intensities. Using the expressions of the fields given by Eqs. (7.2) and (7.3), the interference term (without the factor 2) can be written as

$$\begin{aligned}\mathbf{E}_1 \cdot \mathbf{E}_2 &= \mathbf{E}_{01} \cdot \mathbf{E}_{02} \cos(\mathbf{k}_1 \cdot \mathbf{r} + \phi_1 - \omega t) \cos(\mathbf{k}_2 \cdot \mathbf{r} + \phi_2 - \omega t), \\ &= \mathbf{E}_{01} \cdot \mathbf{E}_{02} [\cos(\mathbf{k}_1 \cdot \mathbf{r} + \phi_1) \cos(\omega t) + \sin(\mathbf{k}_1 \cdot \mathbf{r} + \phi_1) \sin(\omega t)], \\ &\quad \times [\cos(\mathbf{k}_2 \cdot \mathbf{r} + \phi_2) \cos(\omega t) + \sin(\mathbf{k}_2 \cdot \mathbf{r} + \phi_2) \sin(\omega t)]. \end{aligned} \tag{7.9}$$

Keeping in mind the following relations

$$\langle \cos^2 \omega t \rangle = 1/2, \quad \langle \sin^2 \omega t \rangle = 1/2, \quad \langle \sin \omega t \cos \omega t \rangle = 0, \tag{7.10}$$

we have

$$\langle \mathbf{E}_1 \cdot \mathbf{E}_2 \rangle = \frac{1}{2} \mathbf{E}_{01} \cdot \mathbf{E}_{02} \cos(\mathbf{k}_1 \cdot \mathbf{r} - \mathbf{k}_2 \cdot \mathbf{r} + \phi_1 - \phi_2). \tag{7.11}$$

The interference term is then given by

$$I_{12} = \mathbf{E}_{01} \cdot \mathbf{E}_{02} \cos \delta, \tag{7.12}$$

where the phase difference δ is given by

$$\delta = \mathbf{k}_1 \cdot \mathbf{r} - \mathbf{k}_2 \cdot \mathbf{r} + \phi_1 - \phi_2. \tag{7.13}$$

A standard situation corresponds to the case when both the fields have the same linear polarization. In that case, by noting that

$$I_1 = \langle \mathbf{E}_1^2 \rangle = \frac{1}{2} E_{01}^2, \quad I_2 = \langle \mathbf{E}_2^2 \rangle = \frac{1}{2} E_{02}^2, \tag{7.14}$$

Eq. (7.12) can be written in a more meaningful form:

$$I_{12} = 2\sqrt{I_1 I_2} \cos \delta. \tag{7.15}$$

For total irradiance we have

$$I = I_1 + I_2 + 2\sqrt{I_1 I_2} \cos \delta. \tag{7.16}$$

Thus depending on the value of the phase difference, the total irradiance can be more, equal to or less than $I_1 + I_2$. Total constructive interference takes place when $\delta = 0, \pm 2\pi, \pm 4\pi, \cdots$. We say that the disturbances are in phase. A maximum irradiance occurs for $\cos \delta = 1$:

$$I_{max} = I_1 + I_2 + 2\sqrt{I_1 I_2}. \tag{7.17}$$

For $0 < \cos\delta < 1$, the waves are out of phase, resulting in $I_1 + I_2 < I < I_{max}$, which means constructive interference. At $\delta = \pi/2$ $\cos\delta = 0$, the disturbances are 90° out of phase, and $I_1 + I_2 = I$. For $-1 < \cos\delta < 0$, we have destructive interference, resulting in $I_1 + I_2 > I > I_{min}$, where the minimum intensity is given by

$$I_{min} = I_1 + I_2 - 2\sqrt{I_1 I_2}. \tag{7.18}$$

The minimum intensity occurs for odd multiples of π, i.e., for $\delta = \pm\pi, \pm 3\pi, \ldots$. An interesting particular case corresponds to the case when both the wave amplitudes are the same, i.e., $\mathbf{E}_{01} = \mathbf{E}_{02}$. In that case we have $I_1 = I_2 = I_0$ and

$$I = 2I_0(1 + \cos\delta) = 4I_0 \cos^2\frac{\delta}{2}. \tag{7.19}$$

It follows from Eq. (7.19) that $I_{min} = 0$ and $I_{max} = 4I_0$. In fact, one can define the visibility of an interference pattern V as

$$V = \frac{I_{max} - I_{min}}{I_{max} + I_{min}}. \tag{7.20}$$

It is clear that V changes between 0 and 1. For the case corresponding to Eq. (7.19), the visibility is unity.

7.1.1 Conditions for interference

In order to have a stable interference pattern, we need to have very nearly the same frequencies. A large frequency difference would result in a rapidly varying phase, which would lead to a null-averaged value for I_{12} during the detection interval. Nevertheless, if the two sources emit white light, the component red will still interfere with reds and the blue with blues. A great many similar slightly displaced overlapping monochromatic patterns will generate one total white light pattern. It may not be as sharp, as in the case of quasi-monochromatic light. But white light will produce observable interference.

Clearer patterns emerge when the interfering waves have the same or nearly the same amplitudes. The central portions of the dark and light fringes then correspond to complete destructive and constructive interference, yielding maximum contrast.

For an observable fringe pattern, there may be an initial phase difference between the two waves. What is important is that this initial phase difference must remain constant. Such sources that may or may not be in step but always go together are coherent [56].

7.1.2 Temporal and spatial coherence

Coherence plays a very important role in any interference and diffraction phenomena. It is related to the phase correlations between two distinct points in space at the same time or two different moments of time at the same point.

The former defines the spatial coherence while the latter gives the temporal coherence. Consider two points P_1 and P_2 on the same wavefront emitted by a source at time t_0. Let $\mathbf{E}_1(t = t_0)$ and $\mathbf{E}_2(t = t_0)$ be the corresponding electric fields. By definition of the wavefront, the phase difference between these two fields is zero. If the phase difference remains zero at any time t, we say that there is perfect spatial coherence between these two points. If this happens for any two points in the wavefront, the wave has perfect spatial coherence. In practice for any given P_1, the point P_2 must be within some area around P_1 within which there are some good phase correlations. In this case we say that the wave is partially coherent and for any P we can define the suitably introduced coherence area $S(P)$. Analogous notions can be defined in the time domain for temporal coherence. Temporal coherence mostly depends on the monochromaticity of the wave. Partially coherent light will always have a finite spectral width. For a detailed discussion on coherence phenomena, readers are referred to Section 9.2.1 of Ref. [8] and the textbook by Emil Wolf [56].

7.2 Interferometers based on wavefront splitting

It is clear that it is extremely difficult to have two thermal coherent sources. Only modern-day lasers are coherent. Young came up with the brilliant idea of picking a thermal source and using two portions of the same wavefront as the two secondary sources. Since the two secondary sources are on the same wavefront, they are coherent.

7.2.1 Young's double slit interferometer

The schematics of the interferometer are shown in Fig. 7.1. Though the original experiment used pin holes, the modern-day version involves the use

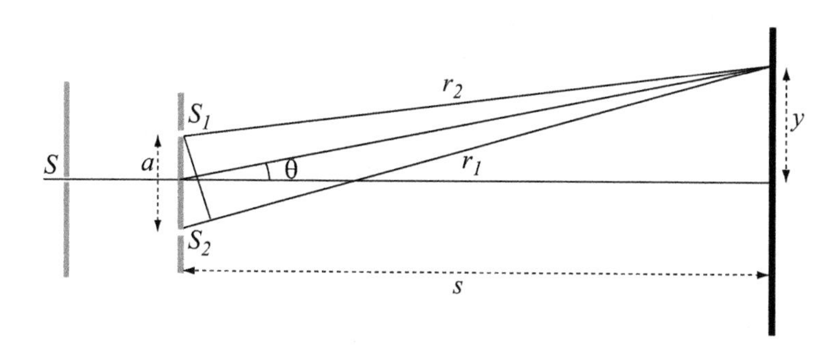

FIGURE 7.1: Schematics of Young's double slit interferometer.

of slits, giving rise to cylindrical waves. A primary wavefront is split into two secondary sources by slits S_1 and S_2. Interference between the waves coming from two slits is expected whenever the optical path difference is less than the coherence length $L_c = c\Delta\tau$. The path difference $r_1 - r_2$ can be expressed as

$$r_1 - r_2 = a\sin\theta \approx a\theta. \tag{7.21}$$

In writing the approximate equality, we used the fact that the point of observation P is far away from the slits so that $\sin\theta \approx \theta$. Using the relation

$$\theta \approx \frac{y}{s}, \tag{7.22}$$

we have

$$r_1 - r_2 \approx \frac{ay}{s}. \tag{7.23}$$

Interference maxima occur for

$$\frac{2\pi}{\lambda}(r_1 - r_2) = 2m\pi, \tag{7.24}$$

or for

$$r_1 - r_2 = m\lambda. \tag{7.25}$$

Thus for the location of the bright m-th fringe we have

$$y_m = \frac{sm\lambda}{a}. \tag{7.26}$$

Using Eq. (7.22), this occurs at

$$\theta_m = \frac{m\lambda}{a}. \tag{7.27}$$

The spacing between the fringes is given by

$$\Delta y = \frac{s\lambda}{a}. \tag{7.28}$$

Thus the fringes are broader for larger wavelengths: red ones will be broader than the blue ones. Using the expressions for the phase difference $\delta = k(r_1 - r_2)$, for intensity variation we have the expression

$$I = 4I_0\cos^2\frac{k(r_1 - r_2)}{2} = 4I_0\cos^2\left(\frac{\pi ay}{s\lambda}\right). \tag{7.29}$$

The results above hold for slits with infinite length. For slits with finite length, due to diffraction effects, the intensity will fall off on both sides of $y = 0$.

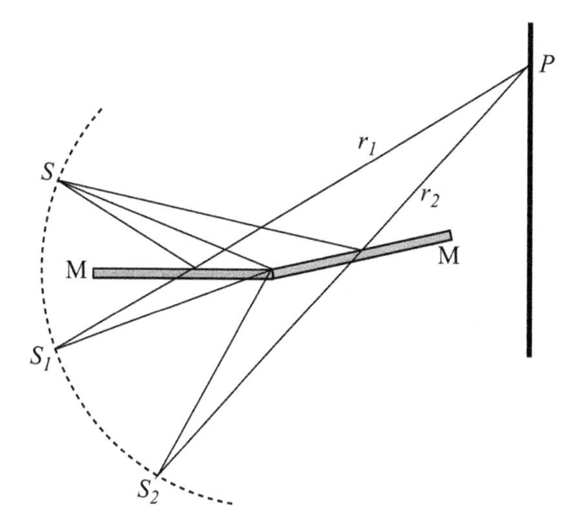

FIGURE 7.2: Schematics of the Fresnel double mirror arrangement.

7.2.2 Fresnel double mirror

The schematics of the interferometer are shown in Fig. 7.2. Two plane front silver mirrors at a small angle are used to create the two virtual sources S_1 and S_2. Denoting the distance of S_1 from point of observation P by r_1 and the same from S_2 by r_2, we will have interference maxima at $r_1 - r_2 = m\lambda$. Thus all the results derived for the case of the double slit experiment will hold here also. The smaller the inclination of the second mirror from the first one, the smaller will be the distance a between the secondary sources.

7.2.3 Fresnel biprism

The schematics of the interferometer are shown in Fig. 7.3. Here a cylindrical wavefront from a single slit S impinges on both prisms. The top portion bends downward while the bottom portion is bent upward, creating the two sources S_1 and S_2. In the region of overlap we have interference. The two virual sources are separated by a, which is determined by the prism acute angle α and the separation d of the plane of the sources from the plane of the prism, as follows:

$$a = 2d(n-1)\alpha, \tag{7.30}$$

where n is the refractive index of the prism material.

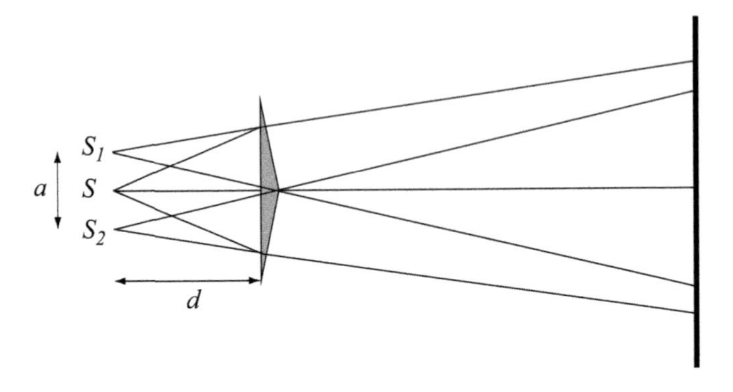

FIGURE 7.3: Schematics of the Fresnel biprism.

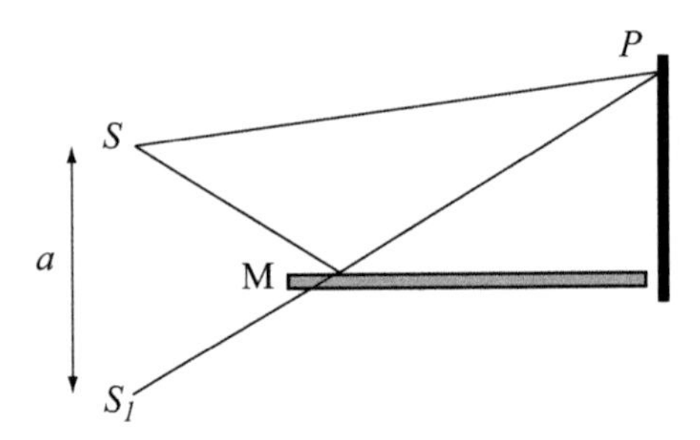

FIGURE 7.4: Schematics of Lloyd's mirror interferometer.

7.2.4 Lloyd's mirror

The schematics of the interferometer are shown in Fig. 7.4. This is perhaps the simplest interferometer with a flat mirror, from which a portion of the wavefront coming from slit S is reflected. Another portion proceeds directly from the slit. Let a be the distance between the slit and the image S_1. Fringe spacing is again $s\lambda/a$. The distinguishing feature of this device is the fact that at glancing angle $\theta_i \sim \pi/2$, the reflected beam undergoes a phase shift of $\pm\pi$. Hence the phase difference now is given by

$$\delta = k(r_1 - r_2) \pm \pi, \tag{7.31}$$

and the irradiance becomes

$$I = 4I_0 \sin^2\left(\frac{\pi a y}{s\lambda}\right). \tag{7.32}$$

Thus the fringe pattern of Lloyd's mirror is complementary to that of Young's interferometer in the sense that the intensity maxima of one will correspond to the minima of the other.

7.3 Interferometers based on amplitude splitting

When a light beam is incident on a beam-splitter, it splits into two parts. One part is transmitted while the other is reflected. Both the beams have lower amplitude than the incident one. In a sense the amplitude has been split. If these two beams can be brought together, interference can result if the original cohence is not destroyed. In other words if the path difference is less than the coherence length, the detector will see the phase correlation and stable fringe patterns can emerge. We restrict ourselves to the case when indeed the path difference is less than the coherence length.

7.3.1 Double beam interference in dielectric films

A film is be said to be thin if its thickness is of the order of the wavelength. Rather spectacular colors of such thin films arise in oil slicks and soap bubbles. With the advent of vacuum-coating technology, manufacturing such films has been rendered possible. There are many diverse applications of such films. Here we investigate the ones based on interference effects in such films.

7.3.2 Fringes of equal inclination

Consider a thin dielectric film of thickness d with very low absorption with refractive index n_t embedded in a medium with refractive index n_i. We also assume the amplitude reflection coefficient to be very low so that only the first two reflected beams need to be considered. Both are assumed to have undergone only one reflection at the top and bottom interfaces, respectively. It is clear that the film serves as an amplitude splitter. The reflected amplitudes E_{1r} and E_{1r} can be considered as coming from the two virtual sources behind the film. The optical path difference Λ between the reflected beams can be written as

$$\Lambda = n_t(AB + BC) - n_i AD. \tag{7.33}$$

Since

$$AB = BC = \frac{d}{\cos\theta_t}, \tag{7.34}$$

$$\Lambda = \frac{2n_t d}{\cos\theta_t} - n_i AD. \tag{7.35}$$

Note that

$$AD = AC\sin\theta_i = AC\frac{n_t}{n_i}\sin\theta_t = 2d\tan\theta_t\frac{n_t}{n_i}\sin\theta_t, \tag{7.36}$$

and hence

$$\Lambda = \frac{2n_t d}{\cos\theta_t}(1 - \sin^2\theta_t) = 2n_t d\cos\theta_t. \tag{7.37}$$

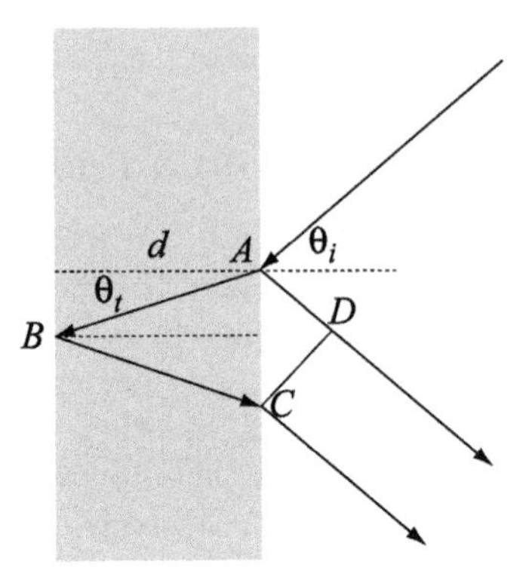

FIGURE 7.5: Explaining fringes of equal inclination.

The phase difference associated with this path difference is $k_0\Lambda$. Noting that there will be an additional phase accumulation in reflection from at least one interface by $\pm\pi$ irrespective of the fact that the refractive index of the film is lower or higher than that of its environment, the phase difference can be written as

$$\delta = \frac{4\pi n_t}{\lambda} d\cos\theta_t \pm \pi \tag{7.38}$$

or

$$\delta = \frac{4\pi d}{\lambda}\left(n_t^2 - n_i^2\sin^2\theta_i\right)^{1/2} \pm \pi. \tag{7.39}$$

We choose the negative sign in Eq. (7.38) to make it look simpler. Interference maxima will occur at $\delta = 2m\pi$, which can be rewritten as

$$d\cos\theta_t = (2m+1)\frac{\lambda_t}{4}, \tag{7.40}$$

where $\lambda_t = \lambda/n_t$. The same conditions correspond to minima in the transmitted light. Interference minima in reflection (maxima in transmission) occur for $\delta = (2m \pm 1)\pi$ and for such cases

$$d\cos\theta_t = 2m\frac{\lambda_t}{4}. \tag{7.41}$$

A comment regarding the refractive indices is in order. If the refractive indices are in increasing or decreasing order, the additional phase shift of $\pm\pi$ would not be there and the formulas for the maxima and minima would be interchanged. Since the phase difference is mainly controlled by θ, such fringes are generally referred to as fringes of equal inclination.

7.3.3 Fringes of equal width

In contrast to the dominating role of θ, a whole class of fringes exists for which the optical thickness of the film plays the most important part. These

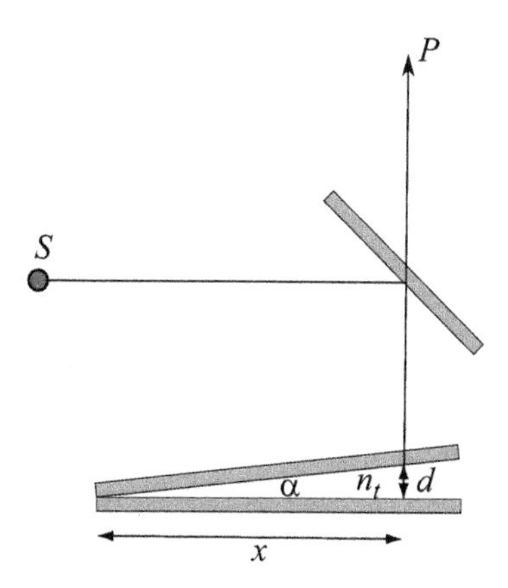

FIGURE 7.6: Explaining fringes of equal thickness.

are referred to as the fringes of equal width. Each fringe is the locus of all points in the film for which the optical thickness is a constant. These fringes are quite useful to determine the surface features. The surface under study can be put in contact with an optical flat (not having a deviation of more than $\lambda/4$). The air gap between the two generates a thin film interference pattern. For flat test surfaces, the fringe will be a series of straight, equally spaced bands. This indicates a wedge-shaped air film. When viewed at nearly normal incidence as in Fig. 7.6, the contours from a nonuniform film are known as Fizeau fringes. For a thin wedge of small angle α, the path difference between the two reflected rays can be given by $\Lambda = 2n_t d \cos\theta_t$, and d can be approximated by $d \approx x\alpha$. For small values of θ_i, the condition for interference maximum can be written as

$$(m + 1/2)\lambda = 2n_t d_m = 2n_t \alpha x_m. \tag{7.42}$$

This yields

$$x_m = \left(\frac{m + 1/2}{2\alpha}\right)\lambda_t, \tag{7.43}$$

where $\lambda_t = \lambda/n_t$ is the wavelength in the film. Thus maxima occur at distances from the apex at $\lambda_t/4\alpha, 3\lambda_t/4\alpha$, etc. The separation between the bright fringes is given by

$$\Delta x = \lambda_t/2\alpha. \tag{7.44}$$

Note that the difference in film thickness for adjacent maxima is given by $\lambda_t/2$. Since the beam reflected by the lower surface traverses the thickness twice, adjacent maxima differ in the optical path by λ_t. In terms of the thickness, the location of maxima is given by

$$d_m = (m + 1/2)\frac{\lambda_t}{2}. \tag{7.45}$$

Traversing the film twice gives a phase shift of π, which when added with additional π phase shift under reflection puts the two rays back in phase. Hence the interference maxima result.

7.3.4 Newton's rings

Two pieces of glass slides, when pressed at a point illuminated by normally incident light, can exhibit concentric fringe patterns. These patterns are known as Newton's rings. These can be studied systematically by the arrangement shown in Fig. 7.7. On top of a glass optical flat, we place a lens. The system is illuminated with normally incident quasi-monochromatic light. The amount of uniformity of the circular pattern is a measure of how perfect the lens is. Let R be the radius of curvature of the convex lens. The relation between the distance x and the film thickness d is given by

$$x^2 = R^2 - (R-d)^2 = 2Rd - d^2. \tag{7.46}$$

Since $d \ll R$, Eq. (7.46) can be rewritten as

$$x^2 = 2Rd. \tag{7.47}$$

Considering only the first two reflected beams, the m-th order interference maximum occurs at

$$2n_t d_m = (m + 1/2)\lambda. \tag{7.48}$$

The radius of the m-th bright ring is then given by

$$x_m = [(m + 1/2)\lambda_t R]^{1/2}. \tag{7.49}$$

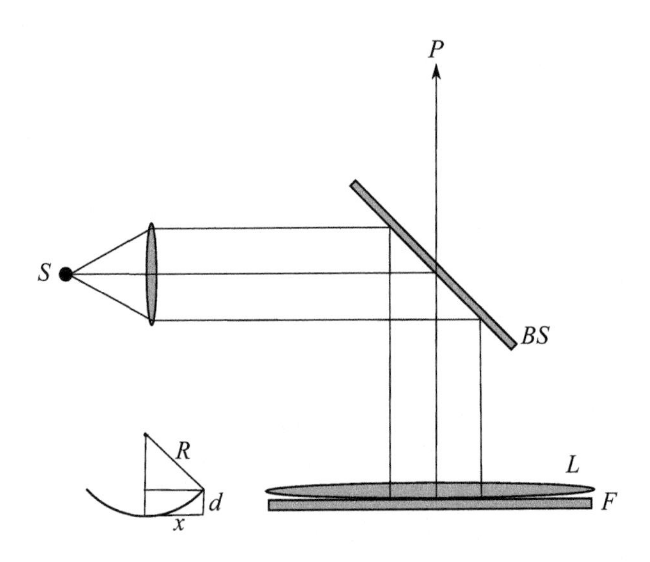

FIGURE 7.7: Setup for observing Newton's rings.

The radius of the m-th dark ring will be

$$x_m = [m\lambda_t R]^{1/2}. \tag{7.50}$$

If there are no dust particles at the center between the lens and the optical flat and the contact is good, then we will have a minimum of intensity at the center (at $x = 0$) since $d = 0$ at that point.

7.3.5 Mirrored interferometers: Michelson interferometer

A variety of amplitude-splitting interferometers are based on multiple mirrors and beam-splitters. Perhaps the best known is the Michelson interferometer. The arrangement of the interferometer is shown in Fig. 7.8. An incident beam is split into two by means of a beam-splitter BS. Both the transmitted and the reflected beams are reflected back onto the beam-splitter by mirrors M' and M''. A compensator C is placed in the path of the transmitted beam in order to compensate for the additional path traversed by the reflected beam in passing through the BS (since it gets reflected by the bottom surface). In order to understand how the fringes are formed, it is better to refer to the equivalent diagram shown in Fig. 7.9. An observer at location D will simultaneously see the two mirrors and the source. Let the mirror separation be d. Thus interference will be observed from light coming from two virtual sources S' and S'' separated by $2d$. Optical path difference between the two rays coming from S' and S'' is given by $2d\cos\theta$ and the condition for interference

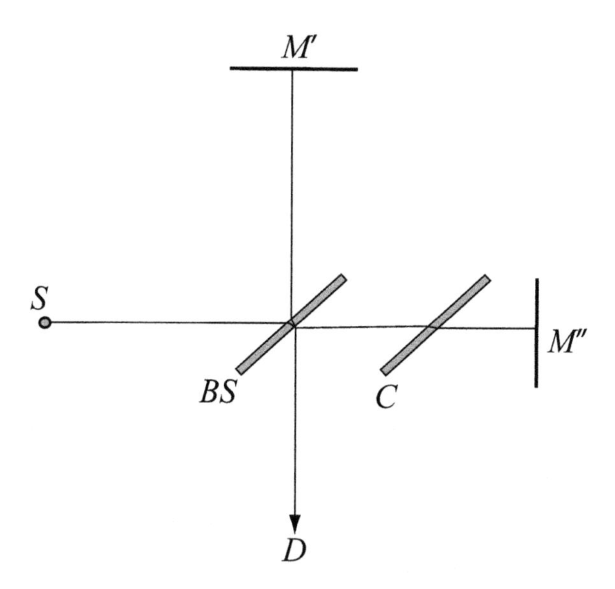

FIGURE 7.8: Michelson interferometer.

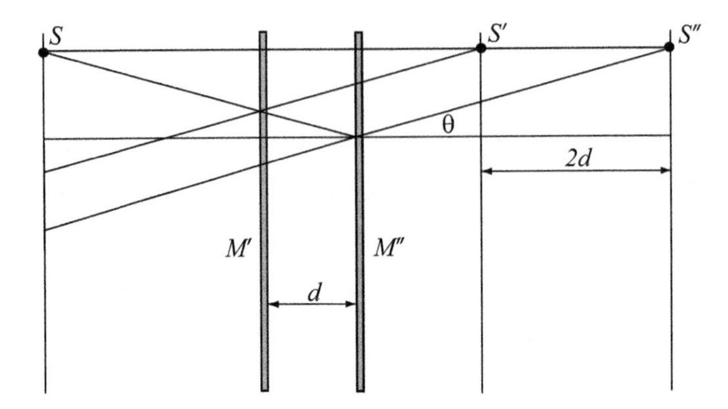

FIGURE 7.9: Equivalent diagram for Michelson interferometer.

maxima can be written as

$$2d\cos\theta_m = m\lambda, \tag{7.51}$$

where m is an integer. An observer will see a circular fringe pattern. Because of the small aperture of the eye, the observer may not see the whole pattern without the use of a large lens.

It is important to note that the path difference $2d\cos\theta$ must be less than the coherence length. This has important implications for sources to be used for Michelson interferometry. For example, for laser sources with large coherence length, d can be as large as 10 cm, while for mercury and other natural sources, the path difference must be very nearly zero. Out of the concentric rings, a particular one corresponds to a given value of order m. As the mirror separation decreases, so does the angle θ_m, and the ring shrinks toward the center. The highest-order one disappears when d is decreased by $\lambda/2$. The central dark fringe occurs for $\theta_m = 0$, and for it we have

$$2d = m_0\lambda. \tag{7.52}$$

For example, for $d = 10$ cm and for $\lambda = 500$ nm, the largest-order fringe is realized with $m = 400{,}000$. For a fixed value of d, the next dark fringes occur at

$$\begin{aligned} 2d\cos\theta_1 &= (m_0 - 1)\lambda, \\ 2d\cos\theta_2 &= (m_0 - 2)\lambda, \\ &\vdots \\ 2d\cos\theta_p &= (m_0 - p)\lambda. \end{aligned} \tag{7.53}$$

Combining the last equation with Eq. (7.52), we have

$$2d(1 - \cos\theta_p) = p\lambda. \tag{7.54}$$

For small θ_p we can use the approximation $\cos\theta_p \approx 1 - \theta_p^2/2$, which reduces Eq. (7.54) to

$$\theta_p = (p\lambda/d)^{1/2}. \tag{7.55}$$

The Michelson interferometer can be used for precise and accurate measurement of distances. This is based on the following fact. As the movable mirror is displaced by $\lambda/2$, each fringe moves to the position previously occupied by the adjacent fringe. We need to count the number of fringes N that have moved through a reference point (say, a crosshair in a microscope objective) in order to determine the displacement Δd of the mirror:

$$\Delta d = N\lambda/2. \tag{7.56}$$

7.4 Multiple beam interference

Upto now have we considered the interference of two coherent beams. There are many cases where interference takes place with participation of more than two coherent beams. For example, for a glass slide and for significant reflections from the surfaces, we need to add up all the reflected rays. If the glass plate is slightly silvered on both sides, it will lead to multiple reflections from both the surfaces. For the time being we will consider transparent dielectrics only to avoid the complications of phase changes at the silvered

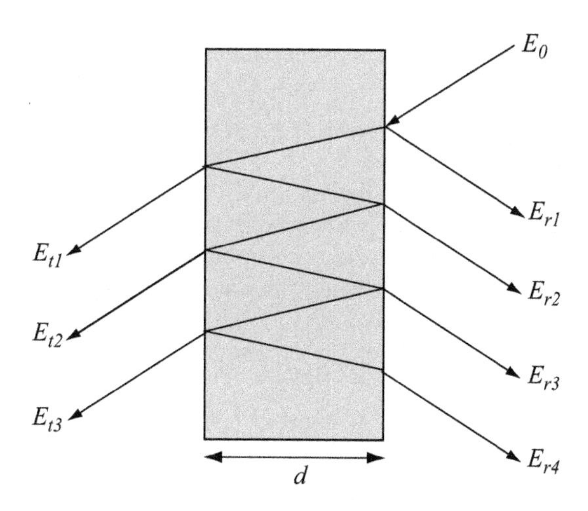

FIGURE 7.10: Multiple beam interference.

surfaces. Again, the path difference Λ between the adjacent rays is given by $\Lambda = 2n_t d\cos\theta$. Initially, we look at the particular case when the slab has specific $\lambda/4$ or $\lambda/2$ thickness, and later we look at the general case.

We first consider the case when the path difference $\Lambda = m\lambda$ ($\lambda/2$ thickness). In this case, the second, third and all higher-order reflected waves will be in phase, but they will be π out of phase with the first reflected wave. Total reflected amplitude is obtained by adding the reflected amplitudes E_{r1}, E_{r2}, etc., as follows:

$$\begin{aligned} E_r &= E_{r1} + E_{r2} + E_{r3} + \cdots, \\ &= E_0 r - (E_0 r t t' + E_0 r^3 t t' + E_0 r^5 t t' + \cdots), \\ &= E_0 r - E_0 r t t'(1 + r^2 + r^4 + \cdots), \\ &= E_0 r - E_0 \frac{r t t'}{1 - r^2} = 0, \end{aligned} \tag{7.57}$$

where r, r' and t, t' are the external and internal reflection and transmission coefficients, respectively. In writing the last row of Eq. (7.57), we used the fact that

$$r' = -r, \quad tt' = 1 - r^2. \tag{7.58}$$

We thus have an interference minimum corresponding to this case.

For $\Lambda = (m + 1/2)\lambda$, the first two rays are in phase while all other consecutive pairs are π out of phase. Hence we have

$$\begin{aligned} E_r &= E_0 r + (E_0 r t t' - E_0 r^3 t t' + E_0 r^5 t t' - \ldots), \\ &= E_0 r + E_0 r t t'(1 - r^2 + r^4 - \ldots), \\ &= E_0 r + E_0 \frac{r t t'}{1 + r^2} = \frac{2r}{1 + r^2} E_0. \end{aligned} \tag{7.59}$$

The corresponding irradiance is proportional to $E_r^2/2$, and hence

$$I_r = \frac{4r^2}{(1 + r^2)^2} I_0, \tag{7.60}$$

where I_0 is the incident irradiance. This case corresponds to the intensity maximum in the reflected light. Since reflection and transmission are complementary, this would correspond to the minima in transmitted light. We now present the rigorous treatment when the phase difference can be arbitrary. We write the reflected fields as

$$\begin{aligned} E_{r1} &= E_0 r \exp[-i(\omega t)], \\ E_{r2} &= E_0 t t' r' \exp[-i(\omega t + \delta)], \\ E_{r3} &= E_0 t t' r'^3 \exp[-i(\omega t + 2\delta)], \\ &\vdots \\ E_{Nr} &= E_0 t t' r'^{(2N-3)} \exp[-i(\omega t + (N - 1)\delta)]. \end{aligned} \tag{7.61}$$

Hence the total reflected field is given by

$$\begin{aligned}
E_r &= E_{r1} + E_{r2} + E_{r3} + ..., \\
&= E_0 e^{-i\omega t}\left[r + tt'r'e^{-i\delta}(1 + r'^2 e^{-i\delta} + (r'^2 e^{-i\delta})^2 + ...)\right], \\
&= E_0 e^{-i\omega t}\left[r + \frac{tt'r'e^{-i\delta}}{1 - r'^2 e^{-i\delta}}\right], \\
&= E_0 e^{-i\omega t}\left[\frac{r(1 - e^{-i\delta})}{1 - r^2 e^{-i\delta}}\right].
\end{aligned} \tag{7.62}$$

In the equations above, we used the fact that $r' = -r$. The expression for reflected irradiance becomes

$$I_r = \frac{E_0^2 r^2 (1 - e^{-i\delta})(1 - e^{i\delta})}{2(1 - r^2 e^{-i\delta})(1 - r^2 e^{i\delta})} \tag{7.63}$$

or

$$\frac{I_r}{I_i} = \frac{2r^2(1 - \cos\delta)}{(1 + r^4) - 2r^2\cos\delta}. \tag{7.64}$$

For the transmitted rays we have

$$\begin{aligned}
E_{t1} &= E_0 tt' \exp[-i(\omega t)], \\
E_{t2} &= E_0 tt'r'^2 \exp[-i(\omega t + \delta)], \\
E_{t3} &= E_0 tt'r'^4 \exp[-i(\omega t + 2\delta)], \\
&\vdots \\
E_{tN} &= E_0 tt'r'^{(2N-2)} \exp[-i(\omega t + (N-1)\delta)].
\end{aligned} \tag{7.65}$$

Hence the total transmitted field is given by

$$\begin{aligned}
E_t &= E_{t1} + E_{t2} + E_{t3} + \cdots, \\
&= E_0 e^{-i\omega t}\left[\frac{tt'}{1 - r^2 e^{-i\delta}}\right].
\end{aligned} \tag{7.66}$$

The expression for reflected irradiance is given by

$$I_t = \frac{I_i (tt')^2}{(1 + r^4) - 2r^2\cos\delta}. \tag{7.67}$$

Finally, introducing the finesse coefficient $F = [2r/(1 - r^2)]^2$, the expressions for both the reflected and transmitted irradiances can be written as

$$I_r = I_i \frac{F\ \sin^2\delta/2}{1 + F\ \sin^2\delta/2}, \tag{7.68}$$

$$I_t = I_i \frac{1}{1 + F\ \sin^2\delta/2}. \tag{7.69}$$

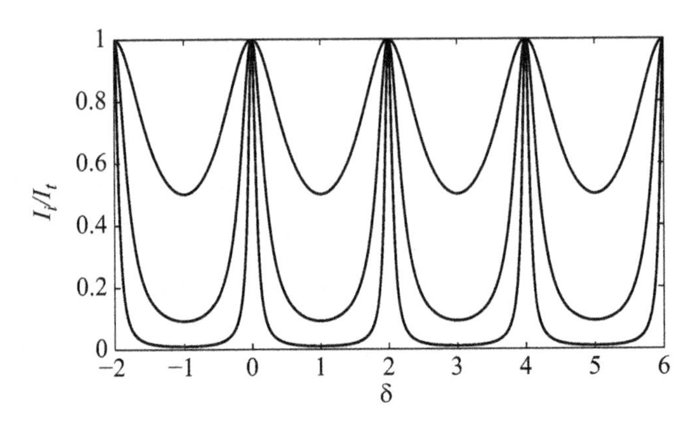

FIGURE 7.11: Transmission resonances. The curves from top to bottom are for F =1, 10 and 100, respectively.

It is clear from the above equations that $I_r + I_t = I_i$ for lossless systems. The minima and maxima of transmission are observed at $\delta = (2m+1)\pi$ and $2m\pi$, respectively, and they are given by

$$I_{t\ min} = I_i \frac{(1-r^2)^2}{(1+r^2)^2}, \qquad I_{t\ max} = I_i. \tag{7.70}$$

It is clear from Eq. (7.69) that transmission resonances (usually referred to as Airy resonances) are evenly spaced with frequency spacing given by $\Delta f = c/(2n_t d)$. The corresponding curves for three different values of F are shown in Fig. 7.11. The half-width at half-maximum $\delta_{1/2}$ can be found from the condition

$$\frac{1}{1+F\ \sin^2 \delta_{1/2}/2} = \frac{1}{2}, \tag{7.71}$$

which yields

$$\delta_{1/2} = \frac{2}{\sqrt{F}} \tag{7.72}$$

so that the full-width at half-maximum (FWHM) is $\gamma = 4/\sqrt{F}$. The finesse of the resonator $\mathcal{F}$ is defined by

$$\mathcal{F} = \frac{2\pi}{\gamma} = \frac{\pi\sqrt{F}}{2}. \tag{7.73}$$

Typical finesse for such resonators is about 30 in the visible range.

7.4.1 Fabry-Pérot interferometer

A Fabry-Pérot interferometer consists of two parallel mirrors. Usually two glass plates with silvered internal surfaces are used. The outer faces are in the form of wedges so that light reflected from those faces need not be accounted for. The working principle is the same as that of a dielectric slab described above. If the gap between the mirrors can be varied, it is referred to as an interferometer, while a fixed-gap system with some transparent material inside is called an *etalon*. The latter is used to resolve the spectral lines in standard spectroscopic devices.

7.5 Diffraction

Deviation of light from rectilinear propagation due to obstacles is usually referred to as diffraction. The phenomenon of diffraction is one of the manifestations of the wave character of light. It is thus characteristic of not only light but also of other waves, like sound waves. The segments of wavefront that propagate beyond the obstacle interfere, causing the particular fringe pattern referred to as the diffraction pattern. In fact, there is no physical distinction between the two phenomena of interference and diffraction. It is customary to call it interference if only a few rays interfere, while in the case of many rays it is called diffraction. Even then, in the context of superposition of multiple beams, it is referred to as interference, while in the context of an array of coherent sources containing the same physics, it is referred to as diffraction.

Let us first look again at the Huygens-Fresnel principle. The basis of the Huygens-Fresnel principle was the Huygens principle, which states: *Each point of the wavefront acts as the source of secondary wavelets. At any later time, the shape of the wavefront is the envelope of the secondary wavelets.* It is clear that the Huygens principle can explain the bending of light but not the intricate diffraction pattern, since it has no reference to the wavelength of light. The difficulty was resolved by Fresnel, who added the important concept of interference: *Every unobstructed point of the wavefront at any given moment serves as a source of secondary wavelets (with the same frequency as that of the primary one). The amplitude of the optical field at any point beyond is the superposition of these wavelets (with consideration of their amplitudes and phases).*

Let us now apply the principle in a situation like in Fig. 7.12, where we have an opaque screen with an opening. The aperture is illuminated with plane waves. Each unobstructed point of the incoming plane wave acts as a coherent secondary source. The maximum optical path difference corresponding to $|AP - BP|$ is $\Lambda_{max} \leq AB$. The equality holds for the point of observation P on the screen. When $\lambda > AB$, it follows that $\lambda \geq \Lambda_{max}$. Since the waves initially were in phase, they interfere constructively (to varying degrees), irrespective of where P is. Thus, if the wavelength is large compared to the linear dimension of the aperture, the waves will spread out at large angles into the region beyond the obstruction. In the opposite case, when $\lambda < AB$, the area

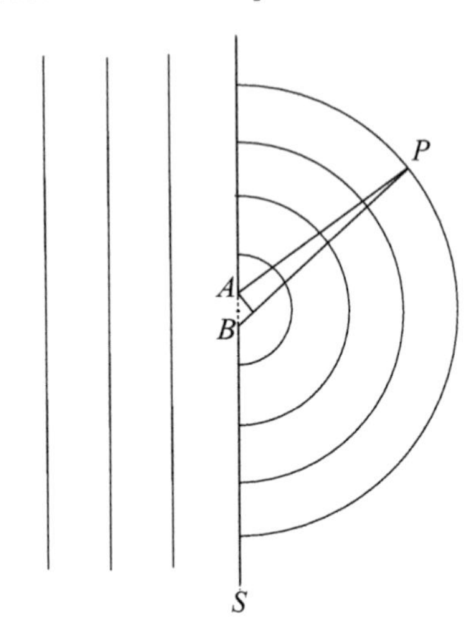

FIGURE 7.12: Plane wave incident on an aperture.

where $\lambda > \Lambda_{max}$ is limited to a small region directly in front of the opening. It is here only that all wavelets will interfere constructively. The idealized geometric shadow corresponds to the case when $\lambda \to 0$.

7.5.1 Fresnel and Fraunhofer diffraction

Consider again an opaque screen with a small aperture, illuminated by a plane wave. We can distinguish three distinct regimes so far as the distances of the screen from the source and the point of observation are concerned.

1. The plane of observation is very close to the screen. An image of the aperture is projected onto the screen. There may be some slight fringing effect at the edges.

2. The observation screen is moved farther apart. The image of the aperture, though recognizable, becomes structured as fringes at the edge become more prominent. This is a Fresnel or near-field diffraction.

3. At large distances, the produced pattern spreads out significantly with no resemblance of the image with the aperture. This is a Fraunhofer or far-field diffraction. If we could decrease the wavelength at this stage, the pattern would revert to Fresnel diffraction. If the wavelength could be reduced to zero, then the fringes would vanish, leading to the geometrical shadow.

The plane wave in Fig. 7.12 can be thought of as coming from a point source very far apart. If the point source is moved to the aperture, the spherical waves would impinge on the aperture, and a Fresnel pattern would exist even at a large distance of the aperture from the observation plane. It is important to note that as long as both incoming and outgoing waves approach being planar (differing therefrom by a small fraction of a wavelength) over the extent of the aperture, Fraunhofer diffraction holds. On the contrary, when the aperture is too close to the source or the observation plane, which results in a curvature of the phase front, Fresnel diffraction prevails. We can write a practical rule of thumb for the region where Fraunhofer diffraction takes place:

$$R > \frac{a^2}{\lambda}, \tag{7.74}$$

where R is the smallest of the two distances of the aperture from the source and from the point of observation, and a is the linear dimension of the aperture. For $R \to \infty$, finite size effects of the aperture are of little consequence. Effectively, this can be achieved by putting two lenses before and after the aperture. In the remainder part of this section, we concentrate on the Fraunhofer diffraction from various sources.

7.5.2 N coherent oscillators

Consider a linear array of N identical oscillators with the same initial phase angle. Also consider a far-off observation point P (Fig. 7.13). If the spatial extent of the oscillators is small, the wave amplitudes arriving at P would be the same, having traveled almost the same distance:

$$E_0(r_1) = E_0(r_2) = E_0(r_3) = .. = E_0(r_{N-1}) = E_0(r_N). \tag{7.75}$$

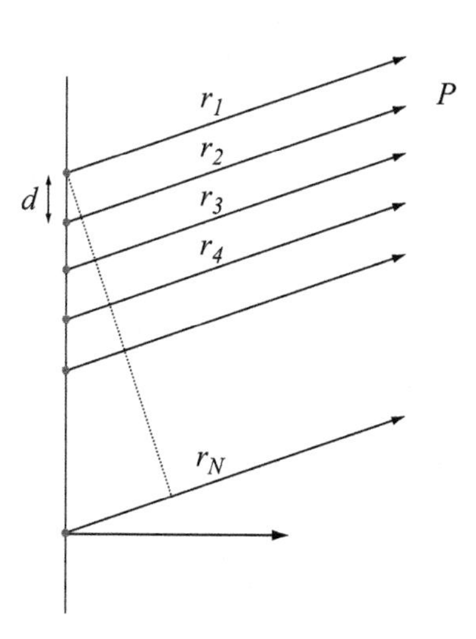

FIGURE 7.13: An array of N equispaced coherent sources.

Then the sum of the interfering wavelets at P is given by the real part of

$$E = E_0(r)e^{i(kr_1-\omega t)} + E_0(r)e^{i(kr_2-\omega t)}) + \cdots + E_0(r)e^{i(kr_N-\omega t)}, \quad (7.76)$$
$$= E_0(r)e^{i(kr_1-\omega t)}\left[1 + e^{ik(r_2-r_1)} + \cdots + e^{ik(r_N-r_1)}\right]. \quad (7.77)$$

In terms of the phase difference between the adjacent sources δ given by

$$\delta = k\Lambda = kd\sin\theta, \quad (7.78)$$

the total field can be written as

$$E = E_0(r)e^{i(kr_1-\omega t)}\left[1 + (e^{i\delta}) + (e^{i\delta})^2 + ... + (e^{i\delta})^{N-1}\right], \quad (7.79)$$
$$= E_0(r)e^{i(kr_1-\omega t)}\left(\frac{e^{i\delta N}-1}{e^{i\delta}-1}\right). \quad (7.80)$$

The last term in the brackets in Eq. (7.80) can be simplified to

$$\left(\frac{e^{i\delta N}-1}{e^{i\delta}-1}\right) = e^{i\frac{\delta}{2}(N-1)}\frac{\sin\left(\frac{\delta N}{2}\right)}{\sin\frac{\delta}{2}}. \quad (7.81)$$

Thus the final expression for the amplitude can be written as

$$E = E_0(r)e^{-i\omega t}\ e^{i[kr_1+(N-1)\frac{\delta}{2}]}\ \frac{\sin\left(\frac{\delta N}{2}\right)}{\sin\frac{\delta}{2}}. \quad (7.82)$$

Denoting the distance from the center of the array to P by R, where

$$R = r_1 + \frac{N-1}{2}d\sin\theta, \quad (7.83)$$

Eq. (7.82) can be rewritten as

$$E = E_0(r)e^{i(kR-\omega t)}\ \frac{\sin\left(\frac{\delta N}{2}\right)}{\sin\frac{\delta}{2}}. \quad (7.84)$$

The corresponding flux density is given by

$$I = I_0\frac{\sin^2\left(\frac{\delta N}{2}\right)}{\sin^2\frac{\delta}{2}} = I_0\frac{\sin^2\left(\frac{Nkd\sin\theta}{2}\right)}{\sin^2\left(\frac{kd\sin\theta}{2}\right)}. \quad (7.85)$$

Here the numerator undergoes rapid oscillation while the denominator varies slowly. The combined expression yields sharp principal peaks separated by small subsidiary maxima. Principal maxima occur at θ_m, satisfying

$$\delta = 2m\pi, \quad m = 0, \ \pm 1, \ \pm 2 \ \cdots \quad (7.86)$$

or for

$$d\sin\theta_m = m\lambda, \quad m = 0, \ \pm 1, \ \pm 2 \ \cdots. \quad (7.87)$$

Note that the ratio of the squares of the sines in Eq. (7.85) $\to N^2$ at principal maxima, and hence the corresponding irradiance is given by $N^2 I_0$. This happens since all the oscillators are in phase. There are maxima in the direction perpendicular to the array for $m = 0$ and $\theta_0 = 0, \pi$. As θ increases I falls off to zero at $N\delta/2 = \pi$ at the first minimum. Note also that if $d < \lambda$, only the principal maximum corresponding to $m = 0$ or the zeroth order exists.

7.5.3 Continuous distribution of sources on a line

Consider an idealized line source along the y-axis with width $\ll \lambda$ as shown in Fig. 7.14. Let D be the entire length of the source. Each point on the source emits a spherical wavefront,

$$E = \frac{\mathcal{E}_0}{r} \exp[i(kr - \omega t)]. \tag{7.88}$$

This case is distinct from the previous one since the sources are weak, their number N is very large and their spacing is vanishingly small. Let the length of the source be divided into M equal segments Δy. Each of these segments will have $\Delta y N/D$ sources. Pick a segment Δy_i $(i = 1 - M)$ at a distance r_i from P. The contribution to the field amplitude from the i-th segment is

$$E_i = \frac{\mathcal{E}_0}{r_i} \exp[i(kr_i - \omega t)]\Delta y_i N/D. \tag{7.89}$$

Transition to a continuous distribution corresponds to the limit $N \to \infty$. Defining a source strength per unit length E_l as the limit of $\mathcal{E}_0 N/D$ as $N \to \infty$, we can write the expression of the net field at P due to all M segments as

$$E = \sum_{i=1}^{M} \frac{\mathcal{E}_l}{r_i} \exp[i(kr_i - \omega t)]\Delta y_i. \tag{7.90}$$

Let the point of observation be far off, i.e., $R \gg D$. Then $r(y)$ never deviates appreciably from the midpoint value R. Thus $\mathcal{E}_l/R$ is essentially constant.

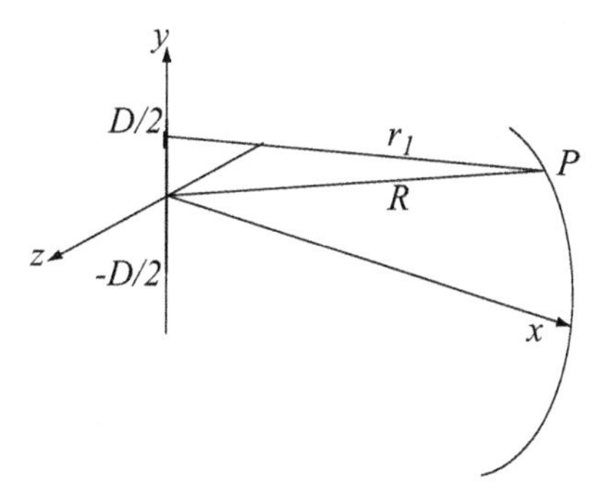

FIGURE 7.14: Single slit.

The field dE due to a length dy of continuous sources can be written as

$$dE = \frac{\mathcal{E}_l}{R} \exp[i(kr(y) - \omega t)]dy. \tag{7.91}$$

Note that phase is much more sensitive to a change in y, and $r(y)$ can be approximated by

$$r(y) \approx R - y \sin\theta. \tag{7.92}$$

The total field can be obtained by integration of Eq. (7.91):

$$E = \frac{\mathcal{E}_l}{R} e^{i(kR-\omega t)} \int_{-D/2}^{D/2} e^{-iky\sin\theta} dy. \tag{7.93}$$

This finally yields

$$E = \frac{\mathcal{E}_l D}{R} \frac{\sin\beta}{\beta} e^{i(kR-\omega t)}. \tag{7.94}$$

The corresponding intensity is given by

$$I(\theta) = I_0 \left(\frac{\sin\beta}{\beta}\right)^2. \tag{7.95}$$

The variable β in Eqs. (7.94) and (7.95) is given by

$$\beta = (kD/2)\sin\theta = (\pi D/\lambda)\sin\theta. \tag{7.96}$$

It is clear that $I(\theta) = I_0$, for $\beta = 0$, and this corresponds to the principal maximum. Note that there is symmetry about the y-axis and this expression holds for θ measured in any plane containing this axis.

Two important points must be noted.

- When $D \gg \lambda$, β can be large and the intensity falls off sharply as θ deviates from zero. The phase as per Eq. (7.94) is equivalent to that of a point source located at the center of the array at a distance R from P. Thus a long line of coherent sources is equivalent to a point emitter radiating predominantly in the forward direction. Its emission resembles a circular wave in the xz plane.
- When $D \ll \lambda$, β can be small, resulting in $I(\theta) = I_0$ since for this case $\sin(\beta)/\beta \approx 1$. Irradiance is constant for all θ and the line source resembles a point source emitting a spherical wave.

7.5.4 Fraunhofer diffraction from a single slit

We can now discuss the diffraction pattern from a single slit based on the understanding of a line source. Let the slit have a width b, which is much less than its length l. The arrangement is shown in Fig. 7.15. The width can be

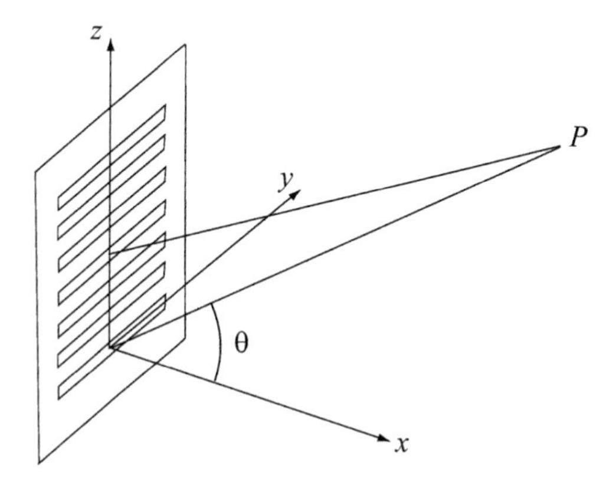

FIGURE 7.15: Diffraction from multiple slits.

several hundred of wavelengths. Usually the length is a few centimeters. The aperture can be thought of as a collection of long differential strips ($dz \times l$). Each strip can be replaced by a point source on the z-axis emitting a circular wave in the xz plane. There will be very little diffraction parallel to the strips. Thus the problem is reduced to one of finding the field in the xz plane due to an infinite number of point sources over the width of the slit b. Thus for the intensity we have Eq. (7.95) with β given by Eq. (7.96) with D replaced by b, i.e.,

$$I(\theta) = I_0 \left(\frac{\sin\beta}{\beta}\right)^2, \quad \beta = (kb/2)\sin\theta = (\pi b/\lambda)\sin\theta. \tag{7.97}$$

Extrema of $I(\theta)$ occur at values of β determined by the equation

$$\frac{dI}{d\beta} = I_0 \frac{2\sin\beta(\beta\cos\beta - \sin\beta)}{\beta^3} = 0. \tag{7.98}$$

Minima occur for $\beta = \pm\pi, \pm 2\pi, \pm 3\pi, \cdots$. When $\beta = \tan\beta$, subsidiary maxima occur in between two consecutive minima.

There is a simple way to understand the diffraction pattern. The expression $\theta = 0$ corresponds to maximum and all rays are in phase. When $b\sin\theta_1 = \lambda$, a ray from the center of the slit is π out of phase from the ray from the top. Another ray slightly below the middle will be out of phase with the one slightly below the top. Thus all the pairs will cancel out and for $\sin\theta_1 = \lambda/b$ there will be perfect cancellation leading to the zero minimum. The same happens when $b\sin\theta_1 = 2\lambda$ and so on. The general zeros occur at $b\sin\theta_m = m\lambda$.

7.5.5 Diffraction from a regular array of N slits

Consider N identical long parallel slits each of width b and center-to-center separation d (see Fig. 7.15). The flux distribution for this case can be written as

$$I(\theta) = I_0 \left(\frac{\sin\beta}{\beta}\right)^2 \left(\frac{\sin N\alpha}{\alpha}\right)^2. \tag{7.99}$$

In Eq. (7.99) I_0 is the intensity along $\theta = 0$ emitted by any of the slits and

$$I(0) = N^2 I_0. \tag{7.100}$$

Eq. (7.100) implies that waves arriving at P are all in phase for $\theta = 0$. If the width of each slit were to shrink to zero, one would recover the expression for an array of coherent sources. Principal maxima occur at $\alpha = 0, \ \pm\pi \pm 2\pi \cdots$, when $(\sin N\alpha)/(\sin\alpha) = N$. Since $\alpha = (ka/2)\sin\theta$, this condition reduces to

$$a \sin\theta_m = m\lambda, \quad m = 0, \pm 1, \pm 2, \cdots. \tag{7.101}$$

Minima occur whenever

$$\alpha = \pm\frac{\pi}{N}, \pm\frac{2\pi}{N} \cdots \pm \frac{(N-1)\pi}{N}, \quad \pm\frac{(N+1)\pi}{N}, \cdots. \tag{7.102}$$

Between consecutive maxima, there will be $N-1$ minima. Between each pair of minima, there will be a subsidiary maximum. The subsidiary maxima are located approximately at points where $\sin N\alpha$ has maximum value:

$$\alpha = \pm\frac{3\pi}{2N}, \pm\frac{5\pi}{2N} \cdots. \tag{7.103}$$

The pattern above is modulated by a single slit diffraction envelope. The pattern is shown in Fig. 7.16 for $b = 25\lambda$, $a = 4b$ and $N = 4$. The top panel shows the variation of $(\frac{\sin(\beta)}{\beta})^2$ and the middle one that of $(\frac{\sin(N\alpha)}{\sin(\alpha)})^2$, while the bottom panel shows the whole pattern modulated by the pattern of a single slit.

7.5.6 Fresnel diffraction

As discussed earlier Fresnel diffraction holds when the source or the observation point is close to the aperture. In that case we need to deviate from the plane wavefront approximation as in Fraunhofer diffraction. The experimental situation here is somewhat simpler since one can avoid the collimating optics. However, the mathematical description is much more complex and one has to resort to several approximations.

7.5.7 Mathematical statement of Huygens-Fresnel principle

Consider the diffraction schematics shown in Fig. 7.17, where the spherical wavefronts from a point source are incident on an aperture, which is not so far from the observation point P. Let the distances from an elemental area da on the aperture be at a distance r' from the source and r from the observer. Compared to the Fraunhofer diffraction, the case under study has several distinguishing features:

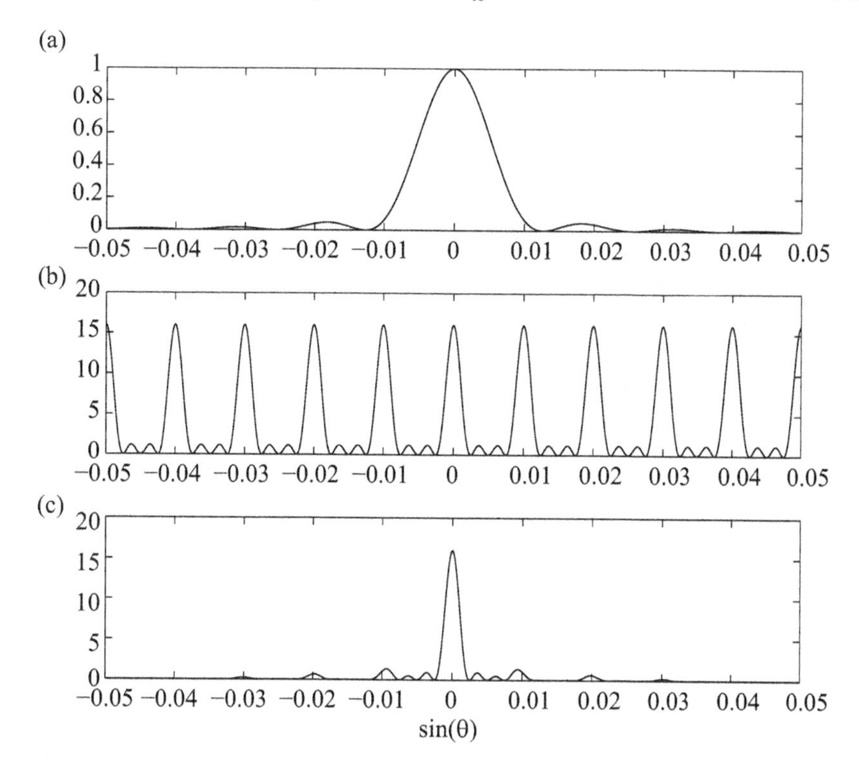

FIGURE 7.16: The diffraction pattern for a grating with $b = 25\lambda$, $a = 4b$ and $N = 4$. (a) shows the variation due to a single slit, (b) that of N coherent sources, while (c) shows the whole pattern due to the grating.

1. Since the approaching waves at the aperture and the observation points are no longer plane, both r and r' enter the relevant diffraction formulas.

2. Since the the direction from various points O on the aperture to a given field point P may no longer be considered approximately constant, the dependence of amplitude on the direction of Huygens wavelets needs be considered. This correction is achieved by the so-called obliquity factor.

Let the contribution to the disturbance at P due to the elemental area da be given by

$$dE_P = \left(\frac{dE_0}{r}\right) e^{ikr}, \tag{7.104}$$

where the amplitude is proportional to the area da, i.e., $dE_0 \sim E_L da$. Since the amplitude E_L arises due to the point source at S, we have

$$E_L = \left(\frac{E_s}{r'}\right) e^{ikr'}. \tag{7.105}$$

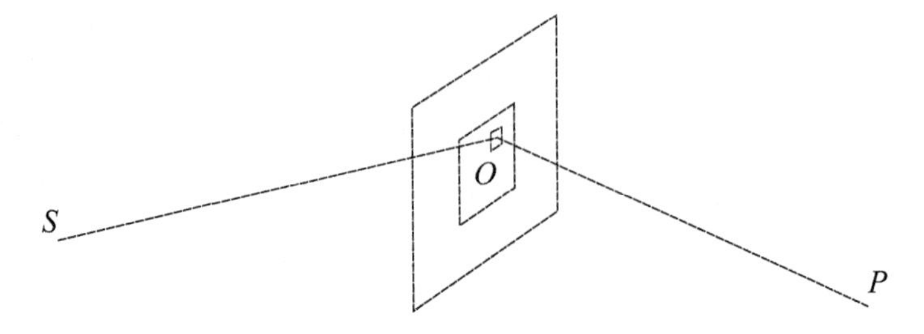

FIGURE 7.17: The schematics of Fresnel diffraction.

Combining Eqs. (7.104) and (7.105), except for a constant we have

$$dE_P = \left(\frac{E_s}{rr'}\right) e^{ik(r+r')} da. \tag{7.106}$$

Thus the field at P due to the entire aperture can be written as

$$E_P = E_s \int\int \left(\frac{1}{rr'}\right) e^{ik(r+r')} da. \tag{7.107}$$

Eq. (7.107) is incomplete on two counts: (i) it does not incorporate the obliquity factor $F(\theta)$, and (ii) it does not have the $\pi/2$ phase change of the diffracted wave with respect to the incoming wave. After incorporating these changes the formula now reads

$$E_P = \frac{-ikE_s}{2\pi} \int\int F(\theta) \left(\frac{e^{ik(r+r')}}{rr'}\right) da, \tag{7.108}$$

where $-i = \exp(-i\pi/2)$ accounts for the phase shift and $F(\theta)$ is given by

$$F(\theta) = \frac{1+\cos(\theta)}{2}. \tag{7.109}$$

Eq. (7.109) holds still under the approximation $\lambda < b < r, r'$. The integration is to be performed over a closed surface, including the aperture. Kirchhoff's approximations amount to the fact that the wave function and the derivative vanish right behind the opaque part of the screen. The vector field $\mathbf{E}$ is approximated by a scalar having the same value at the aperture as in the case of its absence.

7.6 Scalar diffraction theory

We first present a qualitative comparison of the scalar theory of diffraction and the rigorous vector theory in order to have a feel for the domain of validity of the scalar theory. For a monochromatic electromagnetic field in a linear medium, the vector fields **E** or **H** satisfy the vector Helmholtz equations. Further, for a homogeneous isotropic medium, any of the Cartesian components of the electric or the magnetic field satisfy the same Helmholtz equation, albeit for the scalar component, However, for light propagating through a step-index medium (interface between two dielectric media) or any localized or distributed inhomogeneity (as in diffraction problems), the situation is not so simple. The assumptions *isotropic* and *homogeneous* break down and there is mixing of the various components of **E** and **H**, and even coupling between them via the boundary conditions. In such cases, if the mixing and couplings are strong, the rigorous theory must incorporate the inherent vector character of the fields. In fact, a comparison of the rigorous vector theory with the scalar counterpart reveals ripples (in the step-index example) in both amplitude and phase in the rigorous treatment, while they are absent in the scalar theory [57]. The differences are noticeable only in the immediate vicinity of the interface. In the case of apertures, in typical diffraction problems the differences show up near the edge of the apertures. After several wavelengths away from the inhomogeneity both the rigorous and the scalar approximation produce similar results and the mixing effects can be ignored. Diffraction from sub-wavelength structures thus may need full vectorial treatment (see Chapter 14). Except for such cases, a scalar theory is a widely accepted tool for diffraction problems. The scalar theory starts with Green's second identity and leads to Kirchhoff's integral theorem. Applied to a specific problem of diffraction from an open aperture in an otherwise opaque screen, this leads to the Fresnel-Kirchhoff diffraction integral. As we go along we will briefly mention the limitations of Kirchhoff's approximation in the boundary conditions, which led to Bethe's theory [58] and the recent developments on extraordinary transmission (see Chapter 14).

7.6.1 Helmholtz-Kirchhoff integral theorem

Consider a volume V enclosed by a surface S (see Fig. 7.18). Let the scalar field U satisfy the Helmholtz equation in V. the Kirchhoff integral relates the field at a point P to the value of the field and its first derivative on the boundary. In order to arrive at the integral theorem we invoke the second Green's identity, applicable to two functions U and G, which are continuous along with their first and second derivatives in V as well as on S. Green's second identity can be written as

$$\iiint_V (U\nabla^2 G - G\nabla^2 U)dV = -\iint_S \left(U\frac{\partial G}{\partial n} - G\frac{\partial U}{\partial n} \right) da, \qquad (7.110)$$

where $\frac{\partial F}{\partial n} = \nabla F \cdot \mathbf{n}$ is the directional derivative along unit inward normal $\mathbf{n}$. In the context of diffraction problems it is more convenient to use the inward normal, though a standard form of Green's identity involves the outward normal [31]. Let G also satisfy the Helmholtz equation so that for both the functions similar relations hold: $\nabla^2 G = -k^2 G$ and $\nabla^2 U = -k^2 U$. Thus the

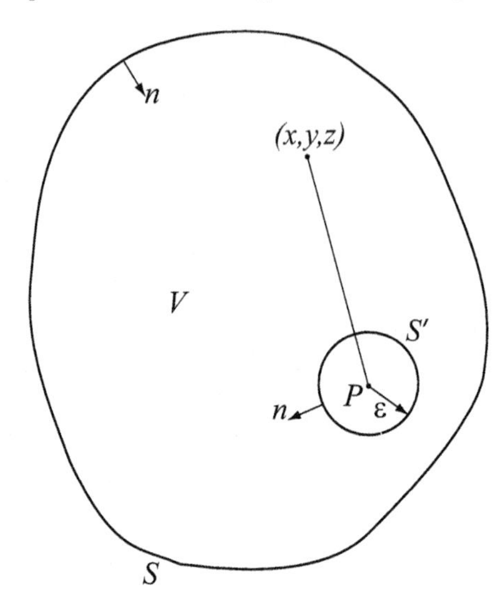

FIGURE 7.18: Schematics of the domain of integration for the derivation of the Helmholtz-Kirchhoff integral theorem.

integrand on the left-hand side of Eq. (7.110) reduces to zero and we have

$$\iint_S \left(U\frac{\partial G}{\partial n} - G\frac{\partial U}{\partial n} \right) da = 0. \tag{7.111}$$

Let the auxiliary Green's function G (sometimes referred to as the point function) be given by

$$G(x, y, z) = \frac{e^{iks}}{s}, \tag{7.112}$$

where s is the distance from point P to (x, y, z). The field given by Eq. (7.112) represents a spherical wave emanating from a point source at P and thus has a singularity at P ($s = 0$). The closed surface in Eq. (7.111) must exclude any singularity. To this end we exclude a spherical region with surface S' with center at P and with radius ϵ (see Fig. 7.18). The integration in Eq. (7.111) can now be performed on $S \cup S'$, where the enclosed volume does not have any singularity. We get

$$\iint_S \left(U\frac{\partial G}{\partial n} - G\frac{\partial U}{\partial n} \right) da = -\iint_{S'} \left(U\frac{\partial}{\partial n}\left(\frac{e^{iks}}{s}\right) - \frac{e^{iks}}{s}\frac{\partial U}{\partial n} \right) da'. \tag{7.113}$$

Since $\mathbf{n}$ is the outward normal to S' and s is along the same direction, the directional derivative $\frac{\partial}{\partial n}$ can be replaced by $\frac{\partial}{\partial s}$, and for the directional derivative of the auxiliary Green's function we have

$$\frac{\partial G}{\partial n} = \frac{\partial}{\partial s}\left(\frac{e^{iks}}{s}\right) = \frac{e^{iks}}{s}\left(ik - \frac{1}{s}\right). \tag{7.114}$$

In order to evaluate the integral on the right-hand side of Eq. (7.113), we replace the surface element da' by the element of the solid angle $d\Omega$ and rewrite it in the form

$$-\iint_{\Omega}\left(U\left(ik-\frac{1}{\epsilon}\right)\left(\frac{e^{ik\epsilon}}{\epsilon}\right)-\frac{e^{ik\epsilon}}{\epsilon}\frac{\partial U}{\partial s}\right)\epsilon^2 d\Omega. \tag{7.115}$$

Finally, taking the limit $\epsilon \to 0$ on the right-hand side of Eq. (7.115), we have finite contribution only from the second term, reducing the integral over S' to $4\pi U(P)$. Thus Eq. (7.113) reduces to

$$U(P)=\frac{1}{4\pi}\iint_{S}\left(U\frac{\partial}{\partial n}\left(\frac{e^{iks}}{s}\right)-\frac{e^{iks}}{s}\frac{\partial U}{\partial n}\right)da. \tag{7.116}$$

It is clear from Eq. (7.116) that a knowledge of the field and its first derivative on the surface is adequate for its evaluation at an interior point.

7.6.2 Fresnel-Kirchhoff diffraction integral

Consider now a typical diffraction scenario as depicted in Fig. 7.19. A point source is placed at P_0 and let the observation point be at P. The source and the observation points are separated by an opaque screen $\mathcal{B}$ with an opening (aperture) $\mathcal{A}$. Let the distance between an arbitrary point P_A on the aperture to the observer (source) be denoted by s (r). We assume the linear dimension of the opening to be larger than the wavelength λ, though much smaller than both r and s (see Chapter 14 for near and far-field definitions). Let the closed surface be formed by the part of a spherical surface S_R with center at P with

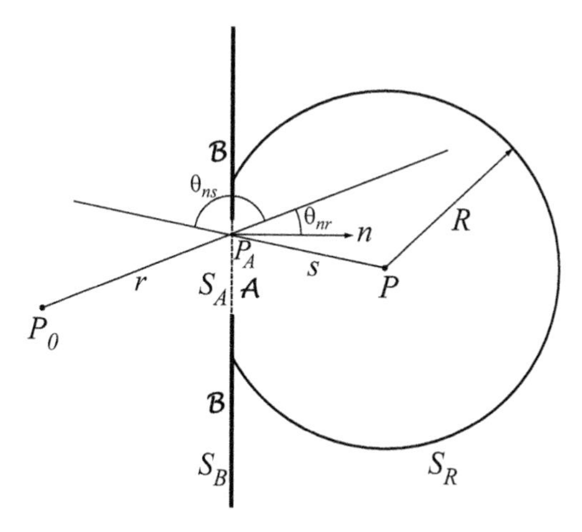

FIGURE 7.19: Schematics of the domain of integration for the derivation of the Fresnel-Kirchhoff diffraction integral.

a large radius R, opaque screen S_B and the opening S_A. Thus the Kirchhoff integral given by Eq. (7.116) can be broken up as follows:

$$\iint_S = \iint_{S_A} + \iint_{S_B} + \iint_{S_R}. \tag{7.117}$$

It can be shown that for sufficiently large R, the contribution from S_R is negligible [31]. In order to evaluate the contributions from S_A and S_B, Kirchhoff made the following approximations.

- On the opaque screen S_B, both the function and its derivatives vanish:

$$U = 0, \qquad \frac{\partial U}{\partial n} = 0. \tag{7.118}$$

- On the opening S_A, both the function and its derivatives are the same as created by the source, as if no screen were there:

$$U = U_i, \qquad \frac{\partial U}{\partial n} = \frac{\partial U_i}{\partial n}, \tag{7.119}$$

where U_i is the field produced by the source at P_A in absence of any obstacles.

There are several mathematical and physical inconsistencies in Kirchhoff approximations in the context of realistic systems. The mathematical inconsistency stems from the fact that in standard boundary value problems, we use the Dirichlet or the Neumann boundary conditions but not both in general. Kirchhoff's approximations given by Eq. (7.118) lead to the mathematical conclusion that the field must be zero everywhere in space. Rayleigh showed that either $U = 0$ or $\frac{\partial U}{\partial n} = 0$ is enough to derive another integral, namely, the Rayleigh-Sommerfield diffraction integral [31]. Moreover, in reality most of the opaque screens in diffraction optics are made of metals with finite thickness and conductivity. Such screens, along with the holes, can support surface plasmons and localized plasmons. Also as discussed earlier, scalar theory can break down in the near vicinity of the aperture. A deeper understanding of these limitations led to the recently discovered area of extraordinary transmission (see Chapter 14).

Referring back to Fig. 7.19 and assuming an amplitude A for the spherical wave, we have

$$U_i = A\frac{e^{ikr}}{r}, \qquad \frac{\partial U_i}{\partial n} = A\frac{e^{ikr}}{r}\left(ik - \frac{1}{r}\right)\cos\theta_{nr}. \tag{7.120}$$

For the auxiliary Green's function we have similar expressions:

$$G = \frac{e^{iks}}{s}, \qquad \frac{\partial G}{\partial n} = A\frac{e^{iks}}{s}\left(ik - \frac{1}{s}\right)\cos\theta_{ns}. \tag{7.121}$$

In Eqs. (7.120) and (7.121), θ_{nr} (θ_{ns}) is the angle between the inward normal and r (s). In view of the boundary conditions given by Eq. (7.118), there is no contribution from S_B. Further, in the limit $kr \gg 1,\ ks \gg 1$, the field at P can be written as

$$U(P) = -\frac{iA}{2\lambda}\iint_{S_A}\frac{e^{ik(r+s)}}{rs}(\cos\theta_{nr} - \cos\theta_{ns}))da. \tag{7.122}$$

Eq. (7.122) is known as the Fresnel-Kirchhoff diffraction integral.

For symmetric (with respect to the aperture) illumination when $\cos\theta_{nr} = 0$, Eq. (7.122) simplifies to

$$U(P) = -\frac{iA}{2\lambda}\frac{e^{ikr_0}}{r_0}\iint_{S_A}\frac{e^{iks}}{s}(1+\cos\chi))da, \tag{7.123}$$

where $\chi = \pi - \theta_{ns}$ and r_0 is the radius of the spherical wavefront reaching the aperture. In Eq. (7.123) we can easily see the manifestation of the Huygens-Fresnel principle (see Eq. (7.108)) by recognizing the obliquity/inclination factor $F(\chi)$ given by Eq. (7.109). We can further distinguish between the Fraunhofer and Fresnel regimes of diffraction, which are covered extensively in the standard optics literature [31]. We covered the scalar theory mainly in order to understand the inconsistencies in the Kirchhoff diffraction theory.

7.7 Rayleigh criterion

In different parts of the book, we touch upon the notion of resolution and its limits in order to explore how to beat these limits. In the context of the far-field patterns, Rayleigh proposed a criterion that now bears his name. Let us understand the Rayleigh criterion in the context of diffraction from a single slit, though it is applicable to many other situations and instruments. The normalized intensity pattern for a slit (see also Eq. (7.97)) is given by

$$\frac{I(\beta)}{I_0} = \left(\frac{\sin\beta}{\beta}\right)^2. \tag{7.124}$$

According to Rayleigh the two lines with equal intensity $\lambda_0 \pm \Delta\lambda_0/2$ separated by $\Delta\lambda_0$ are resolved if the principal intensity peak of one corresponds to the first intensity minimum of the other. Translating $\Delta\lambda_0$ into $\Delta\beta$, this implies that $\Delta\beta = \pi$ as shown in Fig. 7.20. The solid lines show the individual intensity patterns due to the two lines, while the dash-dotted curve shows the superposition. Note that for the resultant intensity, the ratio of the midpoint intensity to the peak value is given by 0.811. We also define the resolving power of an instrument by the ratio $\lambda_0/\Delta\lambda_0$.

Note that the Rayleigh criterion, based on formulas like Eq. (7.97), holds for the far-field and propagating waves, where interference is the dominant

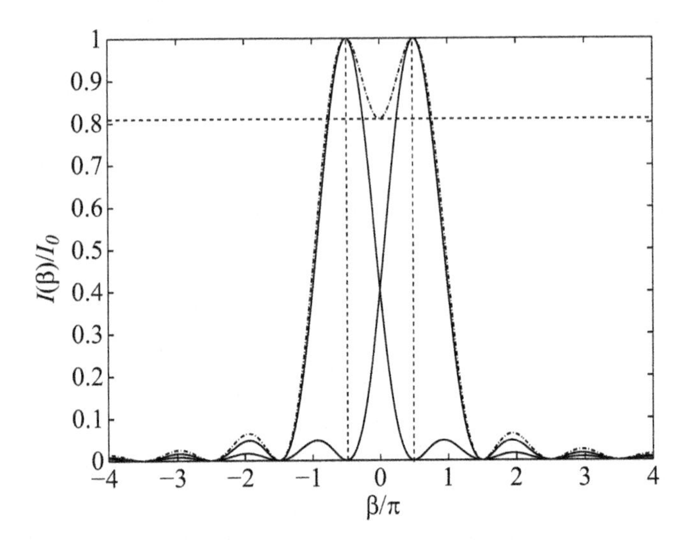

FIGURE 7.20: Single slit diffraction pattern for the explanation of the Rayleigh criterion.

physical process. The natural length-scale for such processes is the wavelength of light. It is thus not difficult to figure out that in the near-field we can possibly beat the Rayleigh limit in order to resolve sub-wavelength structures. These will correspond to nanostructures with visible light to probe them.

Chapter 8

Rays and beams

In our daily lives, we are accustomed to the formation of a shadow when an object is illuminated with light. From a broad perspective, our experience with the shadow of an object confirms our basic understanding of light as having rectilinear propagation. A closer inspection of the boundary between the illuminated and the shadow region reveals that the transition from one to another is never abrupt. A more detailed investigation reveals the oscillatory nature (in space) of the light-shadow boundary, which is a consequence of the wave nature of the light giving rise to diffraction, studied in Chapter 7. In our daily lives, we are mostly concerned about the visual appeal of objects much larger than the wavelength of light used to probe them. We thus consider a regime of optics to explain our daily experiences with objects around us and some devices we often use (telescopes, binoculars, camera, etc.). In reality many of these devices were developed long before a satisfactory understanding of wave and physical optics.

Geometrical optics deals with the limiting case when $\lambda \to 0$, and it can explain most of the optical phenomena in terms of geometrical constructions. In the first part of this chapter we will focus on some aspects of geometric optics or ray optics, particularly paraxial rays. We start with the derivation of the ray trajectories from the first principles, i.e., from the Helmholtz equation. We then consider the propagation of paraxial rays through typical optical elements and how this can be described by a 2×2 $ABCD$ matrix formulation. We apply this formulation to derive criteria for the stable propagation of rays. In the second part of the chapter we consider optical beams, which are the solution of the paraxial wave equation. We show how the propagation of the beam can be described by the same $ABCD$ matrices developed for rays. We define a complex beam parameter, the evolution of which can be described by a linear fractional transform (Möbius transformation), which is well known in standard complex analysis.

8.1 Eikonal equation and rays

We start from the scalar wave equation for a monochromatic wave leading to the Helmholtz equation

$$[\nabla^2 + k^2]E(\mathbf{r}) = 0, \quad k = k_0 n(\mathbf{r}) = \omega n(\mathbf{r})/c. \tag{8.1}$$

Eq. (8.1) is a generalization for a typical isotropic inhomogeneous medium, where n could be a function of the space coordinates. We seek a solution of the form

$$E(\mathbf{r}) = A(\mathbf{r})e^{ik_0\mathcal{S}(\mathbf{r})}, \tag{8.2}$$

where A and $\mathcal{S}$ in Eq. (8.2) are assumed to be smooth functions of their argument. Using Eq. (8.2) we calculate the partial derivatives, say, with respect to x in order to arrive at the expression of $\nabla^2 E$ and thus for the left-hand side of Eq. (8.1):

$$\frac{\partial E}{\partial x} = \left[\frac{\partial A}{\partial x} + ik_0\frac{\partial \mathcal{S}}{\partial x}A\right]e^{ik_0\mathcal{S}}, \tag{8.3}$$

$$\frac{\partial^2 E}{\partial^2 x} = \left[\frac{\partial^2 A}{\partial x^2} - Ak_0^2\left(\frac{\partial \mathcal{S}}{\partial x}\right)^2 + i\left(2k_0\frac{\partial \mathcal{S}}{\partial x}\frac{\partial A}{\partial x} + Ak_0\frac{\partial^2 \mathcal{S}}{\partial x^2}\right)\right]e^{ik_0\mathcal{S}}, \tag{8.4}$$

$$\left[\nabla^2 + k^2\right]E = \left[\left(\nabla^2 A + Ak^2 - Ak_0^2\left|\nabla\mathcal{S}\right|^2\right) + ik_0\left(A\nabla^2\mathcal{S} + 2\nabla A\cdot\nabla\mathcal{S}\right)\right]e^{ik_0\mathcal{S}}. \tag{8.5}$$

Separating the real and imaginary parts of Eq. (8.5), we have

$$n^2 - |\nabla\mathcal{S}|^2 + \frac{1}{k_0^2}\frac{\nabla^2 A}{A} = 0, \tag{8.6}$$

$$A\nabla^2\mathcal{S} + 2\nabla A\cdot\nabla\mathcal{S} = 0. \tag{8.7}$$

The limit of geometrical optics corresponds to $\lambda \to 0$ and $k_0 \to \infty$. Assuming vanishingly small wavelength, we rewrite Eq. (8.6):

$$n^2 - |\nabla\mathcal{S}|^2 = 0. \tag{8.8}$$

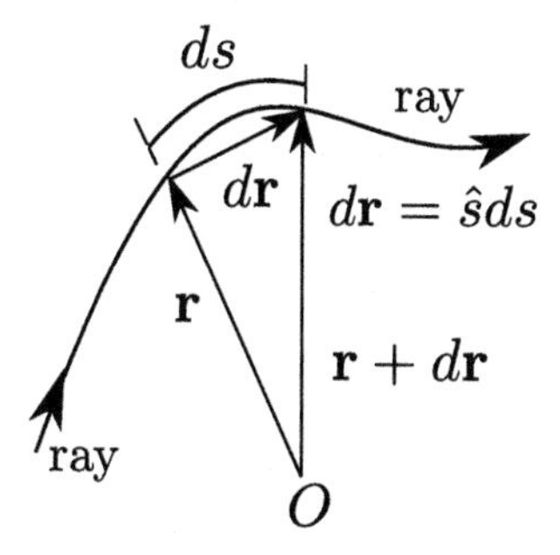

FIGURE 8.1: Schematics of ray propagation with the unit vector $\hat{\mathbf{s}}$ along the direction of propagation.

This is known as the *eikonal equation*, and $\mathcal{S}$ is called the *eikonal*. The geometrical rays are thus defined as the normals ($\nabla\mathcal{S}$) to the surface $\mathcal{S}(\mathbf{r})$ =constant. Further, Eq. (8.8) implies that the magnitude of $\nabla\mathcal{S}/n$ is unity and we can introduce a unit vector $\hat{\mathbf{s}}$, normal to the wavefront surface $\mathcal{S}$ as

$$\frac{\nabla\mathcal{S}}{n} = \hat{\mathbf{s}}. \tag{8.9}$$

Let us now relate $\hat{\mathbf{s}}$ to a given ray as shown in Fig. 8.1. Consider the point P on the ray whose position is given by $\mathbf{r}(s)$, a function of arc length s measured along the ray. It is clear from the figure that $d\mathbf{r}/ds = \hat{\mathbf{s}}$; combining this with Eq. (8.9) we have

$$\nabla\mathcal{S} = n\frac{d\mathbf{r}}{ds}. \tag{8.10}$$

Following Born and Wolf [31] we derive the trajectory of rays as follows. Differentiating both sides of Eq. (8.10) with respect to s gives us

$$\frac{d}{ds}\left(n\frac{d\mathbf{r}}{ds}\right) = \frac{d}{ds}\nabla\mathcal{S}, \tag{8.11a}$$

$$= \left(\frac{d\mathbf{r}}{ds}\cdot\nabla\right)\nabla\mathcal{S}, \tag{8.11b}$$

$$= \left(\frac{1}{n}\nabla\mathcal{S}\cdot\nabla\right)\nabla\mathcal{S}, \tag{8.11c}$$

$$= \left(\frac{1}{2n}\nabla(|\nabla\mathcal{S}|^2)\right), \tag{8.11d}$$

$$= \left(\frac{1}{2n}\nabla(n^2)\right), \tag{8.11e}$$

$$= \nabla n. \tag{8.11f}$$

In writing Eq. (8.11c) and Eq. (8.11e), we have made use of Eq. (8.10) and Eq. (8.8), respectively. Hence from Eq. (8.11f) we have

$$\frac{d}{ds}\left(n\frac{d\mathbf{r}}{ds}\right) = \nabla n, \tag{8.12}$$

which governs the propagation of rays.

In life we are quite used to rectilinear propagation of rays. Such rectilinear propagation takes place in a homogenous medium (n = constant), and the corresponding trajectory can be derived from Eq. (8.12) by setting $\nabla n = 0$. We then have

$$\frac{d^2\mathbf{r}}{ds^2} = 0, \tag{8.13}$$

whose solution corresponds to a straight line (with $\mathbf{a}$, $\mathbf{b}$ as arbitrary constants):

$$\mathbf{r} = s\mathbf{a} + \mathbf{b}. \tag{8.14}$$

The rectilinear propagation in a homogeneous medium can be shown from a different angle from Eq. (8.9). In fact, $\mathcal{S}$ can be expressed as

$$\mathcal{S} = n(\alpha x + \beta y + \gamma z), \tag{8.15}$$

with α, β, γ giving the direction cosines of the ray satisfying $\alpha^2+\beta^2+\gamma^2 = 1$. Eq. (8.15) represents a plane and the rays are aligned along straight lines perpendicular to this surface. It is clear that $\mathcal{S}$ defined by Eq. (8.15) satisfies

$$\frac{\nabla\mathcal{S}}{n} = \left(\alpha\hat{\mathbf{i}} + \beta\hat{\mathbf{j}} + \gamma\hat{\mathbf{k}}\right) = \hat{\mathbf{s}} \tag{8.16}$$

and thus represents rectilinear propagation. We now focus our attention on Eq. (8.7) in order to get insight into the physical interpretation of $\mathcal{S}$ in the context of energy propagation. To that goal we parametrize $\mathcal{S}$ and A and treat them as functions of s. Thus, for example, $\nabla A \cdot \hat{\mathbf{s}} = \frac{\partial A}{\partial s}$ and Eq. (8.7) can be written as

$$A\nabla^2\mathcal{S} + 2n\frac{\partial A}{\partial s} = 0. \tag{8.17}$$

Eq. (8.17) gives the physical basis of this definition of a ray. These would be the trajectories normal to the wavefront surface (the equiphase surface). Integration of Eq. (8.17) leads to

$$A(s) = A_0 \exp\left[-\int_0^s \frac{\nabla^2\mathcal{S}}{2n}ds\right], \tag{8.18}$$

where A_0 is the amplitude at the initial point $s = 0$. Eq. (8.18) clearly implies that in order to get a ray trajectory, it is sufficient to know it at any arbitrary point on that ray. But geometric optics is inadequate to predict the behavior of any adjacent ray, though it can give the change in the amplitude along a separate ray. Thus in geometrical optics, optical fields along one ray are independent of the same along another ray. In order to draw a parallel with the hydrodynamics of incompressible fluid, we multiply Eq. (8.7) by A yielding

$$A^2\nabla \cdot (n\hat{\mathbf{s}}) + 2A\nabla A \cdot (n\hat{\mathbf{s}}) = 0. \tag{8.19}$$

In writing Eq. (8.19) we have made use of the relation $\nabla^2 \mathcal{S} = \nabla \cdot \nabla \mathcal{S} = \nabla \cdot (n\hat{\mathbf{s}})$. In compact form Eq. (8.19) reads as

$$\nabla \cdot (nA^2\hat{\mathbf{s}}) = \nabla \cdot \mathbf{J} = 0, \tag{8.20}$$

where we introduced the current density $\mathbf{J} = nA^2\hat{\mathbf{s}}$. Eq. (8.20) represents the well-known continuity equation for a stationary incompressible fluid. In our case $\mathbf{J}$ gives the light flux density. It is proportional to the (time)-averaged Poynting vector $\langle \mathbf{S} \rangle$ given by [31]

$$\langle \mathbf{S} \rangle = v\langle w \rangle \hat{\mathbf{s}}, \tag{8.21}$$

where v, $\langle w \rangle$ are velocity and the time-averaged energy density, respectively. Thus light flows along the narrow light tubes formed by rays.

8.2 Ray propagation through linear optical elements

As discussed earlier ray propagation in a homogeneous isotropic medium is rectilinear. With respect to a given axis of the optical system, the ray can be characterized by two parameters, namely, the distance r of the ray from the optical axis and its slope $r' = \frac{dr}{dz}$ (see Fig. 8.2).

$$\mathcal{R}(z) = \begin{pmatrix} r(z) \\ r'(z) \end{pmatrix}. \tag{8.22}$$

According to some literature, instead of the slope $r'(z)$, the reduced slope defined by

$$r'(z) = n_0(z)\frac{dr(z)}{dz} \tag{8.23}$$

is used with $n_0(z)$ as the refractive index of the medium. Throughout this chapter we use only the ordinary slope and not the reduced slope. Furthermore, we consider only paraxial rays for which the angle θ between the ray and

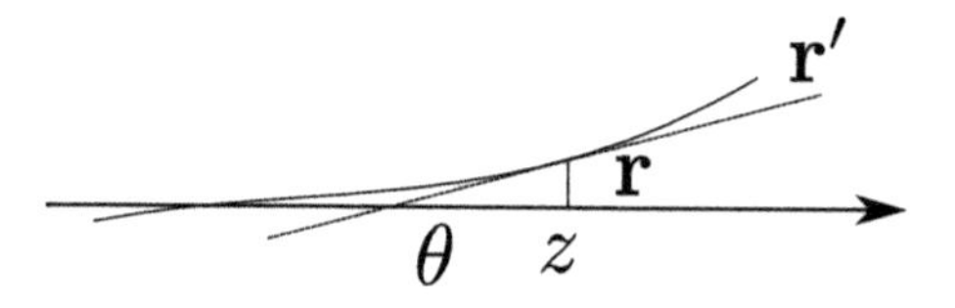

FIGURE 8.2: Schematics of ray parameters (distance and slope) at z.

the axis is small enough so that $\tan\theta \sim \sin\theta \sim \theta$. We now describe a method to evaluate the changes in $\mathcal{R}(z)$ for passage through linear optical elements. Let the initial ray state vector (before it enters the optical element) be given by $\mathcal{R}_1$ and the ray state after it by $\mathcal{R}_2$. Using a 2×2 matrix to characterize the optical element/elements, $\mathcal{R}_1$ and $\mathcal{R}_2$ can be related as

$$\mathcal{R}_2 = M\mathcal{R}_1, \quad M = \begin{pmatrix} A & B \\ C & D \end{pmatrix}. \tag{8.24}$$

Matrix M is known as the $ABCD$ matrix for the particular optical element/elements. For non-lossy elements, we further have $det(M) = AD - BC = 1$. $ABCD$ matrix elements for several linear optical elements are given in Table 8.1. They are also listed in many other standard textbooks [59, 60].

TABLE 8.1: $ABCD$ matrices for typical optical elements.

Homogeneous medium of length d and refractive index n	$\begin{pmatrix} 1 & \frac{d}{n} \\ 0 & 1 \end{pmatrix}$
Dielectric interface with $n_1(n_2)$ as the entry (exit) medium refractive index	$\begin{pmatrix} 1 & 0 \\ 0 & \frac{n_1}{n_2} \end{pmatrix}$
Spherical mirror of radius R	$\begin{pmatrix} 1 & 0 \\ -\frac{2}{R} & 1 \end{pmatrix}$
Spherical dielectric interface of radius $R(R > 0)$ with $n_1(n_2)$ as the entry (exit) medium refractive index	$\begin{pmatrix} 1 & 0 \\ -\frac{n_2-n_1}{n_1 R} & \frac{n_1}{n_2} \end{pmatrix}$
Thin lens of focal length $f(f > 0)$	$\begin{pmatrix} 1 & 0 \\ -\frac{1}{f} & 1 \end{pmatrix}$

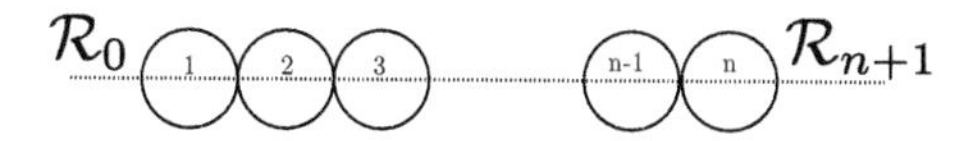

FIGURE 8.3: Ray propagation through a sequence of optical elements, each represented by circles.

8.2.1 Sequence of optical elements

Consider the sequence of optical elements arranged one after the other as shown in Fig. 8.3. The $ABCD$ matrix for this system can be evaluated as follows. The relation between the ray states before and after n optical elements is given by

$$\mathcal{R}_{n+1} = M_n \mathcal{R}_n = M_n M_{n-1} \mathcal{R}_{n-1} = M_{total} \mathcal{R}_0, \tag{8.25}$$

where M_{total} is the $ABCD$ matrix for the entire system

$$M_{total} = M_n M_{n-1} \cdots M_2 M_1. \tag{8.26}$$

Proper attention is to be paid to the order in which the product of matrices is taken in Eq. (8.26), as the matrix product is not commutative in general. Note also that $\det(M_{total}) = 1$ since the product of unimodular matrices is again unimodular.

8.2.2 Propagation in a periodic system: An eigenvalue problem

Consider a periodic system of linear optical elements, each represented by its own $ABCD$ matrix. We pose a general eigenvalue problem to find the stability of the ray propagation. Ray propagation is understood to be stable if r is finite for any z along the axis. On the contrary, diverging r implies an unstable system where the ray is no longer confined near the axis. Let the $ABCD$ matrix for one period be denoted by M. We show below that the eigenvalues of M determine the character of ray propagation and whether the ray is confined near the z-axis. The eigenvalue problem can be stated as [59]

$$M\mathcal{R} = \lambda \mathcal{R}, \tag{8.27}$$

where the eigenvalue λ satisfies the algebraic equation

$$\lambda^2 - 2m\lambda + AD - BC = 0, \tag{8.28}$$

with $m = \frac{A+D}{2}$. As our system is loss-free, we have $AD - BC = 1$. Solving for λ yields

$$\lambda_\pm = m \pm \sqrt{m^2 - 1}. \tag{8.29}$$

Let $\mathcal{R}_\pm$ represent the eigenvectors corresponding to $\lambda_\pm$, which are orthogonal (for distinct roots). Any ray $\mathcal{R}_0$ at the input and $\mathcal{R}_n$ after passage through n

periods can be written as the linear combination of $\mathcal{R}_{\pm}$:

$$\mathcal{R}_0 = c_+\mathcal{R}_+ + c_-\mathcal{R}_-, \tag{8.30}$$

$$\mathcal{R}_n = c_+\lambda_+^n\mathcal{R}_+ + c_-\lambda_-^n\mathcal{R}_-. \tag{8.31}$$

Eq. (8.31) is obtained by the n-fold application of M on Eq. (8.30). It is now clear why the evolution of the ray will depend on the nature of $\lambda_{\pm}$ (as $\lambda_{\pm}^n$ signifies converging or diverging solutions as $n \to \infty$). Based on the value of m, two cases are possible, namely, $m^2 \geq 1$, and $m^2 < 1$ (see Eq. (8.29)). We now consider these two cases separately and draw necessary conclusions.

- Case (a)

$$m^2 \leq 1, \quad \text{or} \quad -1 \leq \left(\frac{A+D}{2}\right) \leq 1. \tag{8.32}$$

We replace m by $\cos\theta$ since $|m| \leq 1$. Eigenvalues $\lambda_{\pm}$ now take the form

$$\lambda_{\pm} = \cos\theta \pm i\sqrt{1-\cos^2\theta} = e^{\pm i\theta}. \tag{8.33}$$

The eigenvector $\mathcal{R}_n$ in Eq. (8.31) can be written as

$$\mathcal{R}_n = c_+e^{in\theta}\mathcal{R}_+ + c_-e^{-in\theta}\mathcal{R}_-, \tag{8.34a}$$

$$= c_+(\cos n\theta + i\sin n\theta)\mathcal{R}_+ + c_-(\cos n\theta - i\sin n\theta)\mathcal{R}_-, \tag{8.34b}$$

$$= a_0\cos n\theta + b_0\sin n\theta, \tag{8.34c}$$

with $a_0 = c_+\mathcal{R}_+ + c_-\mathcal{R}_-$ and $b_0 = i(c_+\mathcal{R}_+ - c_-\mathcal{R}_-)$. It can be seen from Eq. (8.34c) that the ray state oscillates about the axis and, as mentioned earlier, this kind of system is called a geometrically stable system.

- Case (b)

$$|m| > 1, \quad \left|\frac{A+D}{2}\right| > 1. \tag{8.35}$$

As before we parametrize m by $\cosh\theta$ ($\theta \neq 0$). Eq. (8.29) can be rewritten as

$$\lambda_{\pm} = \cosh\theta \pm \sqrt{\cosh^2\theta - 1} = e^{\pm\theta}. \tag{8.36}$$

Let the eigenvectors corresponding to $\lambda_{\pm}$ be $\tilde{\mathcal{R}}_{\pm}$. The ray state after n periods $\tilde{\mathcal{R}}_n$, in terms of the new basis vectors, can be written as

$$\tilde{\mathcal{R}}_n = c_+e^{n\theta}\tilde{\mathcal{R}}_+ + c_-e^{-n\theta}\tilde{\mathcal{R}}_-, \tag{8.37a}$$

$$= c_+(\cosh n\theta + \sinh n\theta)\tilde{\mathcal{R}}_+ + c_-(\cosh n\theta - \sinh n\theta)\tilde{\mathcal{R}}_-, \tag{8.37b}$$

$$= \tilde{a}_0\cosh n\theta + \tilde{b}_0\sinh n\theta, \tag{8.37c}$$

with $\tilde{a}_0 = c_+\tilde{\mathcal{R}}_+ + c_-\tilde{\mathcal{R}}_-$ and $\tilde{b}_0 = (c_+\tilde{\mathcal{R}}_+ - c_-\tilde{\mathcal{R}}_-)$. The ray state after traversing n periods grows exponentially as given by Eq. (8.37c). As a consequence such systems are referred to as unstable systems.

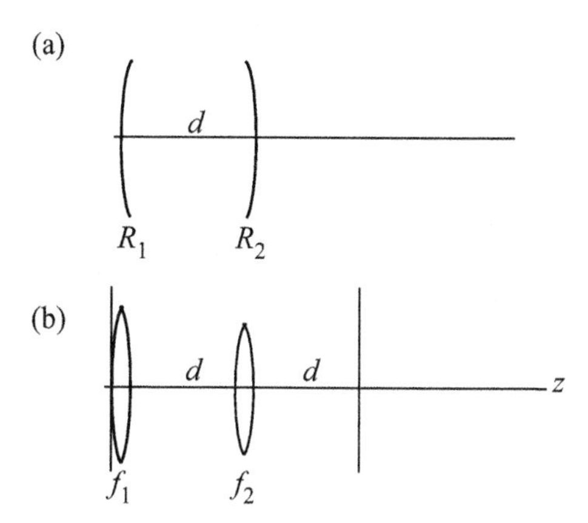

FIGURE 8.4: Schematics of (a) a spherical mirror cavity and (b) its equivalent lens waveguide.

We now present an example to understand the stability of rays in a resonator. Consider the system shown in Fig. 8.4(a) consisting of two mirrors (having radii R_1 and R_2) separated by a distance d. This system is periodic since the rays in the cavity can retrace their path, being bounced by the mirrors repeatedly. We evaluate the conditions under which the ray propagation is stable. Noting the equivalence of a thin lens and a spherical mirror, the system can be modeled by a periodic arrangement of lenses (also known as a lens waveguide) as shown in Fig. 8.4(b) with the focal lengths $f_1 = R_1/2$ and $f_2 = R_2/2$. The $ABCD$ matrix for one period (see Fig. 8.4(b)) is given by

$$M = \begin{pmatrix} 1 & d \\ 0 & 1 \end{pmatrix} \begin{pmatrix} 1 & 0 \\ -\frac{2}{R_2} & 1 \end{pmatrix} \begin{pmatrix} 1 & d \\ 0 & 1 \end{pmatrix} \begin{pmatrix} 1 & 0 \\ -\frac{2}{R_1} & 1 \end{pmatrix}, \tag{8.38}$$

$$= \begin{pmatrix} \left(1 - \frac{2d}{R_1}\right)\left(1 - \frac{2d}{R_2}\right) - \frac{2d}{R_1} & d\left(2 - \frac{2d}{R_2}\right) \\ -\frac{2}{R_2}\left(1 - \frac{2d}{R_1}\right) - \frac{2}{R_1} & 1 - \frac{2d}{R_2} \end{pmatrix}. \tag{8.39}$$

In writing Eq. (8.39) we have made use of Eq. (8.26) and Table (8.1). For the half-trace of the $ABCD$ matrix we have

$$\frac{A+D}{2} = 2\left(1 - \frac{d}{R_1}\right)\left(1 - \frac{d}{R_2}\right) - 1, \tag{8.40}$$

and using Eq. (8.32) we find that the resonator will be stable if

$$0 \leq \left(1 - \frac{d}{R_1}\right)\left(1 - \frac{d}{R_2}\right) \leq 1 \text{ or} \tag{8.41}$$

$$0 \leq g_1 g_2 \leq 1, \tag{8.42}$$

where $g_{1,2} = (1 - d/R_{1,2})$.

8.3 Beam characteristics

In this section we focus on optical beams in a homogeneous medium and then consider propagation of such beams through different optical elements. Since a beam is localized near an axis, we make the paraxial wave approximation to obtain one particular solution of the problem in the form of a fundamental Gaussian beam. Note that the paraxial wave equation allows for higher-order Gaussian beam modes (Hermite-Gaussian or the Lagurre-Gaussian) depending on the spatial symmetry of the problem.

8.3.1 Paraxial wave equation and its solutions

Assuming predominant propagation along the z-axis, we use the ansatz

$$E(x, y, z, t) = \mathcal{U}(x, y, z)e^{i(kz-\omega t)} + \text{c.c}, \tag{8.43}$$

where $\mathcal{U}(x, y, z)$ is assumed to be a slowly varying function of z, since we filtered out the quickly oscillating part (e^{ikz}). Substituting this ansatz in the wave equation leads to

$$\frac{\partial^2 \mathcal{U}}{\partial x^2} + \frac{\partial^2 \mathcal{U}}{\partial y^2} + \frac{\partial^2 \mathcal{U}}{\partial z^2} + 2ik\frac{\partial \mathcal{U}}{\partial z} = 0. \tag{8.44}$$

SVEA (see Section 1.5.4) implies

$$\left|\frac{\partial^2 \mathcal{U}}{\partial z^2}\right| \ll \left|k\frac{\partial \mathcal{U}}{\partial z}\right| \ll \left|k^2\mathcal{U}\right| \tag{8.45}$$

and simplifies Eq. (8.44) as

$$\Delta_\perp \mathcal{U} + 2ik\frac{\partial \mathcal{U}}{\partial z} = 0, \tag{8.46}$$

where $\Delta_\perp$ denotes the transverse Laplacian and Eq. (8.46) is known as the *paraxial wave equation*. We seek a solution of the form

$$\mathcal{U} = A(z)\exp\left(ik\frac{x^2+y^2}{2q(z)}\right). \tag{8.47}$$

Substituting Eq. (8.47) in Eq. (8.46), we have

$$\frac{A(z)k^2}{q^2(z)}\left(\frac{dq}{dz} - 1\right)(x^2+y^2) + 2ik\left(\frac{A(z)}{q(z)} + \frac{dA(z)}{dz}\right) = 0. \tag{8.48}$$

Separating the real and imaginary parts and demanding a nontrivial solution, we arrive at the following set of coupled equations:

$$\frac{dq(z)}{dz} = 1, \tag{8.49}$$

$$\frac{dA(z)}{dz} = -\frac{A(z)}{q(z)}, \tag{8.50}$$

whose solutions are

$$q(z) = q_0 + z, \tag{8.51}$$

$$\frac{A(z)}{A(0)} = \frac{q_0}{q(z)}. \tag{8.52}$$

In writing Eqs. (8.51) and (8.52), we assumed that at $z = 0$, $A(z) = A_0$ and $q(z) = q_0$. The complex function q can be related to the relevant physical beam parameters as

$$\frac{1}{q(z)} = \frac{1}{R(z)} + i\frac{\lambda}{\pi w^2(z)}, \tag{8.53}$$

where $R(z)$, $w(z)$ are the real functions of z. The physical meaning of R and w will be transparent after we derive the expression of the fundamental Gaussian beam. We measure z from $z = 0$ where $R \to \infty$. We show below that this amounts to saying that the beam wavefront is planar at $z = 0$. Similarly we have also assumed that $w(z)$ at $z = 0$ is given by w_0. We can introduce the so-called Rayleigh range z_R as

$$z_R = iq_0 = \frac{\pi w_0^2}{\lambda}. \tag{8.54}$$

With this definition we find that $q(z)$ can be written as

$$\frac{1}{q(z)} = \frac{1}{z - iz_R} = \frac{z + iz_R}{z^2 + z_R^2}. \tag{8.55}$$

Now comparing the real and imaginary parts of Eq. (8.53) and Eq. (8.55), we find that

$$R(z) = z\left[1 + \left(\frac{z_\mathrm{R}}{z}\right)^2\right], \tag{8.56}$$

$$w^2(z) = w_0^2\left[1 + \left(\frac{z}{z_\mathrm{R}}\right)^2\right]. \tag{8.57}$$

Similarly, we also find that

$$\frac{A(z)}{A(0)} = \frac{q_0}{q_0 + z} = \frac{1 - i(z/z_R)}{1 + (z/z_R)^2}, \tag{8.58}$$

which can be rewritten as

$$\frac{A(z)}{A(0)} = \sqrt{\frac{1}{1+(z/z_R)^2}}\, e^{i\psi} = \frac{w_0}{w(z)} e^{i\psi}, \tag{8.59}$$

where

$$\psi = -\tan^{-1}\left(\frac{z}{z_R}\right) \tag{8.60}$$

is known as the Guoy phase. Substituting Eqs. (8.56)–(8.59) in Eq. (8.47), we get

$$\mathcal{U}(x,y,z) = \mathcal{U}_0 \frac{w_0}{w(z)} e^{i\psi} \left[\exp\left(ik\frac{x^2+y^2}{2R(z)}\right)\right] \exp\left(-\frac{x^2+y^2}{w^2(z)}\right), \tag{8.61}$$

where $\mathcal{U}_0$ is the normalization factor. This is known as the fundamental mode and labeled by TEM_{00}. Eqs. (8.56) and (8.57) in Eq. (8.61) provide a clear physical meaning to $R(z)$ and $w(z)$ as the radius of curvature of the phase front and the beam spot size $w(z)$ at z, with $w(0) = w_0$ being the beam waist. In fact, the knowledge of the complex beam parameter $q(z)$ at any z makes it possible to recognize all the physical parameters (R, w, ψ) of the fundamental Gaussian beam. As mentioned earlier there are infinite solutions to paraxial equations. When solved in Cartesian coordinates, any general TEM_{mn} is given by [59, 60]

$$\mathcal{U}_{mn}(x,y,z) = \left(\frac{1}{\pi 2^{n+m-1} n! m!}\right)^{1/2} \frac{w_0}{w(z)} \times \left[H_m\left(\frac{\sqrt{2}x}{w(z)}\right) H_n\left(\frac{\sqrt{2}y}{w(z)}\right) \exp\left(ik\frac{x^2+y^2}{2R(z)} + i(n+m+1)\psi(z)\right)\right] \exp\left(-\frac{x^2+y^2}{w^2(z)}\right), \tag{8.62}$$

where $H_{n,m}$ are the Hermite polynomials. Degeneracy of the modes with the same $m+n$ value is clear from Eq. (8.62). In the cylindrical basis we have the Laguerre-Gaussian modes, which can have vortex character with angular momentum (see Chapter 12).

8.3.2 *ABCD* matrix formulation for fundamental Gaussian beam

The passage of a beam through optical elements can again be derived using the 2×2 matrix formulation, as with rays. We now propagate the complex beam parameter. The input and output beam parameters (q_2 and q_1, respectively) are related by the following linear fractional transform:

$$q_2 = \frac{Aq_1 + B}{Cq_1 + D}, \tag{8.63}$$

and for a sequence of optical elements, we must take the matrix product in reverse order, as in Eq. (8.26). The proof of Eq. (8.63) can be found in Ref. [59]; we do not include it here.

8.3.3 Stability of beam propagation

In the case of rays, stability implied that rays were confined to the axis. The corresponding eigenvalue problem in Eq. (8.27) had complex eigenvalues assuring oscillations about the axis. In the case of beams, a similar picture holds, albeit with necessary changes. For a stable cavity, the beam (to be precise, the complex beam parameter) has to replicate itself after each period, requiring the following equation to be valid:

$$q = \frac{Aq + B}{Cq + D}. \tag{8.64}$$

Eq. (8.64) represents a quadratic equation for q. It is easier to deal with $1/q$, whose roots are given by

$$\frac{1}{q} = \frac{D - A}{2B} \pm \frac{1}{B}\sqrt{\left(\frac{A + D}{2}\right)^2 - 1}. \tag{8.65}$$

In order to qualify as a Gaussian beam with a finite spot size, $1/q$ must be complex, requiring

$$m^2 = \left(\frac{A + D}{2}\right)^2 \leq 1, \tag{8.66}$$

leading to

$$\frac{1}{q} = \frac{D - A}{2B} \pm \frac{i}{|B|}\sqrt{1 - m^2}. \tag{8.67}$$

We thus arrive at the same stability condition for beams as for rays (compare with Eq. (8.32)). Eq. (8.67) easily leads to the stable beam parameters R and w as per Eq. (8.53).

Chapter 9

Optical waves in stratified media

Often in optics we must deal with layered media when the optical properties change only in one specific direction (say, along the z-axis), while in any plane transverse to this direction, the optical properties are invariant. Such media with dielectric and magnetic responses given by $\varepsilon(z)$ and $\mu(z)$ as functions of only z are also referred to as stratified media. There are many examples of such media. The most useful one refers to the optical interference coatings that are essential for most optical instruments. Reflection and transmission through these structures can be handled in terms of simple 2×2 matrices when the constituent layers are homogeneous and isotropic. In the case of an anisotropic layered medium, we can develop a 4×4 matrix formulation for uniaxial materials [61, 62]. In this chapter we deal with isotropic homogeneous layers and develop the characteristics matrix formalism to obtain the refection and transmission coefficients (see also Ref. [31]). We discuss how the dispersion in such structures can be engineered to lead to slow and fast light [7]. We apply the technique to investigate the modes of a structure. We probe the effects of finite temporal width of a pulse leading to the Wigner delay [63]. The space equivalent of Wigner delay, also known as the Goos-Hänchen shift [64], for a spatially finite beam is then discussed. Finally, we show how someone using such structures can realize perfect transmission and coherent perfect absorption. We will define all the necessary notions and concepts as we go along.

9.1 Characteristics matrix approach

Characteristic matrices relate the tangential field components at two interfaces of the medium. We consider a stratified medium consisting of isotropic homogeneous media, as shown in Fig. 9.1. We obtain the matrix for a particular j-th layer with width d_j occupying the space between planes $z = z_j$ and $z = z_{j+1}$. Let the material properties, namely, the relative dielectric permittivity ε_{rj} and permeability μ_{rj}, be given. Since our focus is mostly on plasmonic phenomena, we present results for the TM- or p-polarized monochromatic plane waves (with only nonvanishing components H_y, E_x and E_z), while the case for the TE- or s-polarization can be worked out in an analogous manner.

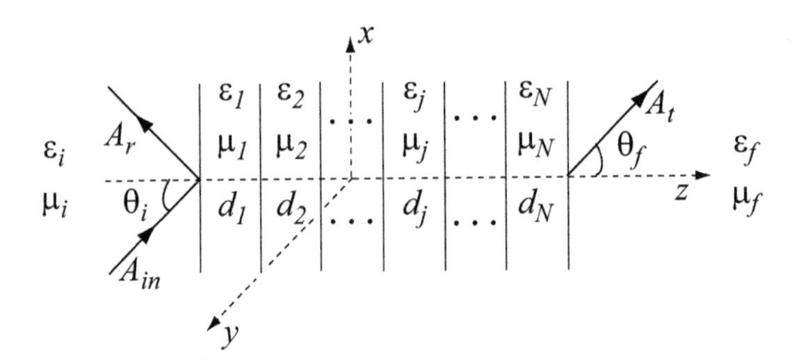

FIGURE 9.1: Schematic of a layered structure.

The medium stratification leads to forward and backward propagating waves and thus the magnetic field H_y can be written as a superposition of forward and backward waves:

$$H_{jy} = [A_{j+}e^{ik_{jz}(z-z_j)} + A_{j-}e^{-ik_{jz}(z-z_j)}]e^{ik_x x}e^{-i\omega t}, \tag{9.1}$$

while the expression for the corresponding tangential component of the electric field E_{jx} from Maxwell's equation (see Eq. (2.28)) is given by

$$\sqrt{\varepsilon_0}E_{jx} = [p_{jz}\sqrt{\mu_0}(A_{j+}e^{ik_{jz}(z-z_j)} - A_{j-}e^{-ik_{jz}(z-z_j)})]e^{ik_x x}e^{-i\omega t}. \tag{9.2}$$

In Eqs. (9.1) and (9.2), $A_{j\pm}$ are the forward and backward wave amplitudes, while k_{jz} and p_{jz} are expressed through the x-component of the propagation constant k_x:

$$k_{jz} = \sqrt{k_0^2 n_j^2 - k_x^2}, \quad k_x = k_0 n_i \sin\theta_i, \quad p_{jz} = \frac{k_{jz}}{k_0 \varepsilon_{rj}}, \tag{9.3}$$

where $k_0 = \omega/c$, $n_j = \sqrt{\varepsilon_{rj}\mu_{rj}}$ and $n_i = \sqrt{\varepsilon_{ri}\mu_{ri}}$ (medium of incidence). In order to ensure causality, the square root computed in Eq. (9.3) is chosen such that the imaginary part of the z component of the wave vector is positive. Writing Eqs. (9.1) and (9.2) in terms of a matrix,

$$\begin{pmatrix} \sqrt{\mu_0}H_{jy} \\ \sqrt{\epsilon_0}E_{jx} \end{pmatrix} = \begin{pmatrix} e^{ik_{jz}z} & e^{-ik_{jz}z} \\ p_{jz}e^{ik_{jz}z} & -p_{jz}e^{-ik_{jz}z} \end{pmatrix} \begin{pmatrix} \sqrt{\mu_0}A_{j+} \\ \sqrt{\mu_0}A_{j-} \end{pmatrix}, \tag{9.4}$$

and relating the tangential components of the fields at the left and right faces of the j-th layer, we can relate the corresponding tangential field components by the matrix relation:

$$\begin{pmatrix} H_y \\ E_x \end{pmatrix}_j = M_j \begin{pmatrix} H_y \\ E_x \end{pmatrix}_{j+1}, \tag{9.5}$$

where we have used the notation

$$\bar{E}_{jx} = \sqrt{\epsilon_0}E_{jx}, \qquad \bar{H}_{jy} = \sqrt{\mu_0}H_{jy}. \tag{9.6}$$

Henceforth, we drop the overbars in fields like in Eq. (9.5). The subscript j in Eq. 9.5 refers to the interface $z = z_j$ and the characteristic matrix M_j is given by [31]

$$M_j = \begin{pmatrix} \cos(k_{jz}d_j) & -(i/p_{jz})\sin(k_{jz}d_j) \\ -ip_{jz}\sin(k_{jz}d_j) & \cos(k_{jz}d_j) \end{pmatrix}. \tag{9.7}$$

For incidence of TE-polarized (s-polarized) light, the characteristics matrix remains the same except that now $p_{jz} = k_{jz}/(k_0\mu_{rj})$. For a layered medium with N layers, as in Fig. 9.1, the total characteristic matrix is given by

$$M_{total} = M_1 M_2 .... M_N. \tag{9.8}$$

Using the total characteristic matrix given by Eq. (9.8), we can relate the tangential components of electric and magnetic fields at the extreme ends of the layered structure as

$$\begin{pmatrix} H_y \\ E_x \end{pmatrix}_{z=0} = M_{total} \begin{pmatrix} H_y \\ E_x \end{pmatrix}_{z=d_N}. \tag{9.9}$$

Later, the characteristics matrix was generalized to Kerr nonlinear stratified media and applied to explore various nonlinear optical effects and photon localization in nonlinear systems [65].

9.2 Amplitude reflection, transmission coefficients and dispersion relation

In this section we present the results for the reflection and transmission features of a layered structure (see Fig. 9.1). Let the structure be illuminated by a TM-polarized plane monochromatic wave at an angle θ_i. Then from Eqs. (9.1) and (9.2) we can write the magnetic and electric fields at $z = 0$ (at the left-most interface) as

$$\begin{pmatrix} H_y \\ E_x \end{pmatrix}_{z=0} = \begin{pmatrix} 1 & 1 \\ p_{iz} & -p_{iz} \end{pmatrix} \begin{pmatrix} \sqrt{\mu_0} A_{in} \\ \sqrt{\mu_0} A_r \end{pmatrix}, \tag{9.10}$$

where $p_{iz} = (\sqrt{k_0^2 n_i^2 - k_x^2})/k_0 \varepsilon_{ri}$ is the normalized z component of the wave vector. A_{in} and A_r are the incident and the reflected amplitudes in the medium of incidence, respectively. We can also write the analogous expression for the fields in the final medium as

$$\begin{pmatrix} H_y \\ E_x \end{pmatrix}_{z=d_N} = \begin{pmatrix} 1 \\ p_{fz} \end{pmatrix} \sqrt{\mu_0} A_t, \tag{9.11}$$

with $p_{fz} = (\sqrt{k_0^2 n_f^2 - k_x^2})/k_0 \varepsilon_{rf}$ as the scaled z component of the wave vector and A_t is the normalized transmitted amplitude in the final medium. Substituting Eqs. (9.10) and (9.11) into Eq. (9.9), we get

$$\begin{pmatrix} 1 & 1 \\ p_{zi} & -p_{zi} \end{pmatrix} \begin{pmatrix} A_{in} \\ A_r \end{pmatrix} = M_{total} \begin{pmatrix} 1 \\ p_{zf} \end{pmatrix} A_t. \tag{9.12}$$

Calculation of the amplitude reflection r and transmission t coefficients is then straightforward:

$$r = \frac{A_r}{A_{in}} = \frac{(m_{11} + m_{12} p_f) p_i - (m_{21} + m_{22} p_f)}{(m_{11} + m_{12} p_f) p_i + (m_{21} + m_{22} p_f)}, \tag{9.13}$$

$$t = \frac{A_t}{A_{in}} = \frac{2 p_i}{(m_{11} + m_{12} p_f) p_i + (m_{21} + m_{22} p_f)}, \tag{9.14}$$

where m_{ij} $(i, j = 1, 2)$ are the elements of the total characteristic matrix of the structure (see Eq. (9.7)), and we have suppressed z from the subscripts of p. This result is also valid for the TE-polarized light with suitable expression for p_{jz}. The intensity reflection (R) and transmission (T) of the structure are given by

$$R = |r|^2, \quad \text{and} \quad T = \frac{p_f}{p_i}|t|^2, \tag{9.15}$$

respectively, and the factor p_f/p_i in T comes from the conservation of light flux across the interface. Note that a common denominator figures in the expressions of both the reflection and transmission coefficients (Eqs. (9.13) and (9.14)). The zeros of the denominator bear the information about the characteristic frequencies (eigenfrequencies) of the system. Physically, this corresponds to the situation when with no input, finite excitations can be sustained in the system. Such specific disturbances with well-defined spatial profiles are referred to as the modes of the structure. The corresponding equation (also known as the dispersion relation) can be written as [65]

$$D = (m_{11} + m_{12}p_f)p_i + (m_{21} + m_{22}p_f) = 0. \tag{9.16}$$

In the context of waveguiding and plasmonic structures, similar dispersion relation plays a very important role since it carries all the information about the modes of the structure. For given system parameters, such equations allow for complex solutions for the eigenfrequencies. The real part of the frequency gives the location of the modes, and the imaginary part defines the lifetime of the specific modes. We can thus define the corresponding quality factors (Q-factor). Usually the excitation of high-Q modes are accompanied by large local field enhancements and they have been exploited for various low threshold optical processes. In most of the cases, Eq. (9.16) is transcendental in nature and it cannot be solved analytically. We have to revert to a graphical or numerical scheme to obtain the distinct branches for the modes. A detailed analysis of the dispersion equation for generic cases will be presented in Chapter 10.

9.3 Periodic media with discrete and continuous variation of refractive index

Periodic structures occupy a very important place in optics because of their various possible applications. For example, interference coatings (antireflective) are used in antiglare screens and in optical instrumentation to increase the light throughput. We may distinguish two broad classes of systems, one with discrete variation of the refractive index, like in the layered media discussed above. The other refers to a continuous variation of the refractive index—for example, a sinusoidal variation along the axis of the optical system. In the first case, in each layer we have forward and backward propagating waves resulting from multiple reflections from each interface, while in the medium with harmonic variation, local inhomogeneity is responsible for the generation of the backward waves. Thus such systems are often referred to as having a distributed feedback. Note that in an infinite homogeneous medium, the forward and backward waves are completely decoupled. The inhomogeneity leads to the coupling in both the discrete and the continuous cases. In the case of weak coupling, we can develop a perturbative approach, which is known in the literature as the *coupled mode theory*, where we retain only the lowest order of scattered waves. More about such systems will be discussed in dealing with the distributed feedback (DFB) systems. However, the case of the discrete variation (layered medium) can be dealt with exactly. In both cases we show the emergence of the band gaps. We calculate the reflection and transmission coefficients for a finite-length periodic medium assuming all the materials to be non-magnetic and non-lossy. The changes for magnetic materials or finite losses can be implemented in a straightforward way.

9.3.1 Discrete variation of refractive index

Consider a periodic layered system like in Fig. 9.1. Let each period consist of two layers with refractive indices n_a, n_b and widths d_a and d_b, respectively. For simplicity we restrict ourselves to the case of wave propagation along the z-axis of the periodic system. Denoting the period of the structure by Λ ($\Lambda = d_a + d_b$) and imposing the periodic boundary conditions, we can relate the output after one period in terms of the input:

$$(e^{i\mu\Lambda}\mathbf{I})\begin{pmatrix} H_y \\ E_x \end{pmatrix}_{z=0} = M_{ab}\begin{pmatrix} H_y \\ E_x \end{pmatrix}_{z=\Lambda}, \tag{9.17}$$

with

$$M_{ab} = M_a M_b. \tag{9.18}$$

The matrix M_a (M_b) denotes the characteristic matrix for an 'a' ('b') type layer. In writing Eq. (9.17) we made use of the Floquet-Bloch theorem. The fields at the output and input faces are the same except for an overall phase accumulation given by the Bloch wave vector μ (see Ref. [61]). Eq. (9.17) represents a homogeneous system and allows for nontrivial solutions if and only if

$$\begin{vmatrix} A - e^{i\mu\Lambda} & B \\ C & D - e^{i\mu\Lambda} \end{vmatrix} = 0, \tag{9.19}$$

where A, B, C and D are the elements of the characteristic matrix for one period M_{ab}:

$$A = \cos\zeta_a \cos\zeta_b - \frac{p_b}{p_a}\sin\zeta_a \sin\zeta_b, \tag{9.20}$$

$$B = -\left(\frac{i}{p_a}\sin\zeta_b \cos\zeta_a + \frac{i}{p_b}\sin\zeta_a \cos\zeta_b\right), \tag{9.21}$$

$$C = -\left(ip_a \sin\zeta_a \cos\zeta_b + ip_b \sin\zeta_b \cos\zeta_a\right), \tag{9.22}$$

$$D = \cos\zeta_a \cos\zeta_b - \frac{p_a}{p_b}\sin\zeta_a \sin\zeta_b, \tag{9.23}$$

with $\zeta_a/k_0 = n_a d_a$ ($\zeta_b/k_0 = k_b d_b$) as the optical width and $p_a = n_a$ ($p_b = n_b$), respectively. The approach and notations we follow here are analogous to the $ABCD$ matrix approach for a lens waveguide system [61]. Eq. (9.19) can be rewritten as

$$e^{i2\mu\Lambda} - (A+D)e^{i\mu\Lambda} + AD - BC = 0. \tag{9.24}$$

We assume all the layers to be lossless, implying that the characteristic matrix of each layer and also their product matrices are unimodular (i.e., $det(M_i) = 1$, with $i = a,\ b,\ ab$). Based on this assumption, Eq. (9.24) can be reduced to

$$e^{i2\mu\Lambda} - (A+D)e^{i\mu\Lambda} + 1 = 0, \tag{9.25}$$

where we have set $AD - BC = 1$. The roots of Eq. (9.25) are given by

$$e^{\pm i\mu\Lambda} = \frac{A+D}{2} \pm \sqrt{\left(\frac{A+D}{2}\right)^2 - 1}. \tag{9.26}$$

The sum of these two roots gives us

$$\frac{e^{i\mu\Lambda} + e^{-i\mu\Lambda}}{2} = \cos\mu\Lambda = \frac{A+D}{2}. \tag{9.27}$$

Making use of Eqs. (9.20), (9.23) and (9.27), we arrive at the dispersion relation for the periodic structure as

$$\cos\mu\Lambda = \cos\zeta_a \cos\zeta_b - \frac{1}{2}\left(\frac{n_b}{n_a} + \frac{n_a}{n_b}\right)\sin\zeta_a \sin\zeta_b. \tag{9.28}$$

We choose the optical pathlengths of the two layers to be same so that $\zeta_a = \zeta_b = \zeta$ and Eq. (9.28) simplifies to

$$\cos\mu\Lambda = 1 - \frac{(n_a+n_b)^2}{2n_a n_b}\sin^2\zeta. \tag{9.29}$$

The character of wave propagation in the periodic medium is governed by the character of the Bloch wave vector μ. The inequality $|\cos\mu\Lambda| \leq 1$ corresponds to real μ, and we have a propagating solution in the structure. In contrast

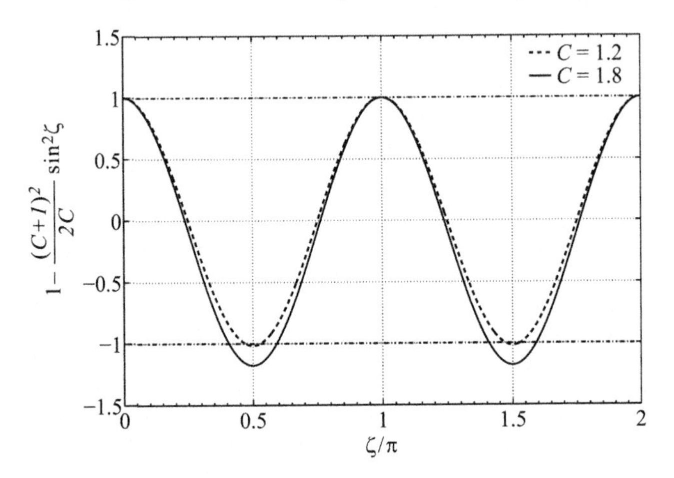

FIGURE 9.2: Right-hand side of Eq. (9.29) as a function of ζ for two different values of C, namely, $C = 1.2$ (dashed) and $C = 1.8$ (solid). The dash-dotted lines represent the maximum and minimum of $\cos\mu\Lambda$ for real arguments.

$|\cos\mu\Lambda| > 1$ would mean a complex μ and corresponding damped waves. We thus have propagating solutions if

$$-1 \le 1 - \frac{(n_a + n_b)^2}{2n_a n_b} \sin^2\zeta \le 1, \tag{9.30}$$

or if

$$0 \le \frac{(C+1)^2}{4C} \sin^2\zeta \le 1, \tag{9.31}$$

where $C = n_a/n_b$ is the refractive index contrast between the two constituent media. We assume that $n_a > n_b$ so that $C > 1$. The band gap occurs in the region where the inequality Eq. (9.31) is violated. The stop gap for two values of C is shown in Fig. 9.2, where we have plotted the right-hand side of Eq. (9.29) as a function of ζ. As can be seen from Fig. (9.2), the stop gap is centered at $\zeta = (2m+1)\pi/2$ and its width is proportional to the contrast. With increasing contrast the band gap broadens. This can be easily seen from the following estimate. For low contrast ($C \approx 1$) the range of ζ (in the principal domain) corresponding to a stop gap is given by

$$\frac{\pi}{2} - \frac{\Delta\zeta}{2} < \zeta < \frac{\pi}{2} + \frac{\Delta\zeta}{2}, \tag{9.32}$$

where $\Delta\zeta$ is the width of the band gap. It can be shown that $\Delta\zeta$ varies as

$$\Delta\zeta = \frac{\delta C}{C}, \tag{9.33}$$

with $\delta C = |C - 1|$.

9.3.2 Continuous variation of refractive index: DFB structures

Consider the continuous periodic variation of refractive index along the direction of propagation z as

$$n(z) = n_0(1 + n_1 \cos Kz), \quad \text{with} \quad K = 2\pi/\Lambda, \tag{9.34}$$

where n_0 is the background refractive index and n_1 is the modulation amplitude while Λ denotes the period. We consider the medium to be nonmagnetic and lossless. Further, the periodic variation is assumed to be a small perturbation to the background (i.e., $n_1 \ll 1$). The dielectric function can then be expressed as

$$\epsilon(z) = n_0^2 + 2n_0n_1 \cos Kz, \tag{9.35}$$

where we have ignored the term containing n_1^2. The solution to the Helmholtz equation can be written as a superposition of forward and backward propagating waves given by

$$E(z) = A_+(z)e^{ikz} + A_-(z)e^{-ikz}, \qquad k = (\omega/c)n_0, \tag{9.36}$$

with $A_+(z)$ ($A_-(z)$) as a forward-propagating (backward-propagating) slowly varying wave amplitude. Note that in absence of modulation both these amplitudes are constants, and the inhomogeneity due to modulation scatters the forward waves into the backward ones. In principle all the scattering events will lead to the spatial harmonics k_s:

$$k_s = k + mK, \quad m = 0, \pm1, \pm2 \cdots, \tag{9.37}$$

with m being the order of the spatial harmonic. These waves exchange energy among themselves as they propagate through the medium. However, for small modulation, significant contribution comes from only the lower-order harmonics. In the spirit of coupled mode theory, we retain only the lowest-order spatial harmonics. Substituting Eq. (9.36) in the Helmholtz equation, we obtain

$$\begin{aligned} &\left[\frac{d^2A_+}{dz^2}e^{ikz} + \frac{d^2A_-}{dz^2}e^{-ikz}\right] \\ +\frac{n_0}{n_1}&\left[A_-e^{ikz}e^{-i(Kz+2kz)} + A_+e^{-ikz}e^{i(Kz+2kz)}\right] \\ &+\left[2ik\frac{\partial A_+}{\partial z} + \frac{n_0}{n_1}k^2A_-e^{i(Kz-2kz)}\right]e^{ikz} \\ &-\left[2ik\frac{\partial A_-}{\partial z} - \frac{n_0}{n_1}k^2A_+e^{-i(Kz-2kz)}\right]e^{-ikz} = 0. \end{aligned} \tag{9.38}$$

The slowly varying envelopes $A_\pm$ satisfy the conditions $|\frac{d^2A_\pm}{dz^2}| \ll k|\frac{dA_\pm}{dz}| \ll k^2|A_\pm|$, and we can neglect the terms in the first square brackets in Eq. (9.38).

The scattering event that is of interest to us corresponds to $2k = K$, and we can rule out the other higher-order events in the second square brackets. Collecting the coefficients of e^{ikz} and e^{-ikz} in Eq. (9.38), we arrive at the coupled mode equations

$$\frac{dA_+}{dz} = i\beta A_- e^{-i\delta z}, \tag{9.39}$$

$$\frac{dA_-}{dz} = -i\beta A_+ e^{i\delta z}, \tag{9.40}$$

where

$$\beta = \frac{\omega n_1}{2c}, \qquad \delta = 2\frac{\omega}{c}n_0 - K. \tag{9.41}$$

Under the transformation

$$\begin{pmatrix} A_+(z) \\ A_-(z) \end{pmatrix} = \mathbf{V}(z) \begin{pmatrix} \bar{A}_+(z) \\ \bar{A}_-(z) \end{pmatrix}, \quad \text{where } \mathbf{V}(z) = \begin{pmatrix} e^{-i\delta z/2} & 0 \\ 0 & e^{i\delta z/2} \end{pmatrix}, \tag{9.42}$$

Eqs. (9.39) and (9.40) can be written in terms of transformed variables as

$$\frac{d}{dz}\begin{pmatrix} \bar{A}_+(z) \\ \bar{A}_-(z) \end{pmatrix} = i\mu\mathbf{M}\begin{pmatrix} \bar{A}_+(z) \\ \bar{A}_-(z) \end{pmatrix}, \tag{9.43}$$

where

$$\mathbf{M} = \frac{1}{\mu}\begin{pmatrix} \delta/2 & \beta \\ -\beta & -\delta/2 \end{pmatrix}, \qquad \mu^2 = -\beta^2 + \delta^2/4. \tag{9.44}$$

The solution to Eq. (9.43) is given by

$$\begin{pmatrix} \bar{A}_+(z) \\ \bar{A}_-(z) \end{pmatrix} = \exp(i\mu\mathbf{M}z)\begin{pmatrix} \bar{A}_+(0) \\ \bar{A}_-(0) \end{pmatrix}. \tag{9.45}$$

It is worth noting that if $-\beta^2 + \delta^2/4 < 0$ in Eq. (9.44), then μ will be purely imaginary (taking only the positive root in order to ensure causality); the waves inside the medium will be evanescent. This means that the forward propagating wave becomes evanescent by giving its energy to the backward propagating wave, resulting in reflection. The interval in which μ becomes imaginary corresponds to the the optical stopgap. For clarity we have plotted real and imaginary parts of μ in Fig. 9.3. It is clear that near the band edge there is considerable dispersion, while inside the gap the Bloch vector becomes purely imaginary. Note also that the gap width is proportional to the coupling strength β ($\Delta\delta = 4\beta$). Noting that $\mathbf{M}^2 = \mathbf{I}$ ($\mathbf{I}$ = identity matrix), we have

$$\begin{aligned} \exp(i\mu\mathbf{M}z) &= \mathbf{I}\left(1 - \frac{(\mu z)^2}{2!} + \frac{(\mu z)^4}{4!} - \cdots\right) \\ &\quad + i\mathbf{M}\left(\mu z - \frac{(\mu z)^3}{3!} + \frac{(\mu z)^5}{5!} - \cdots\right), \\ &= \mathbf{I}\cos(\mu z) + i\mathbf{M}\sin(\mu z). \end{aligned} \tag{9.46}$$

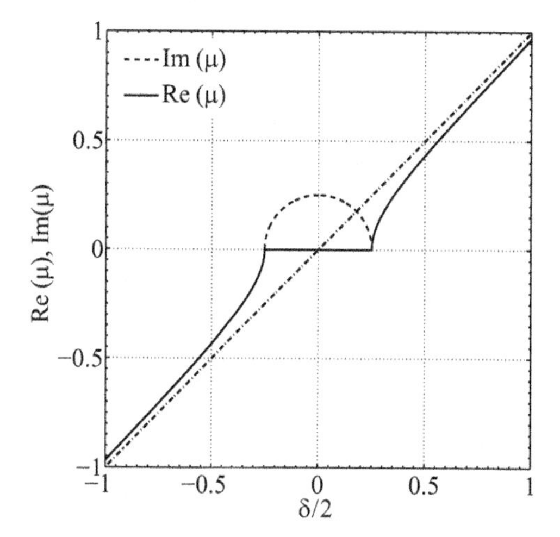

FIGURE 9.3: Re(μ) and Im(μ) as a function of $\delta/2$ for $\beta = 0.25$.

Rewriting the solution (Eq. (9.45)) in terms of $A_{\pm}(z)$, we have

$$\begin{aligned}\begin{pmatrix} A_+(z) \\ A_-(z) \end{pmatrix} &= \mathbf{V}(z)\exp(i\mu\mathbf{M}z)\mathbf{V}^{-1}(0)\begin{pmatrix} A_+(0) \\ A_-(0) \end{pmatrix} \\ &= \mathbf{U}(z)\begin{pmatrix} A_+(0) \\ A_-(0) \end{pmatrix}. \end{aligned} \tag{9.47}$$

Matrix $\mathbf{U}(z)$ gives the spatial evolution of the amplitudes inside the periodic medium. The elements of matrix $\mathbf{U}(L)$ are given by

$$A = \left(\cos(\mu L) + \frac{i\delta}{2\mu}\sin(\mu L)\right)e^{-i\delta L/2}, \tag{9.48}$$

$$B = \frac{i\beta}{\mu}\sin(\mu L)e^{-i\delta L/2}, \tag{9.49}$$

$$C = \frac{-i\beta}{\mu}\sin(\mu L)e^{i\delta L/2}, \tag{9.50}$$

$$D = \left(\cos(\mu L) + \frac{i\delta}{2\mu}\sin(\mu L)\right)e^{i\delta L/2}. \tag{9.51}$$

The $ABCD$ matrix represents the same spirit as in the case of a lens guide or a discrete periodic structure. Like in the earlier cases, we have $AD - BC = 1$ for a lossless system. For a finite segment of such a medium, we can define the amplitude reflection coefficient as

$$r = \frac{A_-(0)}{A_+(0)} = -\frac{C}{D} \tag{9.52}$$

and the amplitude transmission coefficient as

$$t = \frac{A_+(L)}{A_+(0)} = \frac{AD - BC}{D} = \frac{1}{D}. \tag{9.53}$$

Thus, the expressions for reflection and transmission turn out to be

$$r = \frac{i\beta \sinh(sL)}{s\cosh(sL) - i\frac{\delta}{2}\sinh(sL)}, \tag{9.54}$$

$$t = \frac{se^{-i\delta L/2}}{s\cosh(sL) - i\frac{\delta}{2}\sinh(sL)}. \tag{9.55}$$

In Eqs. (9.54) and (9.55) we replaced μ by is in order to make contact with known expressions in the literature [66]. The magnitude of the amplitude reflection coefficient r attains its maximum value at $\delta = 0$, implying $K = 2k$; this is exactly Bragg's condition for reflection from the periodic medium. Considering the ambient media to be same and homogeneous, the intensity reflection and transmission coefficients are given by

$$R = |r|^2, \quad T = |t|^2. \tag{9.56}$$

9.4 Quasi-periodic media and self-similarity

A quasi-periodic system is in between periodic and a random system. There has been a great deal of interest in quasi-periodic (QP) media in the context of weak photon localization. We won't deal with photon localization as such since that goes beyond the scope of this book. Interested readers can find the details in Refs. [67, 68]. Nevertheless, we present results for reflection and transmission for the QP system. Though there are many examples of QP systems, we pick a Fibonacci multilayer as our system. Any $(j+1)$-th generation (S_{j+1}) of a Fibonacci system can be grown from two previous generations starting from two basic ones, say 'A' and 'B,' as follows:

$$S_0 = A, \quad S_1 = B, \tag{9.57}$$

$$S_{j+1} = S_{j-1}S_j. \tag{9.58}$$

Thus, $S_2 = AB$, $S_3 = BAB$, $S_4 = ABBAB$ and so on. The number of elements in each generation S_{j+1} is given by the Fibonacci number F_{j+1}, which is obtained iteratively starting from $F_0 = 1$ and $F_1 = 1$ by the relation

$$F_{j+1} = F_{j-1} + F_j. \tag{9.59}$$

Thus, S_9 and S_{12} would have 55 and 233 layers, respectively. An optical Fibonacci multilayer can be grown by stacking two different dielectric slabs

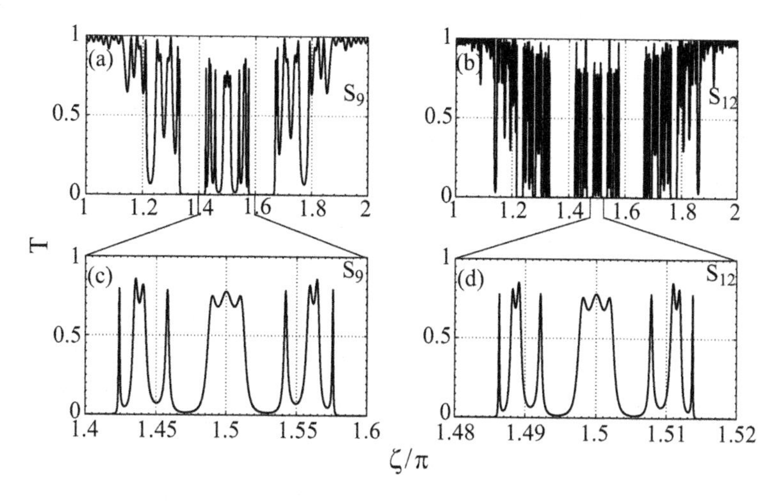

FIGURE 9.4: Transmission through the Fibonacci structure for (a) and (c) S_9 (with 55 layers) and (b) and (d) S_{12} (with 233 layers) as a function of ζ/π with $n_a = 1.5$, $n_b = 2.5$.

arranged in a Fibonacci sequence. The two slabs can be characterized by their refractive indices (n_a and n_b) and widths (d_a and d_b). For simplicity we assume that both the slab materials are nonmagnetic and have the same optical width so that ($k_0 n_a d_a = k_0 n_b d_b = \zeta$). This structure is assumed to be surrounded by a material of refractive index $n_f = n_i = 2.5$. In order to evaluate the reflection and transmission from such structures, we make use of the characteristic matrix method developed in Section 9.1. The transmission as a function of ζ is shown in Fig. 9.4 for both S_9 and S_{12}. The top panels in Fig. 9.4 show the transmission properties of two generations, namely, S_9 and S_{12}, whereas the bottom panel shows the magnified versions corresponding to them. It is clear from Fig. 9.4(b) and Fig. 9.4(d) that the transmission properties look very similar, and this is referred to as self-similarity. Interested readers can find the rigorous mathematical treatment of the multifractal nature and self-similarity aspects in Ref. [67]. The quasi-periodicity has been investigated in a variety of other structures in the past.

9.5 Analogy between quantum and optical systems

In this section we draw an analogy between optical and quantum systems in the context of one-dimensional scattering. The analogy has been known for a very long time (see, for example, [69, 63]) and holds since both the quantum and the optical systems both share common physical ground. Indeed, in view of the inherent wave nature and the deciding role of interference phenomena, both quantum mechanics and optics share a lot of features in common. For example, we encounter propagating and evanescent waves in both areas. In fact, evanescent waves and tunneling have been some of the central issues since they are encountered in the passage of a quantum particle through a barrier (Fig. 9.5(a)). The same phenomenon is there in the tiny air wedge between two prisms in a frustrated total internal reflection geometry (Fig. 9.5b). This analogy has been explored in detail by Kay and Moses in their quest for reflectionless potentials [69]. One of the very first experiments on classical tunneling using microwaves was carried out by an Indian, Sir Jagdish Chandra Bose, more than a century ago [70, 71]. In what follows, we probe this analogy in one-dimensional systems. We start with the time-independent Schrödinger equation given by [72, 73]

$$\frac{d^2\psi}{dz^2} + \frac{2m}{\hbar^2}(E - V)\psi = 0, \tag{9.60}$$

where ψ is the wave function of the electron, m is mass of the electron, V is the potential energy and E is the total energy. On the other hand, in an inhomogeneous, isotropic medium under scalar approximation, the Helmholtz equation is given as

$$\frac{d^2\mathcal{E}}{dz^2} + \frac{n^2\omega^2}{c^2}\mathcal{E} = 0, \tag{9.61}$$

where $\mathcal{E}$ represents scalar amplitude of the electric field and n is the refractive index of the medium. This analogy is complete if we make the following

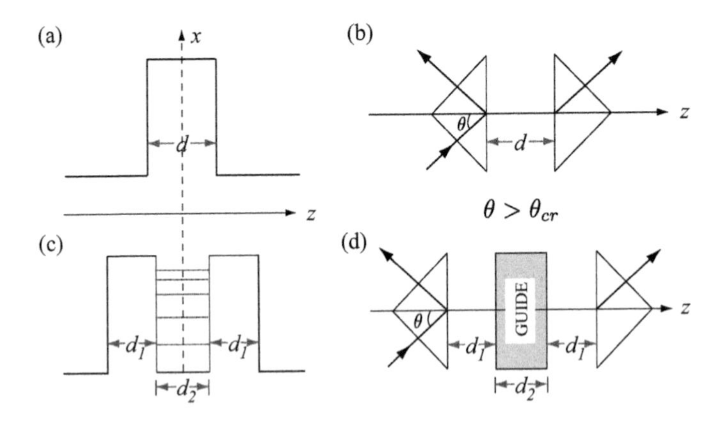

FIGURE 9.5: Analogy between electron and photon tunneling. (a) Electron tunneling through a potential barrier with energy less than the barrier height. (b) Double prism system with light incident at angle $\theta > \theta_c = sin^{-1}(1/n)$. (c) Resonant tunneling and (d) its equivalent optical system.

correspondence:

$$n \Leftrightarrow \frac{[2m(E-V)]^{\frac{1}{2}}\, c}{\hbar\omega}. \tag{9.62}$$

A more detailed scenario for oblique incidence, which can accommodate evanescent waves, will be presented later in the context of reflectionless potentials of Kay and Moses in Section 9.6. The correspondence of Eq. (9.62) allows us to extend the results developed for quantum systems to optical ones and vice versa. Advantages following from this parallel are enormous, since experiments difficult in solid state systems (involving, for example, electron tunneling) can be performed in a much simpler equivalent optical setup. Note also that the optical experiment will be less error-prone since the photons are noninteracting while electrons are not due to Coulomb interaction. It must also be mentioned that there are considerable difficulties in obtaining electron beams with precise control over the flux density in space and time. The typical transit times in the tunneling of an electron wave packet is of the order of 10^{-15}–10^{-18} sec, which requires sophisticated technology to measure such short time scales [74]. Such problems do not exist in optics since the advent of modern lasers and precision detectors. Two broad classes of systems can be identified for studying tunneling phenomena in optical systems, namely, in the absence and in the presence of resonant modes. Both these classes are referred to as the frustrated total internal reflection (FTIR) geometries in optics. The simplest of the first category (see Fig. 9.5(b)) has been discussed above. The same system will be probed later for the saturation of phases leading to the Hartman effect (see Section 9.5.3). Systems of the second class are generally used for exciting guided and surface modes. In fact, without the excitation of the modes, the reflection from the system is close to unity because of total internal reflection. The excitation of the modes can mediate finite transmission through the structure leading to narrow dip/dips in the reflection profile for specific angles of incidence beyond the critical angle. At these specific angles of incidence, energy is transferred to the guided/surface modes from the incident waves and hence the nomenclature of the frustrated (attenuated) total reflection. Such structures (see Fig. 9.5(d)) are also referred to as resonant tunneling geometry, since it mimics the situation of a well enclosed between two barriers as in Fig. 9.5(c). Finite transfer is possible only via the excitation of the quasi-bound states of the well.

9.5.1 Wigner delay: Fast and slow light

In a seminal paper in 1955, Eugene P. Wigner addressed a very important question: how much time a wave packet spends in passing through a sequence of wells and barriers [63]. The time taken was later labeled as the Wigner phase time τ as opposed to the equal time τ_f taken by the packet in absence of the potential. Wigner's derivation was based on bichromatic light. In this section, we generalize the derivation of Wigner to the case of a temporal pulse centered around the carrier frequency ω_c. Let the incident pulse be given by

$$E_I(z,t) = F(t)e^{i(\beta z-\omega_c t)}, \tag{9.63}$$

where the Fourier decomposition of the temporal profile at $z = 0$ is written as

$$F(t) = \int A(\omega)e^{-i(\omega-\omega_c)t}d\omega. \tag{9.64}$$

Let $\tilde{t}(\omega)$ be the complex transmission coefficient of the medium given in terms of the real amplitude and phase as follows:

$$\tilde{t}(\omega) = |\tilde{t}(\omega)|e^{i\phi_T(\omega)}. \tag{9.65}$$

Using Eqs. (9.63)–(9.65), the transmitted pulse (in the region $z > L$) can be written as

$$E_T(z,t) = e^{i(\beta z-\omega_c t)}\int A(\omega)|\tilde{t}(\omega)|e^{i\phi_T(\omega)}e^{-i(\omega-\omega_c)t}d\omega. \tag{9.66}$$

The temporal shift of the transmitted pulse depends on ϕ_T, which bears the signature of the dispersion of the medium. In order to estimate the temporal shift of the pulse, we use the stationary phase approximation. The essence of this approximation is that the phase of the transfer function $\phi_t(\omega)$ is assumed to have a weak dependence on ω so that we can expand ϕ_T in a Taylor series around ω_c and ignore the higher-order derivatives retaining only the lowest-order terms:

$$\phi_t(\omega) = \phi_T(\omega_c) + \left.\frac{\partial\phi_T}{\partial\omega}\right|_{\omega_c}(\omega-\omega_c) + \cdots. \tag{9.67}$$

On substituting the expansion above in Eq. (9.66) and retaining only terms up to the first order, we get

$$E_T(z,t) = e^{i(\beta z-\omega_c t+\phi_T(\omega_c))}\int A(\omega)|\tilde{t}(\omega)|e^{-i\left(t-\left.\frac{\partial\phi_T}{\partial\omega}\right|_{\omega_c}\right)(\omega-\omega_c)}d\omega. \tag{9.68}$$

Assuming a flat (or slowly varying) amplitude response ($|\tilde{t}(\omega)| \sim$ constant) over the spectral spread of $A(\omega)$, Eq. (9.68) reduces to

$$E_T(z,t) = |\tilde{t}(\omega_c)|F\left(t-\left.\frac{\partial\phi_T}{\partial\omega}\right|_{\omega_c}\right)\exp i(\beta z-\omega_c t+\phi_T(\omega_c)). \tag{9.69}$$

Thus, the transmitted pulse arrives at the output end (i.e., $z = L$) at

$$\tau_t = \left.\frac{\partial\phi_T}{\partial\omega}\right|_{\omega_c}. \tag{9.70}$$

Similarly, the reflected pulse will be delayed/advanced by

$$\tau_r = \left.\frac{\partial\phi_R}{\partial\omega}\right|_{\omega_c}, \tag{9.71}$$

where ϕ_R is the phase of the reflection coefficient.

9.5.2 Goos-Hänchen shift

The derivation for the Goos-Hänchen shift is quite analogous, except that the incident field is now assumed to be a monochromatic beam with a spread of wave vectors around (α_0, β), where α_0 and β are the x and z components of the wave vector. The interface between the two dielectrics is assumed to be the plane $z = 0$. Thus the incident field can be written as

$$E_I(x,t) = e^{i(\alpha_0 x + \beta z - \omega_0 t)} \int A(\alpha) e^{i(\alpha - \alpha_0)x} d\alpha. \tag{9.72}$$

As mentioned above, the divergence of the Fourier components is so small that $A(\alpha)$ is sharply peaked around α_0 with the corresponding angle of incidence θ_0 greater than the critical angle θ_c. Thus, the total internal reflection condition is assumed to be valid for all Fourier components and the reflected pulse can then be written as

$$E_R(x,t) = e^{i(\alpha_0 x - \beta z - \omega t)} \int A(\alpha)\, e^{i(\alpha - \alpha_0)x + i\phi_R(\alpha)} d\alpha, \tag{9.73}$$

where $\phi_R(\alpha)$ is the phase of the complex reflection coefficient. An expansion of $\phi_R(\alpha)$ around α_0 just like in Eq. (9.67) reduces Eq. (9.73) to

$$E_R(x,t) = e^{i(\alpha_0 x - \beta z - \omega t + \phi_R(\alpha_0))} \int A(\alpha)\, e^{i(\alpha - \alpha_0)\left(x + \left.\frac{\partial \phi_R}{\partial \alpha}\right|_{\alpha_0}\right)} d\alpha. \tag{9.74}$$

With a definition of the incident beam profile at $z = 0$ as

$$F(x) = \int A(\alpha) e^{i(\alpha - \alpha_0)x} d\alpha, \tag{9.75}$$

the reflected beam takes the final form given by

$$E_R = F\left(x + \left.\frac{\partial \phi_R}{\partial \alpha}\right|_{\alpha_0}\right) e^{i(\alpha_0 x - \beta z - \omega t) + i\phi_R(\alpha_0)}, \tag{9.76}$$

which implies a longitudinal displacement of the beam by an amount

$$d_{GH} = -\left.\frac{\partial \phi_R}{\partial \alpha}\right|_{\alpha_0.}. \tag{9.77}$$

9.5.3 Hartman effect

In barrier-crossing problems a very important issue is the time spent by the wave packet in the barrier region. The very first quantitative assessment in this regard was carried out by Hartman [75]. The theory was based on the time-dependent Schrödinger equation and Hartman calculated the transit time of Gaussian wave packets through a rectangular barrier. The transit time turned out to be longer than equal time (vacuum time) for thin barriers, while it exhibited saturation with increasing barrier width. For the latter, the transit time was shown to be less than the equal time. This effect of saturation of phase time for sufficient barrier width came to be known as the Hartman effect. The first-ever experimental demonstration of the Hartman effect was carried out by Enders and Nimtz [76] in the microwave domain. Phase saturation (which is linked with transit time saturation) was nicely noted in a remarkable experiment by Carniglia and Mandel [77]. Their theory was based on multiple reflections in the low index air gap in a frustrated total internal reflection setup (see Fig. 9.5(b)). A detailed analysis of related physics can be found in the thesis of Manga Rao [78]. Here in Section 9.5.4, we derive the same results using the characteristics matrix approach. A large body of literature exists on the Hartman effect and its generalizations in periodic structures [79, 66, 80]. In view of the widespread interest in the photonic bandgap structures, we follow Winful [66] to present a brief derivation of the Hartman effect for DFB structures in Section 9.5.5.

9.5.4 Precursor to Hartman effect: Saturation of phase shift in optical barrier tunneling

Let a plane monochromatic TM-wave be incident on the dielectric-air interface at an angle θ as shown in Fig. 9.5(b). Using the characteristic matrix approach of Section 9.1, the complex amplitude transmission coefficient is given by (see also Eq. (9.14))

$$\tilde{t} = \frac{\frac{2}{n}\cos\theta}{(\cos\zeta - \frac{i}{np_z}\cos\theta\sin\zeta)\frac{\cos\theta}{n} - ip_z\sin\zeta + \cos\zeta\frac{\cos\theta}{n}}. \tag{9.78}$$

Recall that the z component of the wave vector $k_z = k_0 p_z$, $p_z = (1 - n^2\sin^2\theta)^{1/2}$ and $\zeta = k_z d$. Eq. (9.78) can be rewritten as

$$\tilde{t} = \frac{1}{\cos\zeta - i\left(\cos^2\theta + n^2(1-n^2\sin^2\theta)\right)\frac{\sin\zeta}{2n\cos\theta(1-n^2\sin^2\theta)^{1/2}}}. \tag{9.79}$$

For propagating waves in the air gap, k_z is real, and using Eq. (9.79), we can calculate the phase shift in the transmitted light as follows:

$$\phi_t = \tan^{-1}\left[\frac{\cos^2\theta + n^2(1-n^2\sin^2\theta)}{2n\cos\theta(1-n^2\sin^2\theta)^{1/2}}\tan\zeta\right]. \tag{9.80}$$

For angles of incidence $\theta > \theta_c$, k_z is purely imaginary, leading to evanescent waves in the air gap. The phase shift can then be written as

$$\phi_t = \tan^{-1}\left[\frac{\cos^2\theta + n^2(n^2\sin^2\theta - 1)}{2n\cos\theta(n^2\sin^2\theta - 1)^{1/2}}\tanh\zeta\right]. \tag{9.81}$$

It follows from Eq. (9.81) that for a larger width of the air gap, the phase saturates to

$$\phi_t = \tan^{-1}\left[\frac{\cos^2\theta + n^2(n^2\sin^2\theta - 1)}{2n\cos\theta(n^2\sin^2\theta - 1)^{1/2}}\right]. \tag{9.82}$$

Similarly, for a TE-polarized incident wave, Eq. (9.82) reads as

$$\phi_t = \tan^{-1}\left[\frac{(n^2\cos 2\theta + 1)}{2n\cos\theta(n^2\sin^2\theta - 1)^{1/2}}\right]. \tag{9.83}$$

9.5.5 Hartman effect in distributed feedback structures

As mentioned earlier, the Hartman effect leads to the saturation of phase time for sufficiently wide barriers. The effect was analyzed in great detail by Winful [66] in the context of distributed feedback structures. In order to have a valid coupled mode approach, low modulation depths were assumed. The major aim of the study was to identify the physical origin of the Hartman effect. It was shown that the energy in the barrier region was stored approximately within an attenuation length. The link between the stored energy and the phase time was explored and energy localization led to the independence of the phase time on the barrier width. Another plausible explanation is the evanescent nature of the waves, which implies that there is no accumulation of phase. In what follows, we recall the essential steps of the work of Winful [66] to demonstrate the transit time saturation.

Consider a system with a periodically modulated refractive index given by Eq. (9.34); the reflection and transmission coefficients are given by Eqs. (9.54) and (9.55), respectively. From Eqs. (9.54) and (9.55), the corresponding phases are given by

$$\phi_r = \tan^{-1}\left(-\frac{s\coth(sL)}{\delta/2}\right). \tag{9.84}$$

The overall phase factor $e^{i\delta L/2}$ can be absorbed in the transmission amplitude and we can obtain the phase of the transmission coefficient as

$$\phi_t = \tan^{-1}\left(\frac{\delta/2\tanh(sL)}{s}\right). \tag{9.85}$$

Eqs. (9.84) and (9.85) yield the relation $\phi_r = \phi_t + \frac{\pi}{2}$. Note that the same relation can be derived from a more general consideration of a stratified medium (see Section 9.8.2). It needs to be mentioned that for a lossless symmetric structure, the phase times for reflection and transmission coincide. The phase time, for example, for transmission can be derived from Eq. (9.85) as follows:

$$\tau_p^t = \frac{2n_0}{c}\frac{d\phi_t}{d\delta} = \frac{n_0}{c}\frac{L}{(s^2 + \delta^2\tanh^2(sL))}\left\{\left[\delta^2\tanh^2(sL) - \delta^2\right] + \frac{\beta^2\tanh(sL)}{sL}\right\}. \tag{9.86}$$

For long Bragg grating, i.e., in the limit of $L \to \infty$, the transmission phase time Eq. (9.86) becomes $\tau_p^t = \dfrac{n_0}{cs}$, which is independent of the length of the grating L.

9.6 Optical reflectionless potentials and perfect transmission

In this section we look at a very interesting application of stratified medium. We show that certain given refractive index profiles can lead to total transmission with no backscattered light. The basic idea originated from the notion of reflectionless potentials proposed by Kay and Moses in a seminal paper in 1956 [69]. The question posed by Kay and Moses was the following: Can we design a potential for the Eq. (9.60) such that the wave function $\Psi(z, E)$ satisfies the conditions

$$\Psi(z, E) \sim e^{i\sqrt{E}z} \quad \text{as} \quad z \to -\infty, \tag{9.87}$$

$$\Psi(z, E) \sim t e^{i\sqrt{E}z} \quad \text{as} \quad z \to +\infty, \tag{9.88}$$

meaning thereby that for all E there is no reflected wave. It was shown that it is possible and such potential $V(z)$ satisfies certain specific conditions (see the box on the Kay-Moses theorems). We now present the Kay-Moses prescription for constructing such a potential.

9.6.1 Construction of the Kay-Moses potential

Consider the N parameter family, each with two positive constants A_j and κ_j $(j = 1 \cdots N)$, and construct the linear set of equations for $f_j(z)$ as follows:

$$\sum_{j=1}^{N} M_{ij} f_j(z) = A_i e^{\kappa_i z}, \quad M_{ij} = \delta_{ij} + \frac{A_i e^{(\kappa_i + \kappa_j)z}}{\kappa_i + \kappa_j}. \tag{9.89}$$

Then, the reflectionless potential is given by

$$V(z) = -2\frac{d^2}{dz^2}[\log D], \tag{9.90}$$

where D is the determinant of the $N \times N$ matrix (M_{ij}). The solution of the scattering problem with potential given by Eq. (9.90) can be written as [69]

$$\Psi(z, E) = \left[1 + \sum_{j=1}^{N} \frac{f_j(z) e^{i\kappa_j z}}{\kappa_j + i\sqrt{E}}\right] e^{i\sqrt{E}z}. \tag{9.91}$$

Several important theorems were proved in the seminal work of Kay and Moses:

Kay-Moses theorems on reflectionless potentials

1. The function given by Eq. (9.91) satisfies the Schrödinger equation with potential in Eq. (9.90).
2. $V(z)$ is negative for all finite z.
3. $V(z)$ as in Eq. (9.90) are the only ones for which there is no reflection.

The general form of Eq. (9.90) and the solution Eq. (9.91) may appear to be too complicated. Hence we illustrate with the simplest possible example of a one-parameter family of two constants, A_1 and κ_1. Eq. (9.89) then leads to

$$D(z) = 1 + \frac{A_1}{2\kappa_1} e^{2\kappa_1 z}, \quad V(z) = \frac{4\kappa_1 A_1 e^{2\kappa_1 z}}{\left(1 + \frac{A_1}{2\kappa_1} e^{2\kappa_1 z}\right)^2}. \tag{9.92}$$

Further demanding that the potential $V(z)$ has a minimum at $z = 0$, we have $A_1 = 2\kappa_1$, and

$$V(z) = -2\kappa_1^2 \text{sech}(\kappa_1 z). \tag{9.93}$$

The potential given by Eq. (9.93) is known in the literature as the modified Poschl-Teller potential [81, 82] and has been studied in great detail. The corresponding solution is given by

$$\Psi(z) = \frac{i\sqrt{E} - \kappa_1 \tanh(\kappa_1 z)}{i\sqrt{E} + \kappa_1} e^{i\sqrt{E}z}, \tag{9.94}$$

which evidently obeys the conditions given by Eqs. (9.87) and (9.88).

9.6.2 Optical realization of reflectionless potentials

Realization of reflectionless potentials in optics [83, 84] has certain difficulties associated with the polarization of light. As discussed earlier the stationary wave equation for TE, and TM waves are different. Recall that the TE (TM) wave has only one nonvanishing electric (magnetic) field component perpendicular to the plane of incidence. For a general case of oblique (to z-axis) incidence, the equation for the nonvanishing component of the electric field $\vec{E} = (0, \mathcal{E}e^{ik_x x}, 0)$ for TE waves can be written as

$$\frac{d^2\mathcal{E}}{dz^2} + (k_0^2\epsilon(z) - k_x^2)\mathcal{E} = 0, \tag{9.95}$$

where $\epsilon(z)$ gives the dielectric function variation along the direction of stratification. Here $k_x = k_0\sqrt{\epsilon_s}\sin\theta$ with ϵ_s and θ as the dielectric constant of the substrate/cladding and the angle of incidence. An analogous equation for the TM waves with $\vec{H} = (0, \mathcal{H}e^{ik_x x}, 0)$ is given by

$$\frac{d^2\mathcal{H}}{dz^2} - \frac{d\mathcal{H}}{dz}\frac{d\ln\epsilon(z)}{dz} + (k_0^2\epsilon(z) - k_x^2)\mathcal{H} = 0. \tag{9.96}$$

We can easily draw a parallel of Eq. (9.95) with the stationary Schrödinger equation (Eq. (9.60)) by defining

$$V(z) = k_0^2(\epsilon_s - \epsilon(z)), \quad E = k_0^2\epsilon_s\cos^2\theta. \tag{9.97}$$

However, it is impossible to simultaneously map Eq. (9.96) to the stationary Schrödinger equation. Thus, we cannot have a reflectionless refractive index profile for both TM and TE waves simultaneously. Eq. (9.97) can be rewritten to yield the relation between the dielectric function profile and the reflectionless potential:

$$\epsilon(z) = n^2(z) = n_s^2 - \frac{V(z)}{k_0}, \quad \epsilon_s = n_s^2. \tag{9.98}$$

Note that the design of the profile depends on $k_0 = 2\pi/\lambda$, which in turn depends on the wavelength. Thus for a given λ_c, the profile can be generated and it works well for all wavelengths smaller that λ_c for TE waves. It is rather fortunate that the profile generated for the TE waves also works for TM waves as well.

For the example of a one-parameter family, the reflectionless profile is given by

$$n^2(z) = n_s^2 + \frac{2\kappa\,\mathrm{sech}^2(\kappa z)}{k_0^2}, \tag{9.99}$$

where we have suppressed the subscript of κ. Note that the ideal potential (profile) extends from $-\infty$ to $+\infty$. In reality any profile is bound to be finite, resulting in $n(z)$ differing from n_s only on a finite support. Any deviation from an infinite extent potential has its effect even for TE-polarization; it is no longer purely transmitting over the whole range of angles of incidence. The features above are illustrated in Fig. 9.6. For a given profile, calculation can be performed by making a finite subdivision of profile and considering the refractive index to be constant over each subdivision (see Fig. 9.6(a)).We can invoke the characteristic matrix method for evaluating the reflection and transmission for the whole structure. Results for R as a function of θ are shown in Fig. 9.6(b) for different levels of truncation. It is clear that, the greater the truncation, the greater the loss of perfect antireflection feature. In order to highlight the remarkable conceptual differences from the standard approaches of $\lambda/4$ antireflection coating, we have shown the angle dependence for one standard $\lambda/4$ plate in Fig. 9.6(d). Recall that coating based on the $\lambda/4$ plates usually works for narrow angular intervals and they usually have spectrally narrow bands. The wavelength dependence is shown in Fig. 9.7. It is clear from Fig. 9.7 that up to the design wavelength, the profile works very well.

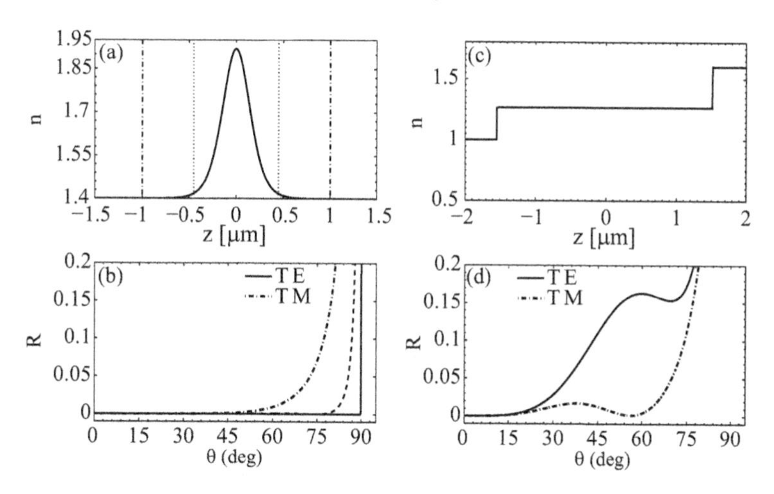

FIGURE 9.6: (a) Refractive index profile (see Eq. (9.98)) for reflectionless potential given in Eq. (9.93) with $\kappa = 5.5$ and $n_s = 1.4$. (b) Intensity reflection coefficient R as a function of θ at $\lambda = 1060$ nm for two truncations of the profile shown in (a), namely, $d = 1$ μm (solid curve), $d = 0.45$ μm (dashed curve) for TE-polarization and the dash-dot curve in (b) is for TM-polarization with $d = 1$ μm. (c) Refractive index profile for a $\lambda/4$ plate with $n_1 = 1, n_2 = 1.2649$, $n_3 = 1.6$ and $d = 1.531$ μm. (d) Intensity reflection coefficient R as a function θ at $\lambda = 1550$ nm for the profile shown in (c).

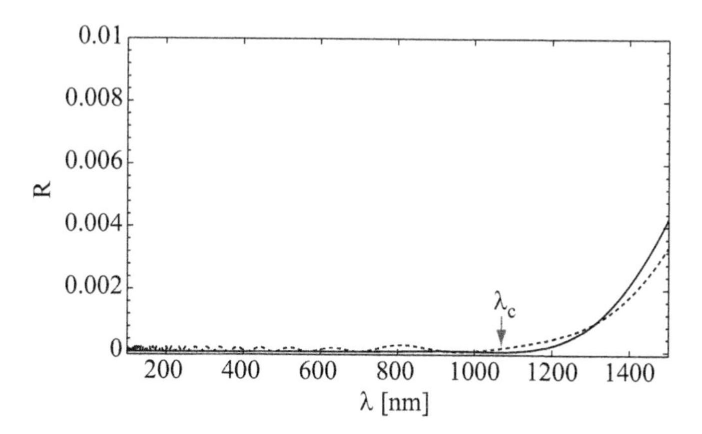

FIGURE 9.7: R as function of λ from the profile shown in Fig. 9.6(a) with design wavelength $\lambda_c = 1060$ μm and $\theta = 0$ for two truncations, namely, $d = 1$ μm (solid curve) and $d = 0.45$ μm (dashed curve).

9.7 Critical coupling (CC) and coherent perfect absorption (CPA)

In this section we reveal yet another interesting application of a stratified medium. This stems from the ability of specific systems to support constructive and destructive interferences to enhance or inhibit effective absorption under coherent illumination. Usually any interference phenomenon depends crucially on the relative phases of the interfering waves. In addition to this, it was highlighted only very recently that the amplitudes of the interfering waves play a very distinctive role in optimizing the interference process [85]. Note that similar ideas were known quite some time ago [86]. For example, two waves with equal amplitudes with a relative phase of π can kill each other due to destructive interference. On the other hand, the two interfering waves in phase can lead to superscattering (SS). The interfering waves can be generated with a single wave or with two incident coherent waves. The possible simplified scenarios for perfect destructive interference or superscattering are shown in Figs. 9.8(a) and 9.8(b). The first (second) is referred to as critical coupling (coherent perfect absorption) configuration. In the first case, we have a thin absorbing layer and a spacer layer on top of a perfect reflector. The perfect reflector prohibits light from passing through. If the reflected waves from all the interfaces happen to satisfy the conditions for destructive interference, then we will end up with null scattering from the structure. For simplicity we assumed the absorbing layer to be thin enough to be represented by a single interface. We thus have a system that neither transmits nor reflects; we say that the incident light has been critically coupled to the system, whereby all the incident light energy has been transferred to the absorbing layer. Critical coupling was first proposed by Yariv [87] and verified in a coupled fiber-microsphere system by Vahala's group [88]. There has been generalization of the scheme

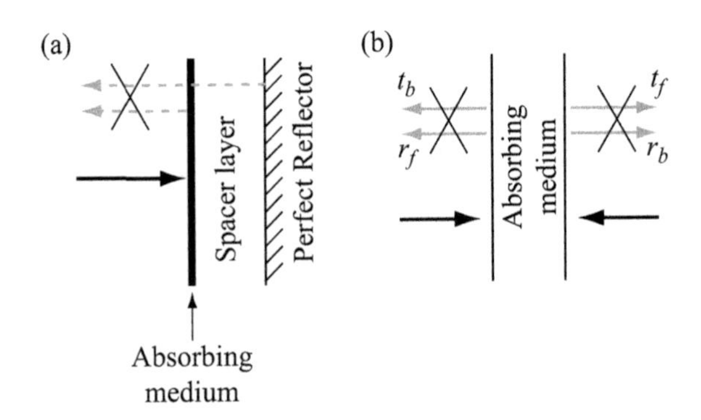

FIGURE 9.8: General schematics of (a) CC and (b) CPA.

(in Fig. 9.8(a)) to oblique incidence [89], where TE and TM results differ as expected. Analogous scenarios can be realized in the context of the system shown in Fig. 9.8(b), where an absorbing slab is illuminated from both sides. In this case, say, on the left, we have reflected light r_f due to incidence from the left and also the light t_b due to transmission from the right. The condition for perfect destructive interference is given by

$$|r_f| = |t_b|, \quad \Delta\varphi = (2m+1)\pi, \tag{9.100}$$

where m is an integer and $\Delta\varphi$ is the phase difference between the two interfering waves. On the right side of the structure, the same interference ensures that there is no scattered wave and that all the incident light is absorbed by the system. A different phase relation, namely, $\Delta\varphi = 2m\pi$, ensures superscattering since now we have constructive interference. These results have been verified experimentally with a thin Si wafer [85]. In what follows, we present some of the results pertaining to CC and CPA.

9.7.1 Critical coupling

Consider the structure shown in Fig 9.9(a) comprising a thin absorbing layer separated from a DFB structure by a spacer layer. A metal (silver) dielectric composite (see Section 5.5) is used as the material for the absorbing layer and its effective dielectric function is estimated by the Maxwell-Garnett formula. The parameters of the system are chosen as follows: $\epsilon_i = 1$, $\epsilon_h = 2.25$, $f = 0.06$, $d_1 = 10$ nm, $\epsilon_2 = 2.6244$, $d_2 = 161$ nm, $N = 12$, $\epsilon_a = 4.84$, $d_a = 46.6$ nm, $\epsilon_b = 2.25$, $d_b = 68.3$ nm and $\epsilon_f = 2.25$. The results for intensity reflection R, transmission T and total scattering $R+T$ are shown in Fig. 9.9(b). It is clear from the figure that at $\lambda \sim 410$ nm both R and T are zero, leading to

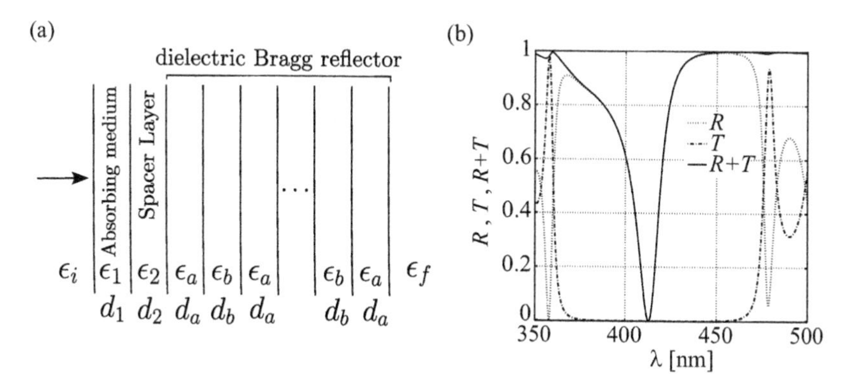

FIGURE 9.9: (a) Schematics and illumination of the critical coupling structure. (b) R, T, $R+T$ from the structure shown in Fig 9.9(a) as a function of the wavelength λ for normal incidence.

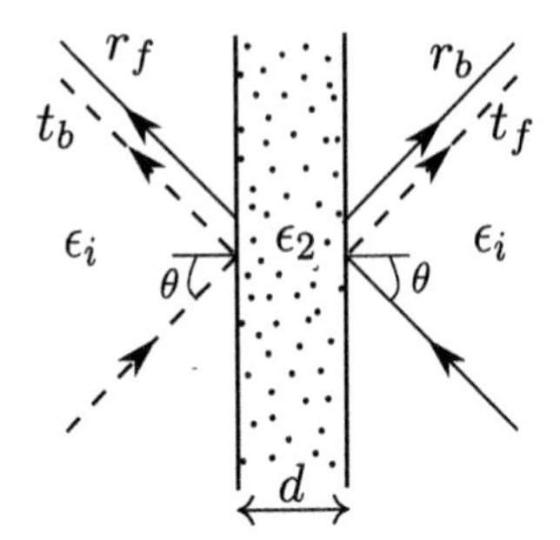

FIGURE 9.10: Schematics of CPA under oblique incidence.

null scattering ($R+T=0$). For incident coherent light at this wavelength, all the light is absorbed in the thin absorbing layer.

9.7.2 Coherent perfect absorption

The general oblique incidence scenario for CPA is shown in Fig. 9.10. For the absorbing layer we used a Bruggeman metal (gold) dielectric composite in order to have a larger range of volume fraction of the metal inclusions. The results for $|r_f|$ and $|t_b|$ and the phase difference between the interfering waves $\Delta\varphi$ (in units of π) along with the absolute value of the scattered amplitude (in log scale) are shown in Fig. 9.11. As can be seen from the figure, we have a CPA dip when the conditions of Eq. (9.100) are satisfied.

9.8 Nonreciprocity in reflection from stratified media

Reciprocity or the lack of it in scattering has intrigued physicists for several decades [90]. A given one-dimensional system is said to exhibit reciprocity if the scattering is insensitive to whether it is excited from the left or from the right. The issues involved are not only fundamental in nature, but they can also lead to many interesting applications. The most important aspect is their generality irrespective of the branch of physics. At various stages, we have stressed the equivalence of one-dimensional optical and quantum mechanical systems. Thus, theorems and conjectures derived for such optical systems can be applied to quantum systems and vice versa. In view of the interesting physics involved and a lack of proper understanding of the underlying physical principles behind the nonreciprocity, Agarwal and Dutta Gupta [91] studied the most general physical principles responsible for the nonreciprocity in reflection from stratified media. The findings were motivated by a beautiful experiment by Armitage et al. involving quantum wells in one compartment

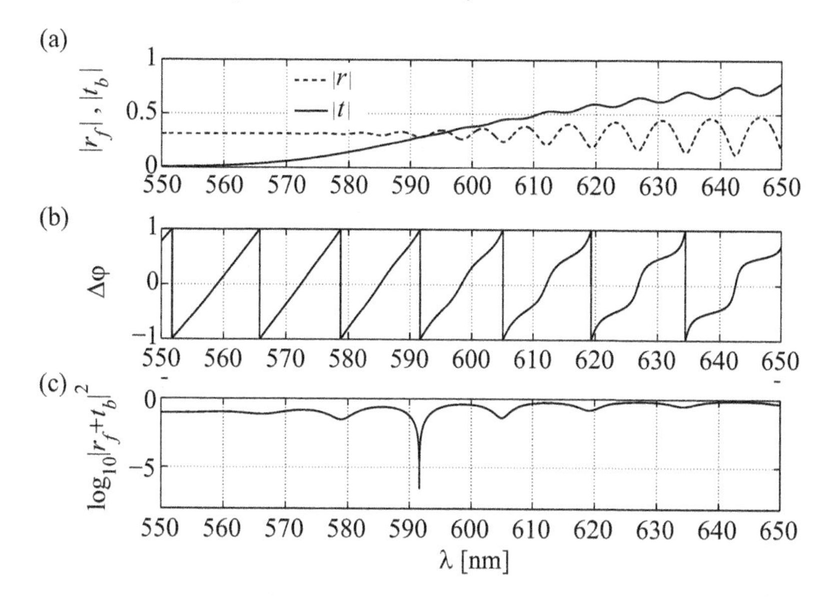

FIGURE 9.11: CPA in a composite layer. (a) Graph of $|r_f|$ and $|t_b|$, (b) the phase difference between the interfering waves $\Delta\varphi$ (in units of π) and (c) the log of the absolute value of the scattered amplitude $|r_f + t_b|$ as functions of λ. Parameters used in the calculations are as follows: $\epsilon_i = 1$, $\epsilon_h = 2.25$, $f = 0.004$ $d = 18.43$ μm and $\theta = 45°$.

of a coupled cavity systems [92]. We discuss the model system for their experiment after we learn about the dispersion relations and their method of analysis in the next chapter. Analogous derivation was carried out for a quantum system by Ahmed [93]. In this section we follow Ref. [91] to derive the general reciprocity relations for both transmitted and reflected light.

9.8.1 General reciprocity relations for an arbitrary linear stratified medium

Consider a stratified medium embedded in a vacuum occupying $-l \le z \le l$ and having a scalar complex dielectric function $\varepsilon(z)$ (see Fig. 9.12). The solution for fields of the Helmholtz equation for incidence from left E_L and right E_R can be written as

$$\begin{aligned} E_L =& E_{Li}e^{ik_0 z} + E_{Lr}e^{-ik_0 z}, \quad z \le -l, \\ =& E_{Lt}e^{ik_0 z}, \quad z \ge l, \\ E_R =& E_{Ri}e^{-ik_0 z} + E_{Rr}e^{ik_0 z}, \quad z \ge l, \\ =& E_{Rt}e^{-ik_0 z}, \quad z \le -l, \end{aligned} \tag{9.101}$$

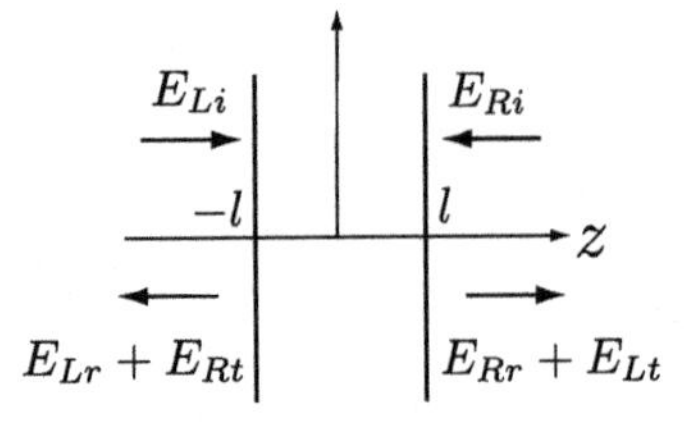

FIGURE 9.12: Schematics of the illumination geometry of the stratified medium with dielectric function $\varepsilon(z)$.

where k_0 is the vacuum wave vector and subscripts i, r, t refer to the incident, reflected and the transmitted waves, respectively. Fields E_L and E_R satisfy the condition

$$\int_{-l}^{l} \left(E_R^* \frac{d^2 E_L}{dz^2} - E_L \frac{d^2 E_R^*}{dz^2} \right) dz + k_0^2 \int_{-l}^{l} E_L E_R^* (\varepsilon - \varepsilon^*) dz = 0, \tag{9.102}$$

which can be easily obtained from the corresponding Helmholtz equations for E_L and E_R^*. Substitution of Eq. (9.101) into Eq. (9.102) leads to the following:

$$E_{Lt} E_{Rr}^* + E_{Rt}^* E_{Lr} + k_0 \int E_L(z) E_R^*(z) \, [\mathrm{Im} \, \varepsilon(z)] dz = 0. \tag{9.103}$$

For identical fields $E_L = E_R = E$, Eq. (9.102), after integration by parts, reduces to a simpler form:

$$\left(E^* \frac{dE}{dz} - E \frac{dE^*}{dz} \right)_{-l}^{l} + 2ik_0^2 \int |E|^2 \mathrm{Im} \, \varepsilon(z) dz = 0. \tag{9.104}$$

Like in Eq. (9.101), writing E as

$$\begin{aligned} E =& E_i e^{ik_0 z} + E_r e^{-ik_0 z}, \; z \le -l, \\ =& E_t e^{ik_0 z}, \; z \ge l, \end{aligned} \tag{9.105}$$

and substituting it into Eq. (9.104), we can recover the standard optical theorem

$$|E_t|^2 + |E_r|^2 + k_0 \int |E(z)|^2 \, \mathrm{Im} \, \varepsilon(z) dz = |E_i|^2. \tag{9.106}$$

In order to get the reciprocity relations for the transmitted amplitudes, we combine the Helmholtz equations for E_L and E_R in the form

$$\int_{-l}^{l} \left(E_R \frac{d^2 E_L}{dz^2} - E_L \frac{d^2 E_R}{dz^2} \right) dz = 0, \tag{9.107}$$

which after integration by parts can be simplified to

$$\left(E_R \frac{dE_L}{dz} - E_L \frac{dE_R}{dz} \right)_{-l}^{l} = 0. \tag{9.108}$$

Making use of Eq. (9.101) in Eq. (9.108), we can obtain the simple relation

$$E_{Ri}E_{Lt} = E_{Li}E_{Rt}. \tag{9.109}$$

Henceforth we will assume the incident fields from left and right to be the same, having unity amplitudes ($E_{Li} = E_{Ri} = 1$), which simplifies Eq. (9.109) further, leading to the reciprocity of the transmitted fields $E_{Lt} = E_{Rt}$. Note that the derivation of Eq. (9.109) never had any reference to the explicit form of the absorption present in the medium and Eq. (9.109) is valid for arbitrary lossy stratified medium. Thus, in linear systems transmission is always reciprocal (though in nonlinear systems such a reciprocal relation does not hold). Equality of the incident fields simplifies Eq. (9.103) and all the main reciprocity relations can be summarized as follows:

$$E_{Lt} = E_{Rt}, \tag{9.110}$$

$$\frac{E^*_{Rr}}{E^*_{Rt}} + \frac{E_{Lr}}{E_{Lt}} + \frac{k_0}{E_{Lt}E^*_{Rt}} \int E_L(z)E^*_R(z) \text{ Im } \varepsilon(z)dz = 0. \tag{9.111}$$

We now look at the effect of absorption and spatial symmetry. In absence of absorption ($\text{Im}(\varepsilon) = 0$), Eqs. (9.111) and (9.106) simplify drastically to

$$E_{Lt} = E_{Rt}, \quad \frac{E^*_{Rr}}{E^*_{Rt}} + \frac{E_{Lr}}{E_{Lt}} = 0, \quad |E_{Lt}|^2 + |E_{Lr}|^2 = 1. \tag{9.112}$$

Eq. (9.112) easily leads to the important result

$$|E_{Rr}|^2 = |E_{Lr}|^2, \quad \text{though} \quad E_{Rr} \neq E_{Lr}. \tag{9.113}$$

Eq. (9.113) implies that there may not be reciprocity in reflected amplitudes since they can differ in phase. Spatial symmetry plays an important role here. In order to highlight the effects of spatial symmetry, we refer back to the system with absorption. Consider a medium with the property $\varepsilon(z) = \varepsilon(-z)$. Then, if $E(z)$ is a solution of the Helmholtz equation, so is $E(-z)$. Let the incident fields also be symmetric, i.e., $E_i(z) = E_i(-z)$. The total field $E(z)$ would then satisfy $E(z) = E(-z)$ everywhere, while for the total fields we have the expressions

$$\begin{aligned} E(z) =& e^{ik_0z} + E_{Lr}e^{-ik_0z} + E_{Rt}e^{-ik_0z}, \quad z \leq -l, \\ =& e^{-ik_0z} + E_{Rr}e^{ik_0z} + E_{Lt}e^{ik_0z}, \quad z \geq l. \end{aligned} \tag{9.114}$$

The symmetry requirement $E(z) = E(-z)$ then implies

$$E_{Lr} + E_{Rt} = E_{Rr} + E_{Lt}. \tag{9.115}$$

Combined with the relation Eq. (9.110), it leads to the reciprocity of the reflected amplitudes

$$E_{Lr} = E_{Rr}. \tag{9.116}$$

Thus spatial symmetry, combined with the identical incident fields from both sides, ensures the reciprocity of the reflected amplitudes irrespective of the absorption in the system. In lossless structures the intensity reflection is reciprocal (Eq. (9.113)), while broken spatial symmetry can lead to nonreciprocity of phases in the reflected waves.

9.8.2 Nonreciprocity in phases in reflected light

In a general scattering event, the quantity of interest is the complex scattered amplitude. In one-dimensional systems as in our case, the situation is simple, namely, we need to extract information about the 'forward' scattered (or the transmitted) and the 'backward' (reflected) amplitudes. Since a complex amplitude involves two real quantities (amplitude and phase), we can ask whether both of them satisfy reciprocity. In absence of absorption ($\mathrm{Im}(\varepsilon(z)) = 0$), assuming a dependence $E_{r,t} \sim |E_{r,t}|e^{i\phi_{r,t}}$, Eqs. (9.110) and (9.111) can be reduced to the relations [78]

$$|E_{Lr}| = |E_{Rr}|, \tag{9.117}$$

$$2\phi_t = \phi_{Lr} + \phi_{Rr} + \pi, \tag{9.118}$$

with the consequence that

$$\tau_t = \frac{(\tau_{Lr} + \tau_{Rr})}{2}. \tag{9.119}$$

Thus, for a lossless structure lacking inversion symmetry, there is no nonreciprocal effect in the intensity reflection coefficient, while studies of the pulse delays can easily reveal the asymmetry of the structure. For example (for vanishingly small transmission delay τ_t), if the reflected pulse acquires a delay (meaning subluminal reflection) for incidence from the left, the same will be superluminal for incidence from the other side. This feature, albeit for a specific system, was noted neatly by Longhi [79]. In addition to the general proof for an arbitrary stratified medium given above, we go one step further to look at a cavity in the presence of absorbers. We show that the change in dispersion resulting from the atom-cavity coupling can result in superluminal propagation for incidence from both sides, indicating (as expected) violation of Eq. (9.119).

The nonreciprocity in the phases can be captured in typical pulse reflection experiments. Recall that the delay/advancements of the reflected/transmitted pulse is given by the frequency derivative of the phase (Wigner phase time) of corresponding reflection/transmission coefficient. Thus, nonreciprocity in phase will translate into nonreciprocity of the phase times, and consequently in different delays of the pulse.

It is always a good practice to keep in mind the limitations of the Wigner phase time. Recall that it was derived under the stationary phase approximation, whereby phase was supposed to be a slowly varying function of frequency.

In resonant structures phase can undergo a jump of π at resonance, and this approximation can prove to be a poor one. It is thus necessary to complement the predictions of the Wigner approach with explicit calculation of the pulse shapes. Since the systems under consideration are linear, we can adopt an approach similar to the one used in filter theory in RF electronics. We start with the Fourier decomposition of the incident pulse into the spectral components; each component is multiplied by the corresponding transfer function (in our case $t(\omega)$ or $r(\omega)$ for the transmitted and reflected light, respectively) to obtain the output spectrum. After the inverse Fourier transform of the output spectrum, we obtain the output pulse shape. For most of our pulse calculations, we use a Gaussian input pulse with the carrier frequency ω_c as follows:

$$E(z,t) = A_0 e^{-(t/\sigma)^2}\, e^{i(kz-\omega_c t)}, \quad A_0 = 1. \tag{9.120}$$

9.9 Pulse transmission and reflection from a symmetric and asymmetric Fabry-Pérot cavities

We present results for symmetric and asymmetric Fabry-Pérot (FP) cavities with or without resonant absorbers with distributed feedback (DFB) mirrors. The DFB mirrors are used so as to lead to high mirror reflections. We first show how the phase symmetry is intact in symmetric systems. In the context of the symmetric cavity, we also focus on resonant atom field coupling to demonstrate normal mode splitting (also known as vacuum field Rabi splitting [94, 95]) and its ability to control group velocity [78, 96]. The resonant absorbers are modeled by a dielectric function $\varepsilon(\omega)$ given by

$$\varepsilon(\omega) = \varepsilon_0 + \frac{\omega_p^2}{\omega_0^2 - \omega^2 - 2i\gamma\omega}. \tag{9.121}$$

In Eq. (9.121), the plasma frequency ω_p can be related to the number density and the dipole matrix elements of the atoms (or excitons in case of quantum wells). We then consider the asymmetric cavity in order to bring out the details of the nonreciprocity aspects in reflection. The nonreciprocity is demonstrated convincingly by the reflected pulse shapes and delays for both the directions for an input Gaussian pulse. We show that broken inversion symmetry alone is not adequate to lead to nonreciprocity in reflected amplitudes, while it is sufficient to lead to different phases for incidence from opposite directions. It is essential for the system to be lossy in order to lead to nonreciprocity in reflected amplitudes.

9.9.1 Symmetric FP cavity with resonant absorbers

Consider the structure shown in Fig. 9.13(a). The DFB mirrors are formed with 13 $\lambda/4$ layers of alternate high ($n_a = 2.4$) and low ($n_b = 1.3$) index materials, and they offer an intensity reflection coefficient ~ 0.99850. For a length of the cavity $d = 6.35\,\mu$m, one of the cavity resonances occurs at $f_c = \omega_c/2\pi = 7.0866 \times 10^{14}$ Hz, resulting in a quality factor $Q \sim 2.24 \times 10^5$. Note that the frequency f_c has been chosen carefully so as to be the central frequency of the bandgap of the DFB structure. All the cavity and pulse features (e.g., cavity resonances, pulse spectral width, etc.) are to be well accommodated within the band gap so as to have high reflection over the whole spectral range of the pulse. Let the cavity be filled with resonant absorber atoms. We calculate the reflection and transmission coefficients as well as their delays for input Gaussian pulse from both ends. For transmitted (reflected) pulse, the output (input) face of the structure is taken as the reference plane. Thus, a transmitted pulse is said to be subluminal (superluminal) if $\tau_t - \tau_f$ is positive (negative). Here $\tau_t = \frac{\partial \phi_t}{\partial \omega}$, $\tau_f = \frac{d}{c}$ with ϕ_t and d representing the phase of the transmission coefficient and total length of the system, respectively. On the other hand, a reflected pulse is subluminal if τ_r is positive ($\tau_r = \frac{\partial \phi_r}{\partial \omega}$, ϕ_r-phase of the amplitude reflection coefficient). The results for the intensity transmission coefficient T, the phase time difference $\tau_t - \tau_f$ and the group index n_g as functions of detuning δ $(= (\omega - \omega_c)/2\pi,\ \omega_c = \omega_0)$ for transmitted light are shown in Fig. 9.14. As mentioned above, the resonant interaction

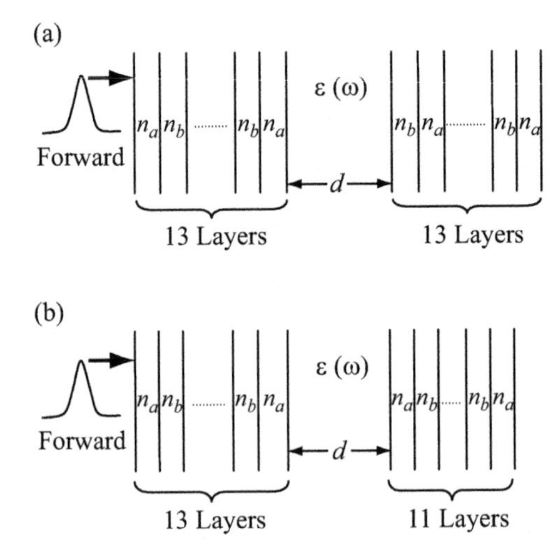

FIGURE 9.13: Schematic view of the (a) symmetric and (b) asymmetric FP cavity. In (a) all the DFB mirrors have 13 alternating high and low index (n_a=2.4, n_b=1.3, respectively) $\lambda/4$ layers. In (b) the left (right) mirror has 13 (11) layers. The length of each cavity is $d = 6.35\,\mu$m. 'Forward' and 'backward' directions imply incidence of the pulse from left or right.

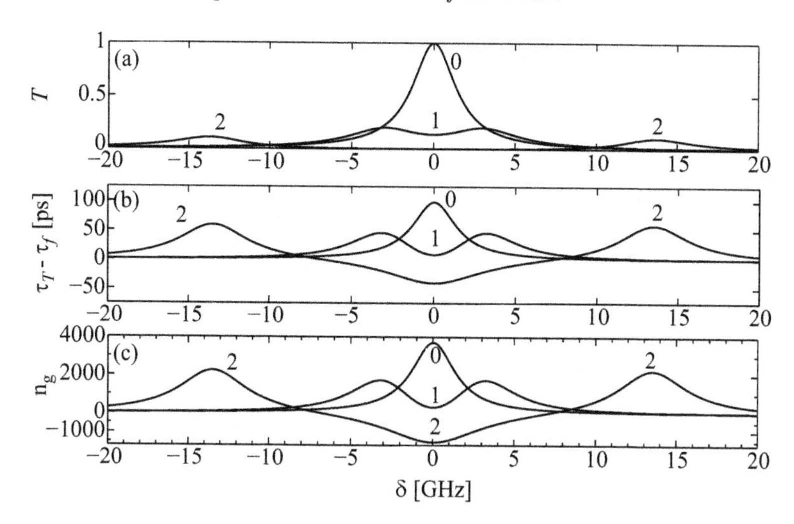

FIGURE 9.14: (a) Intensity transmission coefficient T of symmetric Fabry-Pérot cavity of Fig. 9.13(a), (b) delay/advancement $\tau_t - \tau_f$, (c) group index n_g as functions of detuning δ. The curves with labels 0, 1 and 2 are for $\omega_p^2/\omega_0^2 = 0.0, 0.08 \times 10^{-9}$ and 1.5×10^{-9}, respectively, with $2\gamma/\omega_0 = 1.0 \times 10^{-5}$. The other parameters are as in Fig. 9.13(a).

between the atoms and the cavity leads to the vacuum field Rabi splittings [94, 95], studied in detail for a cavity with metallic mirrors [96]. An increase in the atomic density leads to an increase in the superluminality of a pulse (tuned at $\delta = 0$) to $n_g = -1533$. This increase is mediated by the presence of resonant atoms. For pulses tuned at one of the side bands of the split resonances, the increase in the group index can be as large as $n_g = 2240$. It is thus clear that the cavity can enhance the control over the manipulation of the group velocities. The same control is possible also for reflected pulses (not shown). In contrast to the transmitted pulse, the reflected pulse is superluminal at the side bands of the split resonances. It is thus clear that due to the inherent symmetry of the structure, there is no ground to expect any nonreciprocity. In other words, the system does not distinguish between whether it is illuminated from the left or from the right. The situation changes drastically when the inversion symmetry is broken. In the next section, we consider one such case.

9.9.2 Asymmetric FP cavity

It is now well understood that transmission is reciprocal irrespective of whether the structure under consideration lacks any spatial symmetry. The reciprocity of transmission is insensitive to the presence or absence of absorption as well. As stressed earlier the scenario can be quite different for reflection.

Consider the FP cavity shown in Fig. 9.13(b), which is identical to that in Fig. 9.13(a), except that the right DFB mirror has one period less. Henceforth, we label incidence from left (right) as 'forward' ('backward') directions. The asymmetry leads to a drop of the transmission coefficient from unity (at resonance) resulting in a lower quality factor $Q \sim 9.92 \times 10^4$. The reduced quality factor reflects in lower extremal values of phase time $\tau_t - \tau_f$ and group index n_g for the transmitted pulse. The reflection coefficient (see Fig. 9.15(a)) of an empty asymmetric cavity at resonance is nonzero and finite (curve marked by 0). Note that this curve is the same (see Eq. 9.117) for both forward and backward directions. In contrast, phase times (see Fig. 9.15(b)) are not identical for forward (curve marked by 0) and backward (marked by 0′) directions. In the empty cavity the reflected pulse for forward (backward) incidence is superluminal (subluminal). When atoms are introduced in the cavity, the degeneracy in the reflection coefficient for forward and backward directions is lifted, leading to a more pronounced dip in the backward reflection coefficients for larger densities. This translates into highly superluminal reflection (see Fig. 9.15(b), curve marked by 1′). Note that for forward incidence, the pulse tuned at the side band is also superluminal (curve 2 in Fig. 9.15(b)).

In order to demonstrate these features, we present the reflected pulse profiles for an incident Gaussian profile given by Eq. (9.120) [78]. Results are

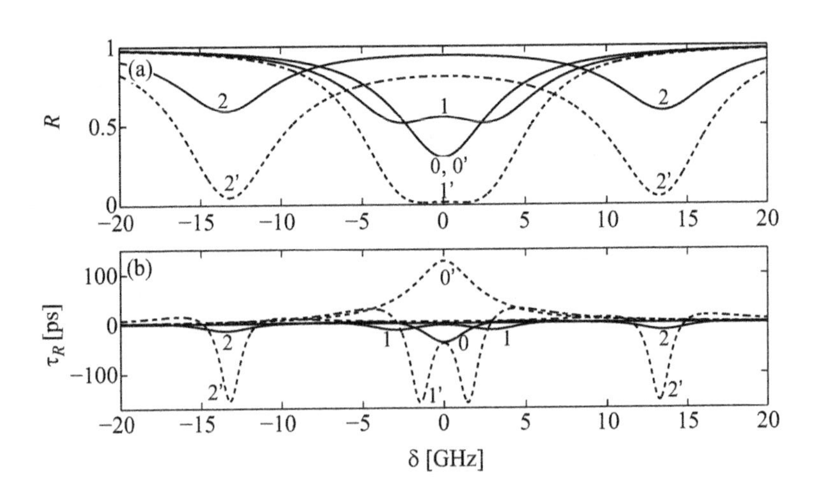

FIGURE 9.15: (a) Forward (solid line) and backward (dashed line) intensity reflection coefficient R, (b) corresponding phase time τ_R as functions of detuning δ. The curves with labels $(0, 0')$, $(1, 1')$ and $(2, 2')$ are for $\omega_p^2/\omega_0^2 = 0.0$, 0.08×10^{-9} and 1.5×10^{-9}, respectively. The other parameters are as in Fig. 9.13(b).

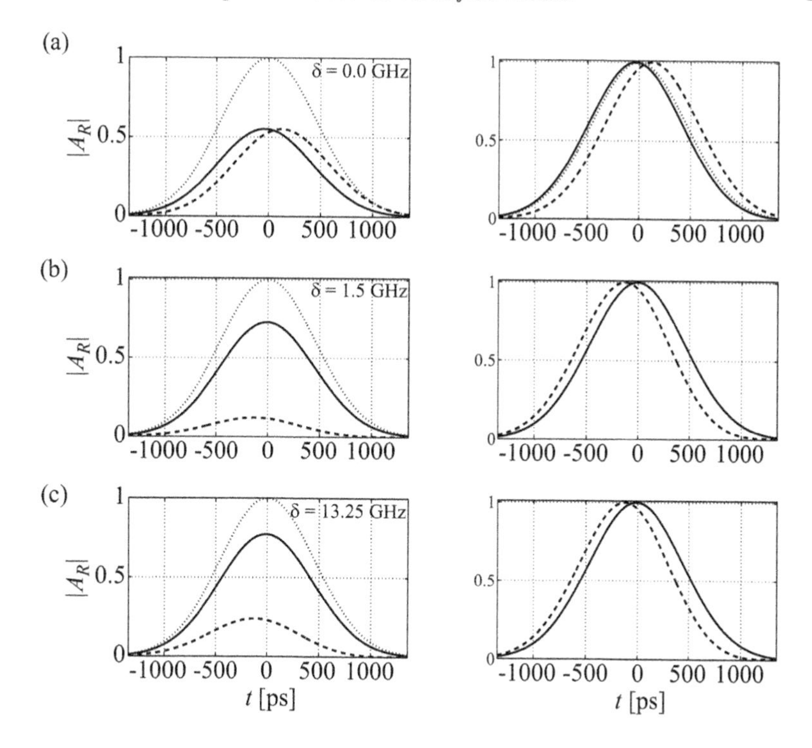

FIGURE 9.16: Reflected pulse shape $|A_R|$ for forward (solid line) and backward (dashed line) incidence as compared to the reference (input) pulse (dotted line) with $\tau = 650$ ps. (a), (b) and (c) are for $(\omega_p/\omega_0)^2 = 0.0$, 0.08×10^{-9} and 1.5×10^{-9}, respectively. The corresponding detuning values are shown on the panels. The left panel highlights the relative strength while the right one (normalized to the peak values) highlights the delay/advancement.

shown in Fig 9.16 for different cases, namely, when the pulse is tuned at the resonance of the empty cavity (i.e., $\delta = 0$) or at one of the side bands, say, at the right one, at $\delta = 1.5$ GHz and the other at $\delta = 13.25$ GHz. In Fig 9.16, we have also plotted the incident pulse profile for reference. The left panels in Fig 9.16 show the actual pulse shapes for reference (dotted line), forward (solid line) and backward (dashed line) incidence. The right panels show the same, albeit with peak normalization so that we can easily read out the delay/advancement. The top to bottom rows are for $\omega_p^2/\omega_0^2 = 0.0, 0.08 \times 10^{-9}$ and 1.5×10^{-9}, respectively. Apart from the opposite delay behavior noted earlier for a pulse tuned at $\delta = 0$ for lower densities of atoms, the interesting feature of the narrowing of the pulse can be seen from the right panel of Fig. 9.16(b). Thus, along with slowing down, we can also temporally squeeze the reflected pulse for backward incidence. The reflected pulse for the input tuned at the side band can be highly superluminal again in backward incidence (Fig. 9.16(c)). However, due to the sharpness of the spectral feature of the side band, there may be distortion at the trailing edge of the pulse.

It must be noted that Eqs. (9.110) and (9.111) and Eqs. (9.117) and (9.118) hold for each frequency component and thus the nonreciprocity in phases as depicted by Eq. (9.118) may be probed in a continuous wave (CW) experiment as well. The reflected fields for CW inputs from both ends can be made to interfere, leading to an interference term proportional to $cos(\phi_{1R} - \phi_{2R})$. The emergence of the interference pattern will be the direct evidence of the asymmetry of the phases. It is also important to note that the atomic damping has a major role to play in determining the delay response. In order to demonstrate this, let us consider a situation (an atomic density) identical to that of curve $2, 2'$ of Fig. 9.15, except that $2\gamma/\omega_0$ was decreased tenfold to 1.0×10^{-6}. The results are shown in Fig. 9.17. It is clear from Fig. 9.17(b) that the pulse tuned at the side band will exhibit superluminality (subluminality) for incidence from the left (right). Note that for the previous value of $2\gamma/\omega_0$ for both cases, pulse reflection was superluminal. These features are demonstrated clearly in terms of reflection of single and twin-hump pulses in Figs. 9.18(a) and 9.18(b), respectively. The twin-hump input was made of a superposition of two closely spaced Gaussians as follows:

$$A(t) = e^{-\frac{(t-\sigma_{sep})^2}{\sigma^2}} + e^{-\frac{(t+\sigma_{sep})^2}{\sigma^2}}, \tag{9.122}$$

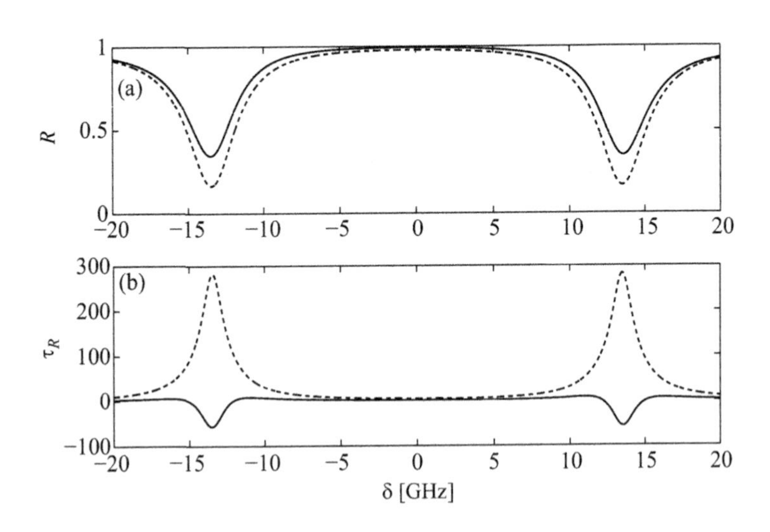

FIGURE 9.17: (a) Intensity reflection coefficient R, for forward incidence (solid curve) and for backward incidence (dashed curve), (b) corresponding delays τ_R as a function of detuning δ for $2\gamma/\omega_0 = 1.0 \times 10^{-6}$ and $\omega_p^2/\omega_0^2 = 1.5 \times 10^{-9}$. The other parameters are the same as in Fig 9.13(b).

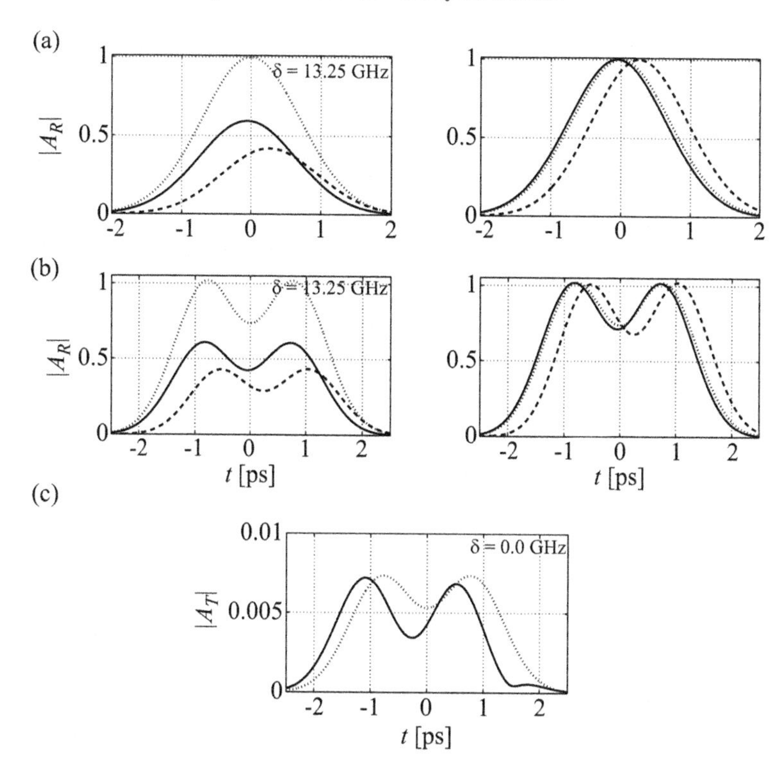

FIGURE 9.18: Reflected pulse shape $|A_R|$ for forward (solid line) and backward (dashed line) incidence as compared to the reference pulse (dotted line) for (a) single Gaussian with $\sigma = 1$ ns, and (b) twin Gaussian with $\sigma = 0.8$ ns and $\sigma_{sep} = 0.8$ ns. The left panels highlight the relative strength while the right ones (normalized to the peak values) highlight the delay/advancement. The corresponding detuning values are shown on the panels. (c) Transmitted pulse shape $|A_T|$ (solid line) as compared to the reference (dotted line) with $\sigma = 0.8$ ns and $\sigma_{sep} = 0.8$ ns. Here the reference pulse is normalized to the peak value of the transmitted pulse.

where $2\sigma_{sep}$ gives approximately the distance between the two peaks. For sake of completeness, we present the transmission of the twin-hump pulse at $\delta = 0$ where the transmission is highly superluminal. The latter is shown in Fig. 9.18(c). The almost distortionless propagation of the twin-hump pulse both in reflection and transmission may shed light on yet another issue as regards the true mechanism of superluminal transit [97].

Chapter 10

Surface and guided modes

Plasmonics opens up the gateway to the sub-wavelength world with its inherent ability to beat the Rayleigh limit. What is truly amazing is the fact that most of the related effects can be understood with an undergraduate background in classical optics, and all of it was around for so many years. Indeed, the past two decades have been an eye-opener, with novel effects arising from classical Maxwellian optics. Superlensing and superresolution, invisibility cloaks and extraordinary transmission are just a few examples. We now understand a large variety of effects under the banner of plasmonics with potential applications ranging from enhanced photovoltaics to imaging and surface-enhanced spectroscopy to the precision spectroscopy of single molecules. In fact, there are many more unexplored areas.

In this chapter we focus mostly on the basic entity, namely, the surface plasmons (SPs). Even today the best possible source to learn about surface plasmons is Raether's monograph [98]. We will try to show how the SP, being the analog of guided modes of an equivalent dielectric structure, is so very different from them. We will avoid a detailed discussion of the free electron model or the bulk plasmons of noble metals. We discussed them briefly in Section 5.3, and besides, we can find a nice description of related topics in the monograph by Maier [99]. The same monograph also presents highlights of all the important applications (see part II of Ref. [99]). Nevertheless, we will focus on the physics that opens up the possibilities for all such applications.

In standard textbooks or monographs, the surface plasmons and related phenomena are introduced starting from a single metal-dielectric interface. We then move on to two or more interfaces yielding coupled plasmons and gap plasmons. We will adopt a different approach starting from a general layered media and looking at certain specific examples. We will try to highlight the similarities and differences between the guided and the surface modes. We start with a multilayered metal-dielectric stratified medium in order to have a general framework encompassing a wide spectrum of effects. We show how the characteristics of the modes can be calculated through the poles of the reflection/transmission coefficients, which is the same as solving the dispersion relation. We specialize to a simpler case study, where we look at the modes of a general symmetric layered medium. We show how the symmetry simplifies the equations and at the outset distinguishes the symmetric and antisymmetric modes. We then consider various limiting cases yielding the well-known results for waveguide modes, surface plasmons (both long- and short-range) and the gap plasmon modes. While most of this can be carried out in the framework of a textbook, we have borrowed a few results from our earlier research articles involving not-so-simple numerical calculations (multibranch dispersion curves). The corresponding codes will be presented in the Appendix so that interested students can reproduce the results.

We start with the results for the amplitude reflection and transmission coefficients and the dispersion relation presented in Section 9.2. Recall that we defined the modes of the structure through the poles of these coefficients. We take a particular case of a symmetric structure [100], which enables us to arrive at the dispersion relations of the symmetric and the antisymmetric modes. By taking suitable limits we show that these dispersion relations are identical to the standard forms well known in the literature. We solve the dispersion relations for simple cases of planar waveguides and coupled plasmons. We show that while waveguide dispersion can be calculated easily by 'cheating,' the same is not possible for the coupled plasmons. In order to highlight the near-field or the sub-wavelength capabilities of plasmonics, we then consider a gap plasmon (GP) guide where a sub-wavelength dielectric layer is enclosed in between two semi-infinite/finite metal layers. The fundamental plasmonic mode can propagate in this guide while any TE mode is prohibited by the Rayleigh criterion. In this context we demonstrate an avoided crossing phenomenon, which is often encountered in optical and solid state systems. We also show how such guides in resonant tunneling geometry can lead to slow light. We then move on to a coupled waveguide structure modeling the experiment of Armitage et al. [92], first to show the nonreciprocity discussed in Section 9.8 and then to present a demonstration of how to analyze the roots of the corresponding dispersion relation [78]. A few more details about the surface plasmons and coupled surface plasmons are then discussed, concentrating mainly on how to excite them. Of course, methods discussed here are not exhaustive, but they serve the purpose of highlighting the role of momentum matching. A brief comparison is made with the quasi phase matching in nonlinear optics.

10.1 Case study: A symmetric $(2N+1)$-layer structure

Consider the TM modes (with nonvanishing components H_y, E_x and E_z) for the symmetric system shown in Fig. 10.1, comprising $2N+1$ layers with any j-th layer characterized by relative dielectric function $\varepsilon_j(\omega)$ and width d_j. The middle layer is assumed to have a width d_0 and dielectric function ε_0 (this latter symbol is not to be confused with vacuum dielectric permittivity). The bounding media are assumed to be air. Like before we restrict our attention only to the TM-polarized waves. Keeping in mind the symmetry of the structure about the plane $z=0$, we write the magnetic and electric field components in the central layer as

$$H_{0y} = A_0(e^{ik_{0z}z} \pm e^{-ik_{0z}z}), \tag{10.1}$$

$$E_{0x} = p_{0z}A_0(e^{ik_{0z}z} \mp e^{-ik_{0z}z}), \tag{10.2}$$

where $k_{0z} = \sqrt{k_0^2\varepsilon - k_x^2}$ and $p_{0z} = \frac{k_{0z}}{k_0\varepsilon}$. Hereafter the upper (lower) sign in Eqs. (10.1) and (10.2) will refer to the symmetric (antisymmetric) magnetic modes. Note that symmetry is being judged by the symmetry of the magnetic field distribution across the layers. Making use of the characteristic matrices, we can then relate the tangential field components at the center, i.e., at $z=0$ and at $z=z_N$, which in terms of amplitudes yields the relation

$$\begin{pmatrix} 1 & \pm 1 \\ p_{0z} & \mp p_{0z} \end{pmatrix} \begin{pmatrix} A_0 \\ A_0 \end{pmatrix} = M_T \begin{pmatrix} 1 \\ p_{tz} \end{pmatrix} A_t. \tag{10.3}$$

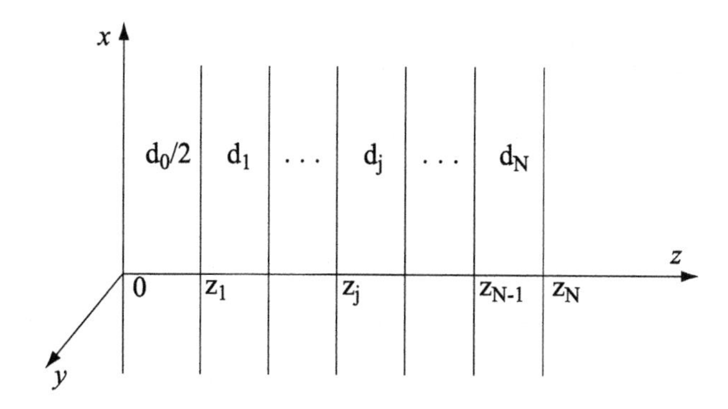

FIGURE 10.1: Schematic view of the symmetric layered medium, where only the right half is shown.

In Eq. (10.3) the subscript t refers to the corresponding quantities in the embedding medium and M_T is given by

$$M_T = M_0(d_0/2)M_1(d_1)\cdots M_j(d_j)\cdots M_N(d_N). \tag{10.4}$$

Referring to the different signs in Eq. (10.3) and demanding the nontriviality of the constant amplitudes, we obtain the dispersion relation for the symmetric and the antisymmetric modes. For the symmetric mode we have

$$m_{21} + m_{22}p_{tz} = 0, \tag{10.5}$$

$$(m_{11} + m_{12}p_{tz})A_t = 2A_0, \tag{10.6}$$

while for the antisymmetric mode we obtain

$$m_{11} + m_{12}p_{tz} = 0, \tag{10.7}$$

$$(m_{21} + m_{22}p_{tz})A_t = 2p_{0z}A_0. \tag{10.8}$$

It is clear that Eqs. (10.5) and (10.7) give the corresponding dispersion relations while Eqs. (10.6) and (10.8) define the amplitudes for the mode functions.

10.1.1 Typical example: Modes of a symmetric waveguide

As an example, we first consider the case of a symmetric planar waveguide, whereby a slab of dielectric constant ε_d and width d_0 is embedded in a dielectric medium with dielectric constant ε_t (see Fig. 10.1 with $d_j = 0,\ j = 1-N$). For guided modes, waves need to be evanescent outside the film. We thus rewrite k_{tz} as

$$k_{tz} = i\sqrt{k_x^2 - k_0^2\varepsilon_t} = i\bar{k}_{tz}. \tag{10.9}$$

Eqs. (10.5) and (10.7) can then be reduced to the well-known dispersion relations for the symmetric and antisymmetric transverse magnetic modes of a

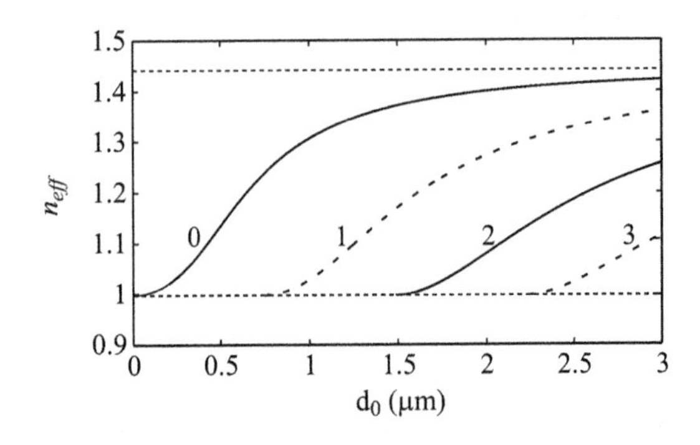

FIGURE 10.2: Transverse magnetic (TM) modes of a symmetric dielectric wave guide for parameters $\lambda = 1.55\ \mu$m, $\varepsilon_d = 2.085$ and $\varepsilon_t = 1.0$.

planar guide [101]:

$$\varepsilon_d \bar{k}_{tz} - \varepsilon_t k_{0z} \tan(k_{0z} d_0/2) = 0, \tag{10.10}$$
$$\varepsilon_d \bar{k}_{tz} + \varepsilon_t k_{0z} \cot(k_{0z} d_0/2) = 0. \tag{10.11}$$

10.1.2 Coherent perfect absorption as antiguiding

Up to now we have studied the relations for waveguiding modes of a structure by demanding finite evanescent output from the structure for null input. An obvious question arises. Is it possible to have null output from the structure for some finite input? CPA (see Section 9.7) corresponds exactly to one such situation, where we require null scattering from the structure for finite input. We now exploit the symmetry principles and show that the nontrivial solutions of CPA can only be either symmetric or antisymmetric. We use this information to find the sufficient condition for CPA.

For CPA we require null scattering: $A_{i-} = 0$ and $A_{f+} = 0$ (see Fig. 10.3) with $A_{i+} = A_{f-} \neq 0$. It immediately follows that the incident field intensities must be the same ($|A_{i+}|^2 = |A_{f-}|^2$) for having CPA. This implies that the forward (A_{0+}) and backward (A_{0-}) propagating amplitudes in the cavity be the same ($A_{0+} = \pm A_{0-}$). The problem then reduces to the symmetric and antisymmetric solutions as before. For symmetric solutions we have $A_{0+} = A_{0-} = A_0$ and $A_{i+} = A_{f-} = A_{in}$. The analogous relations for the antisymmetric solutions are given by $A_{0-} = -A_{0+} = -A_0$ and $A_{i-} = -A_{f+} = -A_{in}$. Consider a case, as before, with only the central slab (d_0, ε_0) and ambient media the same on both sides. Now relating the amplitudes at $z = 0$ to $z = d_0/2$ gives us

$$\begin{pmatrix} 1 & \pm 1 \\ p_{0z} & \mp p_{0z} \end{pmatrix} \begin{pmatrix} A_0 \\ A_0 \end{pmatrix} = M_{d_0/2} \begin{pmatrix} 1 \\ -p_{tz} \end{pmatrix} A_{in}, \tag{10.12}$$

with $M_{d_0/2}$ as the characteristic matrix for the half slab. In writing Eq. (10.12) we have set $A_{i-} = A_{f+} = 0$ and used the symmetry relations. Compar-

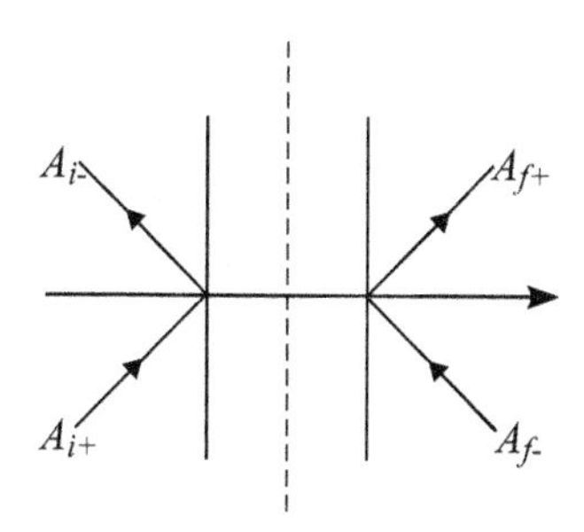

FIGURE 10.3: Schematic view of the incident and scattered amplitudes in CPA configuration.

ing Eq. (10.3) with Eq. (10.12) reveals CPA as an antiguiding phenomenon forming the opposite end of the same scattering problem. Eq. (10.12) can be simplified for symmetric and antisymmetric solutions as [102]

$$p_{tz} + ip_{0z}\tan(k_{0z}d_0/2) = 0, \tag{10.13}$$

$$p_{tz} - ip_{0z}\cot(k_{0z}d_0/2) = 0, \tag{10.14}$$

respectively. The relation given by Eqs. (10.13) and (10.14) is the sufficient condition for CPA and the roots of these transcendental relations give the location and the characteristics of the CPA dips. Note that these relations are in general valid for both TE- and TM-polarizations.

10.1.3 Relevant example: Surface plasmons and coupled surface plasmons

Consider a single metal film with width d_0 and dielectric function ε_m embedded in a dielectric (with constant ε_t) (see Fig. 10.1 with $d_j = 0$, $j = 1 - N$). For surface modes to exist at both the interfaces, waves in all the media need to be evanescent. We thus rewrite the z components of the wave vectors as

$$k_{0z} = i\sqrt{k_x^2 - k_0^2\varepsilon_m} = i\bar{k}_{0z}, \quad k_{tz} = i\sqrt{k_x^2 - k_0^2\varepsilon_t} = i\bar{k}_{tz}. \tag{10.15}$$

Substituting Eqs. (10.15) into Eqs. (10.5) and (10.7) and recalling the structure of the characteristic matrix (Eq. 9.7), we can rewrite the dispersion equations for the symmetric and the antisymmetric modes, respectively, as

$$\varepsilon_m k_{tz} + \varepsilon_t k_{0z}\tanh(x) = 0, \tag{10.16}$$

$$\varepsilon_m k_{tz} + \varepsilon_t k_{0z}\coth(x) = 0, \tag{10.17}$$

where $x = \frac{\bar{k}_{0z}d_0}{2} = \frac{k_{0z}d_0}{2i}$. It is easy to see that Eqs. (10.16) and (10.17) coincide with Eq. (A.20) of Raether's monograph [98]. The case of a single metal-dielectric interface can easily be recovered by taking the limit $d_0 \to \infty$ leading to identical values for tanh and coth (=1) for large arguments. Both the equations then reduce to the same form:

$$\varepsilon_m k_{tz} + \varepsilon_t k_{0z} = 0. \tag{10.18}$$

Eq. (10.18) can easily be reduced to the standard dispersion relation for surface plasmons:

$$k_x = k_0\sqrt{\frac{\varepsilon_t\varepsilon_m}{\varepsilon_t + \varepsilon_m}}. \tag{10.19}$$

We now comment on the decay characteristics of the coupled modes. As can be seen from Fig. 10.4, the antisymmetric (symmetric) modes have a much smaller (larger) decay and thus can propagate a longer (shorter) distance. Therefore, the antisymmetric (symmetric) modes are often referred to as long-range or LR (short-range or SR) modes.

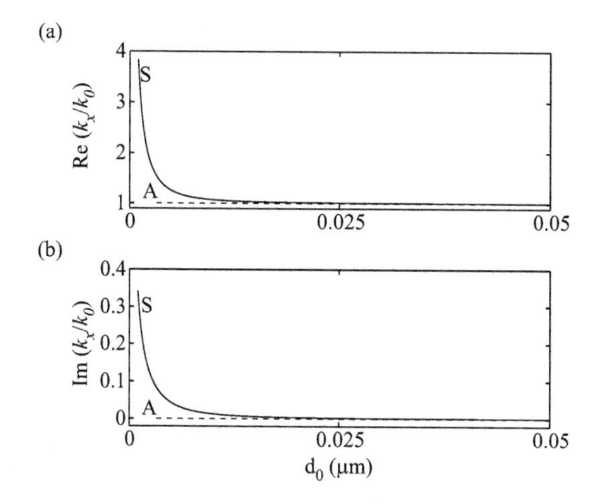

FIGURE 10.4: (a) Real and (b) imaginary parts of the normalized propagation constant k_x/k_0 for the coupled modes. The parameters are $\lambda = 1.55$ μm, $\varepsilon_m = -132 + 12.6i$ (gold) and $\varepsilon_t = 1.0$.

10.1.4 Gap plasmons and avoided crossings

We now restrict our attention to a symmetric metal-clad waveguide with dielectric core thickness d_0, and metal claddings with width d_1. After solving the dispersion equations for complex k_x, the complete spatial dependence of the mode functions can be obtained. For example, for the symmetric modes in the various regions, we have

- for $|z| \leq d_0/2$,

$$H_{0y}(x, z) = 2A_0 \cos(k_{0z}z)e^{ik_x x}, \tag{10.20}$$

- for $d_0/2 < |z| \leq d_0/2 + d_1$,

$$H_{1y}(x, z) = (A_{1+}e^{ik_{1z}(z-d_0/2)} + A_{1-}e^{-ik_{1z}(z-d_0/2)})e^{ik_x x}, \tag{10.21}$$

- for $|z| > d_0/2 + d_1$,

$$H_{ty}(x, z) = A_t e^{ik_{tz}(z-(d_0/2+d_1))}e^{ik_x x}. \tag{10.22}$$

The corresponding equations for the electric field components can be obtained by using Eqs. (10.20)–(10.22) in Maxwell's equations. The constant A_t in Eq. (10.22) is evaluated using Eq. (10.6), while $A_{1\pm}$ in Eq. (10.21) is given by the solution of the following matrix equation:

$$\begin{pmatrix} A_{1+} \\ A_{1-} \end{pmatrix} = \begin{pmatrix} e^{ik_{1z}d_1} & e^{-ik_{1z}d_1} \\ p_{1z}e^{ik_{1z}d_1} & -p_{1z}e^{-ik_{1z}d_1} \end{pmatrix}^{-1} \begin{pmatrix} 1 \\ p_{tz} \end{pmatrix} A_t. \tag{10.23}$$

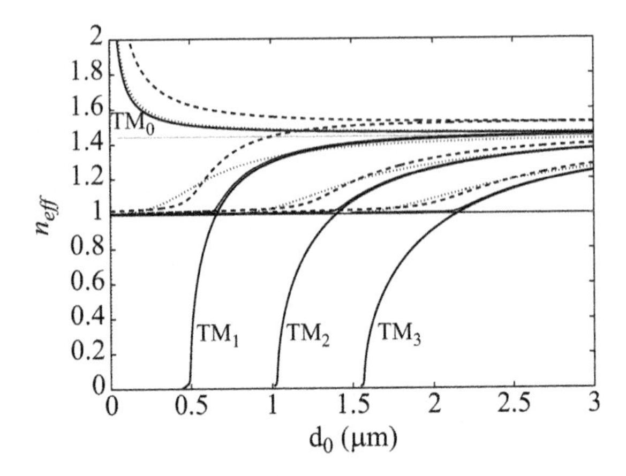

FIGURE 10.5: Solution of the dispersion relation for n_{eff} as functions of the core thickness d_0. The thick lines are for the semi-infinite metal claddings on the silica guide, the dotted lines are for the bare silica guide. The dashed lines are for the metal cladding thickness $d_1 = 0.01$ μm while the thin lines are for $d_1 = 0.03$ μm. The parameters are $\lambda = 1.55$ μm, $\varepsilon_0 = 2.085$, $\varepsilon_1 = -132+12.6i$ and $\varepsilon_t = 1.0$. The leaky and the higher-order branches are not shown. Adapted from Ref. [100].

The arbitrary constant A_0 is fixed by normalization of the modes. With the field profiles known, we can also calculate the time-averaged Poynting vector, giving the power flow along the guide. We do not present the details of the calculation here and refer the reader to Ref. [100] for technical details. We present the results and the related physics here. For numerical calculations the following parameters were chosen: $\lambda = 1.55$ μm, $\varepsilon_0 = 2.085$ (silica), $\varepsilon_1 = -132 + 12.6i$ (gold) [103], $\varepsilon_t = 1.0$ (air). We varied d_0 and d_1. The results for the roots of the dispersion relation are presented in Fig. 10.5, where we have plotted the effective index $n_{eff} = k_x/k_0$ as functions of d_0. We presented the results for both the symmetric and antisymmetric modes as well as the plasmon and oscillating modes. Note that a gap plasmon guide can support modes with fields localized near the metal dielectric interface (plasmonic modes with evanescent field dependence in the core). It can also support the usual oscillatory dielectric guide modes with propagating waves in the core. We label the modes as plasmon (or oscillatory) depending on whether the magnetic field distribution inside the silica guide is expressible as a superposition of hyperbolic sine and cosine (or sines and cosines). For reference we have plotted the cases for (a) the bare silica guide with outside medium as air (dotted lines) and (b) the silica guide with semi-infinite metal claddings on both sides (thick lines) [103]. A comparison of the two cases reveals that with metal cladding we can realize very low effective indices (close to zero) with the oscillatory modes, while the plasmon mode offers very large values. In contrast, the guided mode indices for the silica guide are limited in the range between air and silica refractive indices (i.e., between 1 and $\sqrt{2.085} = 1.44$). In the case of the gap plasmon guide, we have the familiar splitting due to the coupling of the two interface plasmons. The uppermost branch (TM_0 plasmon) corresponds to the symmetric while the lower one corresponds to the antisymmetric oscillatory mode TM_1. Oscillatory modes from left to right are labeled by an increasing integer. It is clear from Fig. 10.5 that for semi-infinite metal claddings, there is a cutoff thickness for the oscillatory modes. For example, for $d < 0.5$ μm, there are no oscillatory modes with the realization of a single-mode operation with just the TM_0 plasmon mode. However, the scenario changes drastically if we restricts the widths of the metal cladding. For example, for $d_1 = 0.01$ μm, the antisymmetric oscillatory mode exists, which has a lower cutoff (see the dashed line). For a slightly larger thickness of the metal films, namely, $d_1 = 0.03$ μm, the behavior almost coincides with the results for the guide with semi-infinite metal claddings. The losses associated with the modes can be studied by looking at the imaginary parts of the roots of the dispersion relation. Mode cutoff is determined by the sudden changes in the losses from small to large values as we reduce the gap width d_0. We also have the avoided crossing phenomenon like in coupled cavity-exciton systems (see Section 10.4). This may lead to the possibility of coupling of the surface plasmons on the two sides of the thin cladding layer when the metal cladding thickness is very small. In other words, the surface plasmon on the metal/air interface can interact with the same on the other metal/silica interface. From a somewhat different angle, this phenomenon can be viewed as the crossing of the dispersion branches of an air/silica/air guide with that of the metal/silica/metal guide. The resulting level repulsion for finite width metal cladding is shown in Fig. 10.6. Indeed, the limiting cases are the bare silica guide and the gap plasmon guide with semi-infinite metal claddings. The case with finite and very low thickness of the metal cladding is in between and has the avoided crossing features. As expected, the avoided crossing effect is stronger for the lower thickness of the metal cladding. Note also that the value of the effective refractive indices for the modes corresponding to the part of the lower branches in Fig. 10.5 is less than unity. Thus these modes are leaky.

10.2 Excitation schemes for overcoming momentum mismatch

Bound modes like the surface and the guided modes localized in the film or near the interface are characterized by an effective index ($=k_x/k_0$) larger than the refractive index of the medium of incidence. Thus such modes can

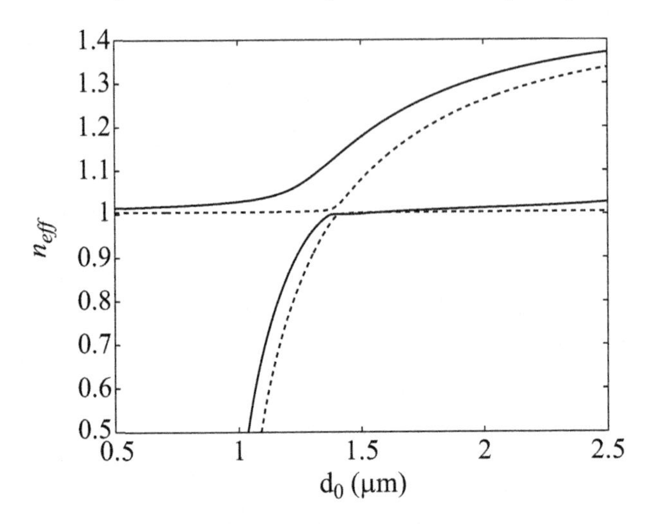

FIGURE 10.6: Avoided crossing phenomenon with the TM_2 mode. The solid (dashed) line is for $d_1 = 0.01$ μm ($d_1 = 0.03$ μm). Other parameters are as in Fig. 10.5. Adapted from Ref. [100].

not be excited just by shining a laser beam on the film or the interface, even for grazing incidence. For plane wave incidence at an angle θ_i, we can never satisfy $k_0 \sin(\theta_i) = k_x = k_{g/sp}$. In order to excite them, we have to compensate for the momentum mismatch. The mismatch can be overcome by a high index prism in attenuated total reflection (ATR) geometry or by periodic engravings on the surface. Such schemes are now widely used and are referred to as the prism and the grating coupling, respectively. Note that a rough surface can also couple the incident light to the relevant mode.

10.2.1 Prism coupling: Otto, Kretschmann and Sarid geometries

In ATR geometry we load the guiding film or the metal-dielectric interface with a high-index prism with dielectric constant ε_p, after a spacer layer, and we operate at angles larger than the critical angle. Thus the waves in the spacer layer are evanescent and we can satisfy the momentum matching as

$$k_0\sqrt{\varepsilon_p}\sin(\theta_i) = k_{g,sp}. \tag{10.24}$$

In the absence of the excitation of any modes, the reflectivity would be unity due to total internal reflection. For specific angles of incidence corresponding to Eq. (10.24), the modes can be excited, leaving sharp dips in reflection. This signifies the channeling of the energy from the incident wave to the specified mode, resulting in a corresponding drop in the reflected light. There can be variations of the ATR geometry (Fig. 10.7). The Otto geometry has a low

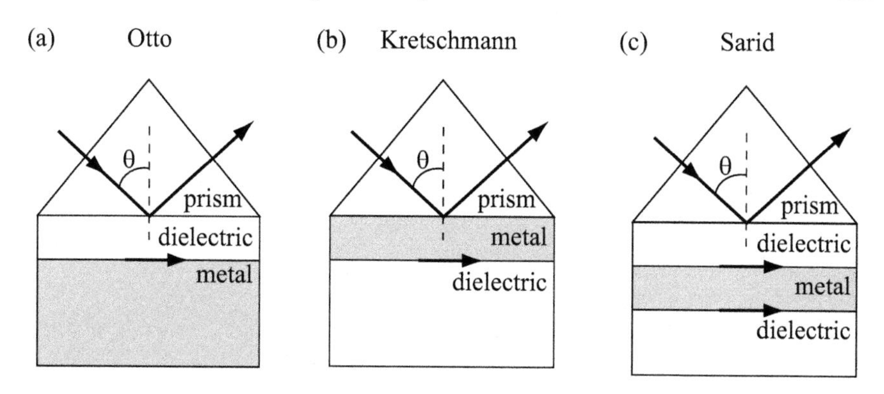

FIGURE 10.7: (a) Otto, (b) Kretschmann and (c) Sarid geometries for ATR.

index spacer layer between the high-index prism and the metal film, whereas in the Kretschmann configuration the metal film is deposited on the base of the high-index prism [98]. A different geometry that can support coupled surface plasmons in very thin metal films was suggested by Sarid [104]. In the Sarid geometry, we can excite both the symmetrical short-range (SR) and the antisymmetrical long-range (LR) surface plasmons.

10.2.2 Grating coupling: Analogy to quasi phase matching

In this scheme the incident light falls on a grating with grating vector K ($= 2\pi/\Lambda$, Λ grating period). The period can be chosen such that one of the diffraction orders, namely, the m-th order matches the guided/surface mode

$$k_{g/sp} = k_0 \sin(\theta) + mK, \quad m = \pm 1, \pm 2, \cdots . \tag{10.25}$$

Momentum matching the +1 diffraction order is shown in Fig. 10.8(b). The dips for the coupled plasmon modes of a free-standing metal film (see Fig. 10.8(a)) are shown in Fig. 10.9 [105, 106]. The results for the specu-

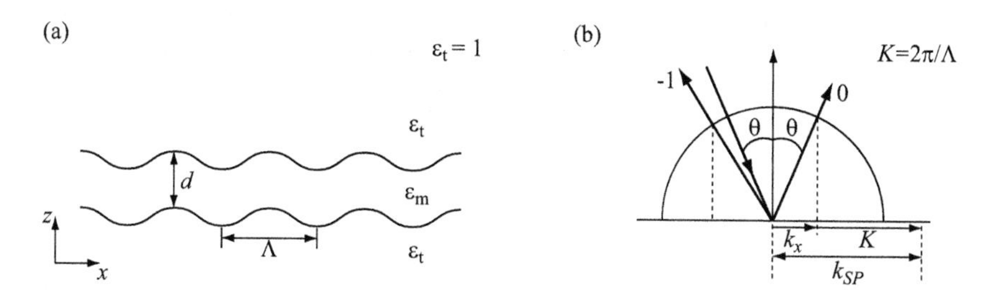

FIGURE 10.8: Schematics of grating coupling. (a) Corrugated film. (b) Wave vectors for different diffracted orders.

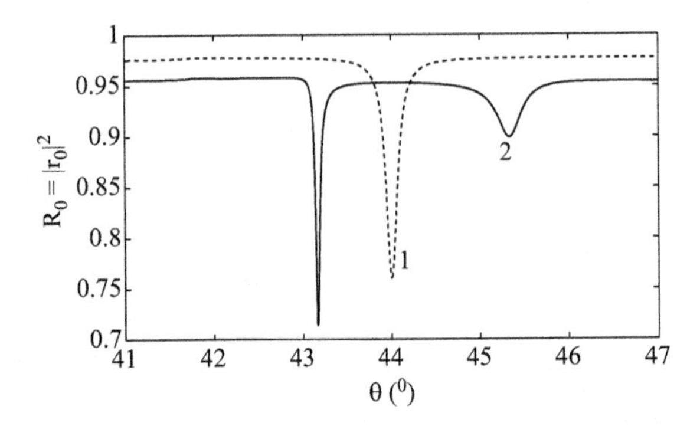

FIGURE 10.9: Reflected intensity as a function of angle of incidence for a corrugated silver film. The parameters are $\lambda = 633$ nm, $\varepsilon_m = -18.0 + 0.51i$, $\varepsilon_t = 1.0$, $d = 150$ nm in (1) and $d = 50$ nm in (2), $a = 8$ nm and $\Lambda = 1.893$ μm.

lar intensity reflection for a silver film of corrugation amplitude a are shown in Fig. 10.9, which clearly shows the splitting for lower film thicknesses.

There is an interesting example where such grating-assisted momentum compensation is employed in nonlinear optics. For example, for efficient second harmonic generation, the momentum mismatch Δk defined as $\Delta k = 2k_\omega - k_{2\omega}$ is compensated by the m-th order of the domain reversal grating

$$\Delta k = mK, \tag{10.26}$$

and for a first-order process, the domain reversal period is chosen as $\Lambda = 2\pi/\Delta k$, which is just double the coherence length beyond which synchronized propagation of fundamental and second harmonic is not possible.

10.2.3 Local field enhancements: Applications

A close inspection of the expressions of the reflection and transmission amplitudes [Eqs. (9.13) and (9.14)] and a mere appreciation of the fact that modes correspond to poles of these coefficients [see Eq. (9.16)] immediately lead to the understanding of the remarkable potentials of these modes. 'Diverging' r and t imply dramatic enhancements of the local fields, which opens up a host of applications. The narrower the mode resonance (lower decay), the tighter it is bound to the surface and the larger the enhancement. In the context of Kretschmann geometry, large transmitted amplitude does not violate any energy conservation, since the transmitted wave is evanescent and does not carry any energy.

The long-range surface plasmons (LRSP) have the added advantage of large local field enhancements associated with them [65]. Various nonlinear optical phenomena exploiting narrow LRSP modes were demonstrated by Sarid's group and others [107, 108]. Optical bistability with surface plasmons at a metal nonlinear dielectric interface was demonstrated by many (see references in [65]). Exact results for optical bistability with surface plasmons in a layered structure on a nonlinear substrate were reported [109] using Leung's solutions [110].

There have been other notable applications of the local field enhancement effect in surface enhanced Raman processes, in high-resolution spectroscopy, single-molecule spectroscopy and many other areas. As mentioned previously, a detailed description of all such applications is far beyond the scope of this book.

10.3 Resonant tunneling through gap plasmon guide and slow light

We have discussed the general resonant tunneling structure in Section 9.5, and here we show how a gap plasmon guide can mimic such a situation. Consider a gap plasmon guide (discussed in Section 10.1.4) enclosed between two high-index prisms so that the dielectric core can form the 'well.' Waves can be evanescent or propagating in the central dielectric layer depending on which modes are excited. For plane wave incidence at an arbitrary angle, very little will be transmitted. There can be significant transmission only via the resonant states. As mentioned earlier, this is referred to as resonant tunneling. One such case is shown in Fig. 10.10(a) for a multimode gap plasmon guide.

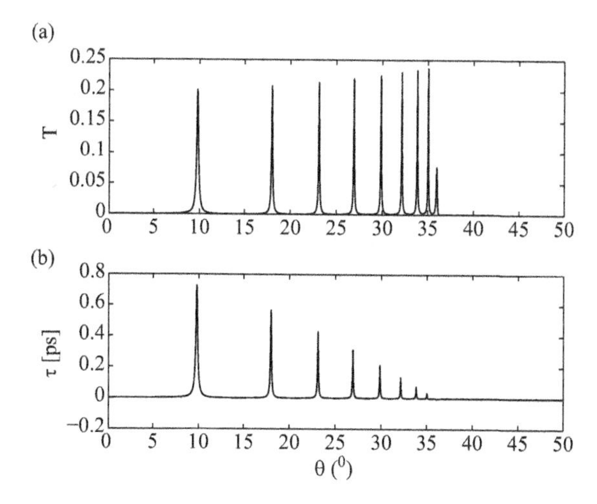

FIGURE 10.10: (a) Intensity transmission coefficient T and the (b) Wigner delay τ as functions of the angle of incidence θ for $d_0 = 3.0\ \mu\text{m}$, $d_1 = 0.03\ \mu\text{m}$, $\varepsilon_t = 6.145$. The other parameters are as in Fig. 10.5. Adapted from Ref. [100].

Different peaks correspond to the different modes. In order to determine the delay characteristics of the structure, we extracted the phase of the transmission coefficient. Recall that the frequency derivative of the phase of amplitude transmission gives the Wigner delay through the structure (see Section 9.5.1). Fig. 10.10(b) clearly shows how light is slowed down when these modes are excited.

Often in the literature, the results for delayed/advanced pulses are presented either in terms of the Wigner phase time τ or in terms of the group index n_g. It is thus useful to relate them for a segment of length d by the simple relation

$$\tau = d/v_g = (d/c)n_g. \tag{10.27}$$

10.4 Nonreciprocity in reflection from coupled microcavities with quantum wells

The work of Armitage et al. [92] was mentioned frequently in the context of nonreciprocity in reflection. In fact, it served as a stimulus for many later theoretical works for the understanding of such nonreciprocity. The system of Armitage et al. consisted of two identical coupled cavities with distributed feedback (DFB) mirrors (see Fig. 10.11). One of the cavities (say, the left one) contained three quantum well (QW) layers. The middle DFB structure was made of fewer layers so that considerable coupling between the adjacent

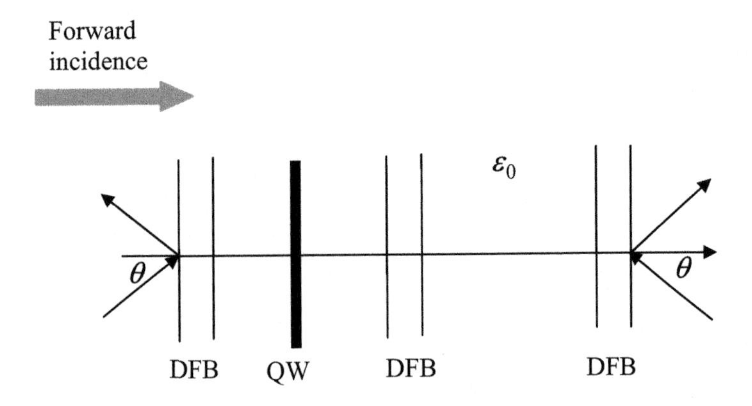

FIGURE 10.11: Schematic view of the coupled cavity system. The quantum well is replaced by resonant absorbers with dielectric function $\epsilon(\omega)$ filling the left cavity.

cavities could be achieved. Under resonance conditions (i.e., QW exciton frequency coinciding with the cavity frequency) they observed that the coupled cavity system behaved differently for illumination from the left and from the right. In the frequency response of the reflected light, two (three) dips were observed when the system was illuminated from the left (right). An intuitive argument based on three coupled oscillators was used in order to explain the absence of one dip for the forward illumination (henceforth, the illumination from the left with the QW in the first cavity will be labeled as forward). However, this explanation is inadequate and suffers from the fact that unavoidable losses near the exciton resonance were totally neglected. In fact, inclusion of absorption in their model accounts for the appearance of the third dip. Later, they came up with a more rigorous analysis [111] and also with additional experimental results whereby excitons were included in both the cavities.

Here we show that a simple model (where the excitons are replaced by resonant absorbers filling the cavity) can capture the essential physics of nonreciprocity [78]. We analyze the coupling-induced mode splittings in the dual cavity system by means of a detailed study of the underlying dispersion equation. We clearly demonstrate the avoided level crossing phenomena in the dispersion for the cavity, the excitonic branches and the exchange of decay behavior. As mentioned above, we use a model similar to the one used in Section 9.9 and use Eq. (9.121) to serve as the macroscopic dielectric function filling the left cavity. From the point of view of cavity quantum electrodynamics, we would like to know what the eigenfrequencies of the resonantly coupled exciton-dual cavity system are. How does these eigenfrequencies react to the change in system parameters? In other words, how many distinct branches do the dispersion relation have? How do the roots of the dispersion relation behave as we change the experimental condition—say, angle of incidence? Is it possible to observe more than three dips in the frequency response as reported in the experiment? We show that all these questions can be addressed by means of the model above. There can be additional branches that can, in principle, lead to additional dips in the frequency scan of the reflection coefficient. We also present a detailed study of the decay behavior of the modes. We show that as we scan through the resonance angle, pairs of modes exchange their decay characteristics.

Consider the structure shown in Fig. 10.11 consisting of two cavities coupled through a DFB structure with the end faces made of DFB mirrors. Let the left cavity be occupied by a medium doped with resonant absorbers leading to a frequency dependent dielectric function given by Eq. (9.121). Let the right cavity contain a homogeneous nondispersive medium with dielectric constant given by ε_0. Thus, in absence of doping ($\omega_p = 0$), both the cavities are identical with degenerate resonance frequencies. Under coupling, there will be a normal mode splitting, resulting in a doublet in the reflection/transmission coefficient. Let the coupled cavity system be illuminated by a plane wave incident from the left at an angle θ. In the presence of doping, reflected or transmitted light will bear the additional signature of the resonant atoms. We have chosen the individual cavities to be λ cavities with a width such that they are at resonance with the atomic frequency for $\theta = 30°$. In what follows we present the numerical results. In all our calculations, we have taken the following system parameters: $\lambda = 0.85$ μm, $2\gamma = 0.001$ μm^{-1}. Reflection coefficient R as a function of frequency for $\omega_p^2 = 0.01$ and 0.05 μm^{-2} and for different values of the angle of incidence was calculated. The results for $\omega_p^2 = 0.01$ μm^{-2} are shown in Fig. 10.12. The left (right) panel shows the results for forward (backward) illumination, while the dashed curve in each panel is for the undoped cavity. It is clear from Fig. 10.12 that with an increase in the angle of incidence, the resonances shift to the right, sweeping over the atomic spectral feature. As observed in the experiment for the resonant case ($\theta = 30°$) we see quite different features in reflection for forward and backward cases. We have two (three) prominent dips in forward (backward) illumination. We do not show the case for $\omega_p^2 = 0.05$ μm^{-2}. Since the atom-photon coupling in this case is stronger, we end up with larger splittings. Moreover, the fourth dip also shows up slightly away from the resonance. Next we study the reflection coefficient for a fixed angle of incidence (say, $\theta = 30°$) for varying atom-photon coupling ω_p^2 (not shown). From a two-dip feature in absence of doping, the response evolves to a three-dip feature bearing the signature of the atom-field coupling.

In order to have a better understanding of the spectral response of the reflection coefficient, we need to look into the roots of the dispersion equation. The dispersion equation for the multilayered medium is given by Eq. (9.16). Recall that the dispersion relation Eq. (9.16) is obtained by setting the denominator of the reflection and the transmission coefficient to zero. The roots of Eq. (9.16) for frequencies are complex in general. The real part of the roots gives the location of the mode while the imaginary part gives the width or the decay rate associated with the mode. The results for the roots of the dispersion equation are shown in Fig. 10.13. The top (bottom) panel shows the real (imaginary) part of the roots as functions of angle of incidence. The dashed curves in Fig. 10.13 show the roots for the coupled cavity in absence of the dopant atoms. It is worth noting that the bare coupled cavities without the dopant atoms are sensitive to the angle of incidence, while the branch for the atomic dispersion does not have any angle dependence. There is a level crossing phenomenon of the bare systems. In a coupled atom dual cavity system, we observe the avoided level crossing signatures. There cannot be any physical crossing in a coupled system since it would then imply different group velocities at the point of intersection. It is clear from Fig. 10.13 that there are five branches of the dispersion curves, implying thereby that in general five distinct roots are possible. For convenience of future discussions, we have labeled the modes by tags from 1 to 5. An important question is whether all these roots can be resolved or not. For overlapping resonances it may be difficult to resolve the roots. In fact, close to the resonance ($\theta = 30°$) the branches 2, 3 and 4 are so close that it may be really difficult to distinguish them because of the individual widths of the modes. In that case we would observe three dips due to the branches 1, (2,3,4) and 5. However, away from the resonance angle,

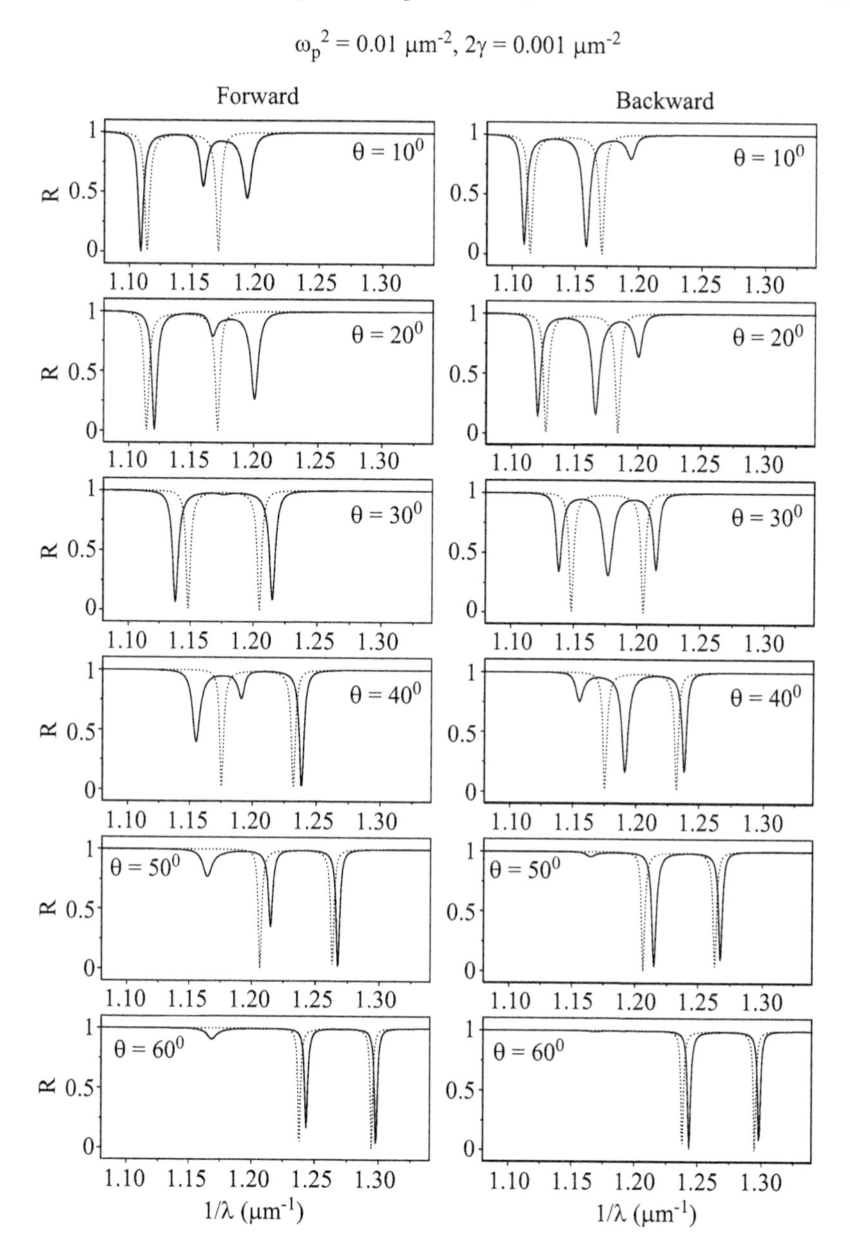

FIGURE 10.12: Reflection coefficient R as a function of frequency for various angles of incidence. Left (right) panels are for forward (backward) illumination. Dashed lines give the results for empty coupled cavities. Adapted from Ref. [78].

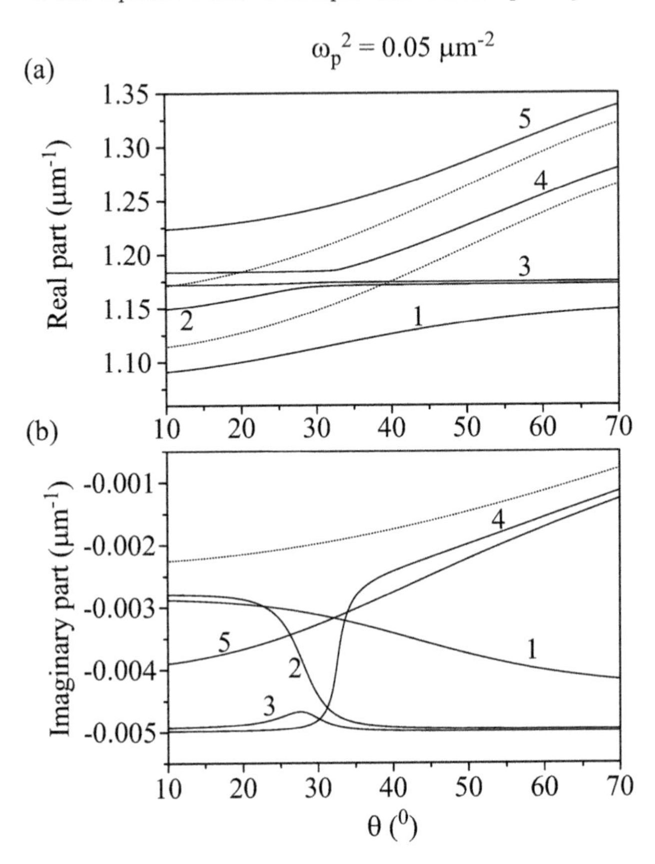

FIGURE 10.13: Roots of the dispersion relation for (a) real and (b) imaginary parts. Dashed lines give the results for empty coupled cavities. Adapted from Ref. [78].

when the spacing between the modes increases, it may be possible to see more than three dips in reflection. We demonstrate this in Fig. 10.14 where the top (bottom) curve shows the resonance dips for $\theta = 45°$ ($\theta = 10°$). The solid (dashed) curve gives the result for the forward (backward) case. It is also interesting to note from Fig. 10.13 (bottom curve) that in passing through the resonance angle, the mode pair 2, 4 exchange their decay behavior. For angles less than the resonance angle, mode 2 is less lossy than mode 4, while for angles beyond the resonance angle, mode 2 is characterized by a higher loss. These features can be easily read from a comparison of the widths of the dips due to the 2-nd and 4-th modes in Fig. 10.14 for the two different angles. There is a similar exchange of decay behaviour, but somewhat less in magnitude, between the mode pairs 1 and 5.

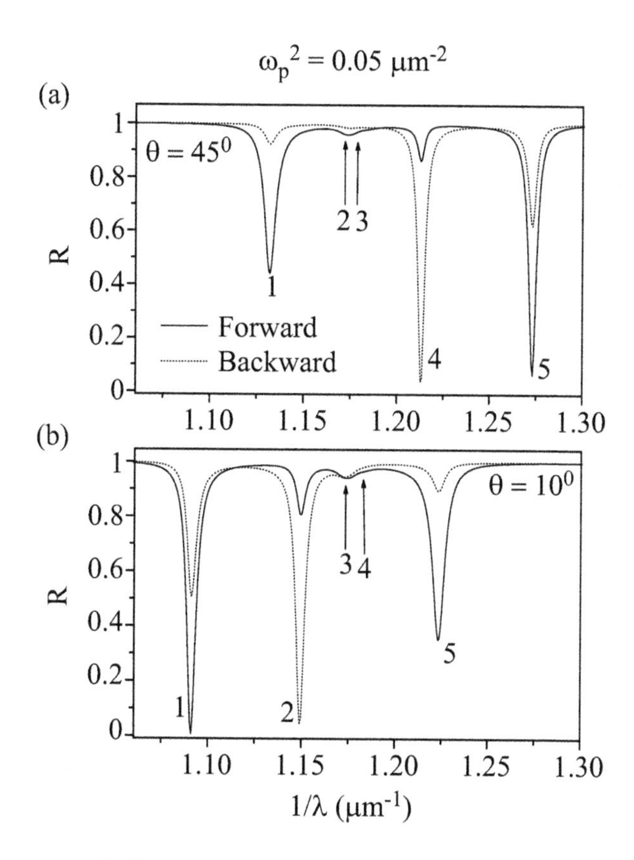

FIGURE 10.14: Reflectivity R for illumination from left (solid) and from right (dashed) for two angles of incidence with the quantum well in the left cavity. (a) $\theta = 45°$ (b) $\theta = 10°$. Adapted from Ref. [78].

Chapter 11

Resonances of small particles

In this chapter we discuss the resonances of small particles. Our focus will be on spherical and cylindrical particles where we can develop a detailed analytical approach. The resonances of such small particles have a great influence on their scattering properties. They also play an important part in modifying the density of states nearby and thus play a dominant role in determining the radiation characteristics of atoms and molecules near them. Moreover, since the excitation of these resonances are always associated with large local field enhancements, they find major applications in nonlinear optics for low threshold phenomena as well as in developing novel types of sensors with high resolution.

In order to have some preliminary ideas of the origin of these resonances, consider a microsphere of radius a with relative refractive index m (with respect to the ambient medium). Let a ray inside the microsphere hit the interface at an angle larger than the critical angle for total internal reflection. The ray can never escape the sphere since it bounces off the surface due to total internal reflection. This may result in a mode of the sphere, provided that the principal quantum number n associated with this mode satisfies the inequality

$$x \leq n \leq mx, \qquad x = \frac{2\pi a}{\lambda}. \tag{11.1}$$

Here x is referred to as the size parameter. The left (right) inequality in Eq. (11.1) reflects the fact that the optical path along the inner (outer) periphery of the sphere (inside and outside) has to be an integer multiple of the wavelength λ (the sphere is assumed to be in a vacuum). Ray optics or eikonal approximation, especially for large spheres ($x \gg 1$) and plane wave illumination, can reveal many of the features of these modes, referred to as the *whispering gallery modes.* The nomenclature owes its origin to an acoustic phenomenon in the gallery of St. Paul's Cathedral in London. A person whispering near the wall can be clearly heard by another standing at the opposite end of the diameter of the gallery, but not by another at the center. The limitations of the ray approach are also evident. A complete analysis valid for all parameter ranges calls for an exact Lorenz-Mie theory. We briefly recall the important steps of the Lorenz-Mie theory [32]. The goal is to reveal the role and importance of the Mie coefficients, which pop up now and again in many problems ranging from simple scattering to QED applications.

11.1 Elements of Mie theory and the whispering gallery modes

Let a microsphere of radius a be illuminated by an x-polarized plane wave propagating along the z direction. The spherical symmetry of the scatterer dictates the decomposition of all the fields in the vector spherical harmonics $\mathbf{M}$ and $\mathbf{N}$ (see Appendix B). Because of the nonstandard choice of the basis, even the simple (in Cartesian geometry) incident plane wave $\mathbf{E}_i$ looks complicated:

$$\mathbf{E}_i = \sum_{n=1}^{\infty} E_n(\mathbf{M}^{(1)}_{o\,1n} - i\mathbf{N}^{(1)}_{e\,1n}), \qquad E_n = \frac{i^n E_0(2n+1)}{n(n+1)}, \tag{11.2}$$

where E_0 is the incident field amplitude. The internal $\mathbf{E}_1$ and scattered fields $\mathbf{E}_s$ can also be expressed in terms of the vector spherical harmonics as follows:

$$\mathbf{E}_1 = \sum_{n=1}^{\infty} E_n(c_n\mathbf{M}^{(1)}_{o\,1n} - id_n\mathbf{N}^{(1)}_{e\,1n}), \tag{11.3}$$

$$\mathbf{E}_s = \sum_{n=1}^{\infty} E_n(ia_n\mathbf{N}^{(3)}_{e\,1n} - b_n\mathbf{M}^{(3)}_{o\,1n}). \tag{11.4}$$

Taking the curl of Eqs. (11.3) and (11.4) leads to the corresponding magnetic fields. All the fields (incident, internal and scattered) need to fulfill the boundary conditions

$$(\mathbf{E}_i + \mathbf{E}_s - \mathbf{E}_1) \times \mathbf{e}_r = 0, \tag{11.5}$$

$$(\mathbf{H}_i + \mathbf{H}_s - \mathbf{H}_1) \times \mathbf{e}_r = 0, \tag{11.6}$$

where $\mathbf{e}_r$ is the unit outward normal to the surface of the sphere. Eqs. (11.5) and (11.6) can be solved for extracting the Mie coefficients a_n, b_n for the scattered light:

$$a_n = [(D_n(mx)/m + n/x)\psi_n(x) - \psi_{n-1}(x)]/A, \tag{11.7}$$

$$b_n = [(mD_n(mx) + n/x)\psi_n(x) - \psi_{n-1}(x)]/B. \tag{11.8}$$

A and B in Eqs. (11.7) and (11.8) are given by

$$A = (D_n(mx)/m + n/x)\xi_n(x) - \xi_{n-1}(x), \tag{11.9}$$

$$B = (mD_n(mx) + n/x)\xi_n(x) - \xi_{n-1}(x). \tag{11.10}$$

Here ψ_n and ξ_n are the Ricatti-Bessel functions and $D_n(x) = \frac{d}{dx} \ln \psi(x)$ is the logarithmic derivative. All these functions can be evaluated recursively using a downward scheme, which is more stable numerically. As in any scattering problem, the zeros of A (B) herald the excitation of the transverse magnetic (electric) or the TM (TE) modes. Here the word *transverse* is used to denote the absence of the radial component of the corresponding field. Thus, for example, TE would mean $E_r = 0$. The dispersion relation for the modes is given by

$$A = 0, \quad \text{for } TM \text{ or } a\text{-modes} \tag{11.11}$$

$$B = 0, \quad \text{for } TE \text{ or } b\text{-modes.} \tag{11.12}$$

In general Eqs. (11.11) and (11.12) allow complex solutions for the size parameter x. The corresponding real part localizes the mode on the frequency/wavelength axis and the imaginary part gives the width of the resonances. In other words they carry information about how lossy the modes are. In general WGMs are extremely narrow resonances with quality factors 10^9 in the visible domain, which is practically impossible to achieve in standard Fabry-Pérot cavities. The roots of Eqs. (11.11) and (11.12) can be labeled by the polarization type and by two integers n and l, where n (l) gives the mode number (order number). The mode number n gives the number of half-waves along the grand perimeter of the sphere while the order number l gives the number of peaks in the radial intensity distribution.

The whispering gallery character of the modes can be easily seen from the radial dependence of the internal fields. For example, for TE modes the radial component of the electric field is zero and the other two $\boldsymbol{e}_\theta$, and $\boldsymbol{e}_\phi$ components inside the sphere are given by

$$\boldsymbol{e}_\theta : \quad \cos(\phi)\pi_n(\cos(\theta))j_n(k_1 r), \tag{11.13}$$

$$\boldsymbol{e}_\phi : \quad -\sin(\phi)\tau_n(\cos(\theta))j_n(k_1 r), \tag{11.14}$$

where k_1 is the wave vector in the sphere, π_n and τ_n are the angle-dependent functions, and they can be expressed in terms of the associated Legendre function of order 1 as follows:

$$\pi_n = \frac{P_n^1}{\sin(\theta)}, \tag{11.15}$$

$$\tau_n = \frac{dP_n^1}{d\theta}. \tag{11.16}$$

The radial intensity distribution for a resonant TE mode (given by j_n^2) has peaks near the inner edge of the sphere and is nearly zero at the center, which explains the whispering gallery character of these modes. Similar results hold for magnetic fields for TM modes. The often-used experimentally measurable quantity is the extinction coefficient Q_{ext}, which is a measure of both absorption and scattering and can be written as follows:

$$Q_{ext} = \frac{2}{k^2 a^2} \sum_{n=1}^{\infty} (2n+1) Re(a_n + b_n). \tag{11.17}$$

It is clear from Eq. (11.17) that resonances in extinction lead to corresponding enhancements in the internal field coefficients c_n and d_n since they have the same resonant denominator. Thus excitation of the modes leads to significant local field enhancements near the periphery of the sphere.

Several important features of the WGMs identify them as potential candidates for diverse applications ranging from nonlinear optics to lasing and cavity QED.

- *Extra high quality factor.* The modes with large n and low l are characterized by extra high quality factors. A quality factor of 10^{10} in the visible range has been reported experimentally [112].
- *Large local field enhancement.* As mentioned earlier, due to the poles of the field coefficients, there can be significant enhancements of local fields. This is of great importance for nonlinear optical applications. In fact, many low threshold nonlinear optical phenomena and lasing have been reported [113].
- *Low mode volume.* In many applications, such as cavity QED, not only the temporal confinement of light (long-lived modes due to large quality factors) but also the spatial confinement (localization) play a very important role. The radial distribution is confined mostly in the region $a/m \leq r \leq a$. The localization of the field near the rim of the microsphere leads to low mode volumes leading to large atom-field coupling.

11.1.1 Excitation and characterization of the WGMs

In the early days, the WGMs were excited basically by coupling in the radiation using a prism coupler, much like in waveguide geometries. The evanescent wave in the gap between the prism and the sphere results from total internal reflection of the incident wave off the base of the prism. For proper angles of incidence and the frequency of the incident radiation, incident light can couple selectively to some of the WGMs. Recent techniques use the evanescent field of single-mode optical fibers. Generally the fiber is mounted on a flat substrate and side-polished to expose the field of the propagating mode. Coupling efficiency is higher if the microsphere is embedded in a liquid that is index-matched to the fiber cladding. The analysis of the fiber-microsphere system requires an extension of the standard Lorenz-Mie theory for plane wave illumination and is referred to as the generalized Lorenz-Mie theory. In essence the fiber-microsphere system has a direct relevance to the problem of off-axis excitation of the sphere by a Gaussian beam. It turns out that a resonant

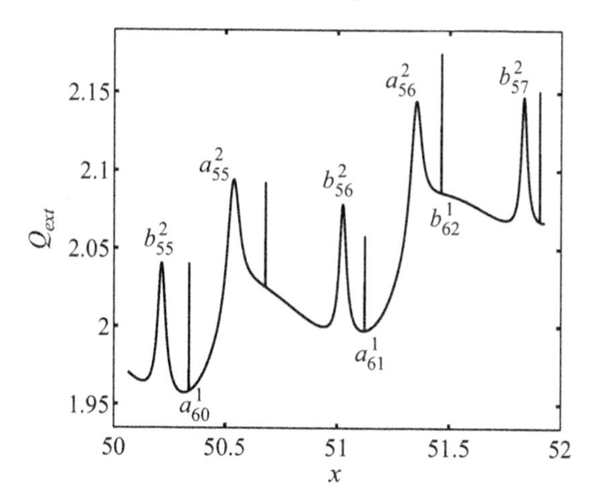

FIGURE 11.1: Extinction coefficient Q_{ext} as a function of size parameter x for a water droplet with refractive index 1.33.

off-axis Gaussian beam excites the WGMs of a microsphere more efficiently than does a plane wave.

11.2 Typical example for a water droplet

A typical extinction feature for a water droplet is shown in Fig. 11.1, where each peak is labeled by the corresponding mode index. The high Q modes are supported on the ripple structure of the microspheres. The polarization (a- or b-mode) and the quantum numbers associated with each mode can be identified by suppressing the term with given n in Eq. (11.17). Subsequently, looking at the peaks of the radial intensity pattern, l can be identified. Fig. 11.2 shows the radial field profile (j_n) for two TE modes, namely, b^2_{56} and b^1_{61}, for a water droplet that demonstrates the localization of the fields near the rim of the microsphere.

11.3 Broken spatial symmetry and its consequences

Up to now we have been discussing the case of perfect spheres. A great deal of research was devoted to the case when the spheres are deformed [114].

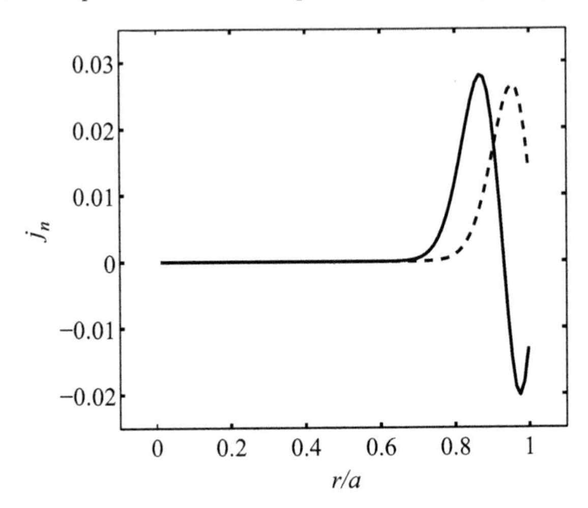

FIGURE 11.2: Radial field profile (j_n) for two TE modes, namely, b_{56}^2 (solid line) and b_{61}^1 (dashed line), for a water droplet with refractive index 1.33.

For deformed spheres a third quantum number, namely, $\tilde{m}$ (not to be confused with the relative refractive index) comes into play. A perfect sphere has all the m-modes degenerate with $(2n+1)$-fold degeneracy. For axisymmetric deformation (e.g., for prolate or oblate spheroids) the degeneracy is partially lifted, leading to distinct frequencies for $\tilde{m}$ values. The plus (minus) refers to the counterclockwise (clockwise) modes. Note that in perfect spheres or spheres with axisymmetric deformation, the degeneracy of the clockwise and counterclockwise modes (also known as Kramers degeneracy) are still present. There have been many experiments demonstrating broken Kramers degeneracy in diverse physical systems. These include a figure-eight ring laser, spheres with impurities, coupled sphere/disc-fiber systems, etc. The broken degeneracy manifests itself in the normal mode splittings. The splitting resulting from the lifting of Kramers degeneracy can be explained in terms of a simple model of coupled oscillators depicting the counter-propagating modes. In a very recent experiment, such normal mode splittings have been exploited to count the nanoparticles deposited on a micro toroid [115]. It is clear that in a perfect sphere, modes differing in $|\tilde{m}|$ will be degenerate since great circles inclined at different angles are the same. The latter does not hold in prolate or oblate spheroids. Thus axisymmetric deformation of a sphere can lift the $|m|$ degeneracy. Usually the mode with $n = |\tilde{m}|$ is referred to as the fundamental mode and it corresponds to motion close to the equatorial plane at an angle $\theta \sim 1/\sqrt{n}$. Assuming the deformed sphere as an ellipsoid of rotation with a profile given by

$$r(\theta) = r_0(1 + \frac{\varepsilon}{3}(3\cos^2\theta - 1)), \tag{11.18}$$

we can estimate the azimuthal mode spacing to be [116]

$$\Delta\omega_{azimuthal} = |\omega_{nl\tilde{m}+1} - \omega_{nl\tilde{m}}| \sim \omega_{nl\tilde{m}}\varepsilon\frac{|\tilde{m}| + 1/2}{n^2}, \tag{11.19}$$

where ε is the eccentricity given by $\varepsilon = (r_p - r_e)/r_0$ and where r_p and r_e are the polar and equatorial radii, respectively. Eq. (11.19) clearly reveals that the spacings decrease with a decrease in $|\tilde{m}|$. The previously mentioned mode splitting due to axisymmetric deformation was studied in yet another beautiful experiment by Haroche's group [117]. The axisymmetric deformation of the dielectric sphere was introduced by the fiber stem, serving the dual purpose of holding and exciting the sphere. The intensity variation with respect to changing the polar angle θ (about the equator) was mapped by a molten fiber tip. This tip was moved on the surface and the radiation collected was sent to the detector. Asymptotic analysis (for large n and $|\tilde{m}| \sim n$) revealed that the mode intensity pattern is proportional to

$$H_{n-|\tilde{m}|}(n^{1/2}\cos\theta)\sin^{|\tilde{m}|}(\theta)e^{i\tilde{m}\phi}. \tag{11.20}$$

It follows from Eq. (11.20) that there will be $n - |\tilde{m}| + 1$ lobes in the angular pattern for the $nl\tilde{m}$-th mode. The (polar) angular intensity distribution is centered about $\theta = \pi$ and extends up to $\theta = \pi \pm \cos^{-1}(\tilde{m}/n)$. It was thus possible to assign quantum number $|\tilde{m}|$ to a given mode by looking at the intensity profile.

11.4 Nanoparticles and quasi-static approximation

In the preceding sections of this chapter, we have briefly outlined the mathematical framework of the Lorenz-Mie theory for scattering of EM waves by spherical particles. The solution of the scattering problem yielded an expression for the scattered field ($\mathbf{E_s}$) via the so-called Mie coefficients a_n and b_n. As mentioned in Section 11.1, for a given n, two distinct types of modes were identified, one for which there is no radial magnetic field component (transverse magnetic TM or a-modes), and another for which there is no radial electric field component (transverse electric TE or b-modes). The condition for excitation of these normal modes was given in Eqs. (11.11) and (11.12)—if for a particular value of size parameter x, the denominator of the a_n (b_n) coefficient approaches a very small value (or tends to vanish), the corresponding mode dominates. The corresponding roots of Eqs. (11.11) and (11.12) were accordingly labeled by the polarization type and by two integers, n and l, where $n(l)$ represents the mode number (order number). These normal modes of the sphere were identified as the *whispering gallery modes* (WGMs). Selective excitation of these modes leads to huge enhancement of the local field near the periphery of the sphere. It was subsequently illustrated that large-sized dielectric spheres (radius $a \gg \lambda$, large value of size parameter x) typically exhibit extremely high quality factor of these modes. In this section, we turn to the other regime of the scattering problem, i.e., scattering by small particles (small compared to wavelength $a \ll \lambda$, $x \ll 1$), and we discuss the normal modes of sub-wavelength metal particles in this context. These so-called *localized surface plasmon* modes of small metallic particles have evoked intensive investigations in the recent past. Detailed accounts of localized plasmons and their various intriguing properties and manifestations are beyond the scope of this book. Here, we shall briefly introduce the concepts of the normal modes of sub-wavelength metal particles and leave the interested reader with appropriate references dealing with relevant development in the growing field of localized plasmon resonance (sometimes referred to as particle plasmons). We shall begin with the Lorenz-Mie theory and show that in the small particle limit, the exact solution of the scattering problem converges to a much simpler one based on the so-called *quasi-static approximation* (QSA). This approximation, simple yet intuitive, helps us acquire useful insights into the normal modes of the sub-wavelength metallic particles. So far, in this chapter, we have dealt with only one set of optical parameters of the scattering particle, namely, the refractive index (n_1 and n_m, the refractive index of the scattering particle and surrounding medium, respectively, or m the relative refractive index $= n_1/n_m$). Here, wherever necessary, we shall also use the other corresponding set of optical properties, namely, the dielectric permittivity (ϵ_1 and ϵ_m of the scatterer and surrounding medium, respectively). As we know, these two sets of quantities are not independent and are connected via standard relations. Nevertheless, the use of the dielectric constant would help us identify useful connections between electrostatics and scattering by small sub-wavelength particles, as we discuss subsequently.

Recall the expressions for the internal and the scattered fields given in Eqs. (11.3) and (11.4) via the a_n and b_n Mie coefficients. The expressions for the fields in the limit of small size parameters ($x \ll 1$) can be obtained by expanding the various functions (the Ricatti-Bessel functions) in the scattering coefficients a_n and b_n and by retaining only the first few terms. If we retain terms up to the order of x^5, the first three coefficients (a_1, b_1 and a_2) of the expansion would only contribute [32]

$$\begin{aligned} a_1 &= -\frac{i2x^3}{3}\frac{m^2-1}{m^2+2} - \frac{i2x^5}{5}\frac{(m^2-2)(m^2-1)}{(m^2+2)^2}, \\ b_1 &= -\frac{ix^5}{45}(m^2-1), \quad a_2 = -\frac{ix^5}{15}\frac{m^2-1}{2m^2+3}. \end{aligned} \tag{11.21}$$

The contributions of a_2- and b_1-modes are much weaker than the a_1-mode ($|b_1| < |a_2| \ll |a_1|$) in the limit of vanishingly small x and $|m|x \ll 1$ (for example, if we consider a scattering particle with radius $a = 20$ nm and $\lambda = 600$ nm, the magnitudes of a_2 and b_1 are two orders weaker than the a_1 mode). Thus, for such nanoparticles, for the moment, we would keep terms up to the order of x^3, wherein the only significant mode is the a_1-mode (the lowest-order TM mode). Within this small particle limit, we shall first try to get some physical insight into this a_1 scattering mode and subsequently we would also inspect the other lower-order modes (a_2 and b_1, according to the hierarchy). With this assumption, the expressions for the scattering, absorption and extinction efficiencies can be obtained using the framework of Mie theory:

$$Q_{sca} = \frac{8}{3}x^4 \left| \frac{m^2-1}{m^2+2} \right|^2, \quad Q_{abs} = 4xIm\left(\frac{m^2-1}{m^2+2}\right), \quad Q_{ext} = Q_{sca} + Q_{abs}. \tag{11.22}$$

If we now replace the relative refractive index m in the expressions for the absorption and scattering efficiencies by the corresponding relative dielectric permittivity, we obtain the very familiar quantity $\frac{\epsilon_1-\epsilon_m}{\epsilon_1+2\epsilon_m}$. This appears in the problem of a sphere in a uniform static electric field and is related to the dipolar polarizability of a sphere (see Eq. (5.45) of Chapter 5). This suggests a connection between the scattering by small particles and electrostatics. Let us now briefly examine the reason for this connection. The problem of a sphere (radius a) in a uniform static electric field can be solved by using the Laplace equation for the potential $\nabla^2\phi = 0$ (let us assume the field is applied along the z direction $\vec{E} = E_0\hat{z}$). From the solution of this equation in spherical polar coordinates (with appropriate boundary conditions), the expressions for the potential at any point (r, θ, ϕ) inside (Φ_1) and outside (Φ_2) the sphere can be obtained as [3]

$$\begin{aligned} \Phi_1 &= -\frac{3\epsilon_m}{\epsilon_1+2\epsilon_m}E_0 r\cos\theta, \\ \Phi_2 &= -E_0 r\cos\theta + a^3 E_0 \frac{\epsilon_1-\epsilon_m}{\epsilon_1+2\epsilon_m}\frac{\cos\theta}{r^2}. \end{aligned} \tag{11.23}$$

Apparently, the potential is independent of the azimuthal angle ϕ due to the azimuthal symmetry of the problem. The field outside the sphere can now be interpreted as a superposition of the incident field and the field due to an ideal dipole kept at the origin with dipole moment

$$\mathbf{p} = 4\pi\epsilon_m a^3 \frac{\epsilon_1-\epsilon_m}{\epsilon_1+2\epsilon_m}\mathbf{E_0}, \tag{11.24}$$

where

$$\begin{aligned} \mathbf{p} &= \epsilon_m \alpha \mathbf{E_0}, \\ \alpha &= 4\pi a^3 \frac{\epsilon_1-\epsilon_m}{\epsilon_1+2\epsilon_m}. \end{aligned} \tag{11.25}$$

Here, α is the dipolar polarizability of the sphere. It follows that a sphere in a uniform static electric field is equivalent to an ideal dipole. We now turn to the problem of scattering an incident plane EM wave by a sphere. Unlike the electrostatic case, the incident field varies in time and space. For the moment, we would assume that replacing the sphere by an ideal dipole is valid for plane wave as well. For an incident x-polarized field, propagating along the z direction with $E = E_0 \exp(ikz - i\omega t)\hat{x}$, an ideal dipole located at $z = 0$ would oscillate with the frequency of the applied field with dipole moment $\mathbf{p} = \epsilon_m \alpha E_0 \exp(-i\omega t)\hat{x}$. The electric field due to this oscillating dipole in the far field ($kr \gg 1$) can be written as [3]

$$\mathbf{E_s} = \frac{e^{ikr}}{-ikr}\frac{ik^3}{4\pi\epsilon_m}\mathbf{e}_r \times (\mathbf{e}_r \times \mathbf{p}). \tag{11.26}$$

We can perform a few simple algebraic steps and show that the expression for the electric field due to the oscillating dipole is identical to that obtained from the Lorenz-Mie theory (Eq. (11.4) with the far-field approximation $kr \gg 1$), retaining only the lowest-order TM (a_1) scattering mode (which was used to obtain the absorption and scattering efficiencies for the sub-wavelength scatterer in Eq. (11.22). In fact, using this approximation of replacing the sphere by an ideal dipole, we obtain the same expressions for the absorption and scattering efficiencies as that of Eq. (11.22) (derived from the Lorenz-Mie theory in the limit of small-size parameter):

$$\begin{aligned} Q_{abs} &= \frac{kIm(\alpha)}{\pi a^2} = 4xIm\left(\frac{\epsilon_1 - \epsilon_m}{\epsilon_1 + 2\epsilon_m}\right), \\ Q_{sca} &= \frac{1}{6\pi}\frac{k^4\left|\alpha\right|^2}{\pi a^2} = \frac{8}{3}x^4\left|\frac{\epsilon_1 - \epsilon_m}{\epsilon_1 + 2\epsilon_m}\right|^2. \end{aligned} \tag{11.27}$$

This approximation of replacing the spherical scatterer by an ideal oscillating dipole amounts to the *quasi-static approximation* mentioned earlier. The reason why this approximation holds for describing the scattering of plane wave by spherical particle with small-size parameter ($x \ll 1$ and $|m|x \ll 1$) can be understood from the following simple argument. For an incident plane wave, the amplitude variation inside the sphere is $E = E_0 \exp(ikz)$; when $x = ka \ll 1$, the field is almost uniform throughout the sphere. Further, note that the characteristic time of changing of the field is $\tau = 1/\omega$. The time required for the signal to propagate across the sphere is $\tau^* = an_1/c$. It is apparent that every point inside the sphere would respond simultaneously (in phase) provided $\tau^* \ll \tau$ or if $2\pi a n_1/\lambda \ll 1$. Thus, when the conditions $x \ll 1$ and $|m|x \ll 1$ are satisfied, the sphere can be treated as an ideal dipole with its moment given by the electrostatic theory.

Having described the scattering problem of small particles in the quasi-static limit, we are now in a position to find out the condition of the resonance of the normal modes of the small particles. It follows from Eq. (11.27) that resonant enhancement of both scattering and absorption is possible under the condition $\epsilon_1 = -2\epsilon_m$. Note that the dielectric permittivity of the scattering particle may have both real and imaginary parts ($\epsilon_1 = \epsilon' + i\epsilon''$) and the condition for resonance is ideally satisfied when $\epsilon' = -2\epsilon_m$ with $\epsilon'' = 0$. This is known as the *Fröhlich condition.* If the surrounding medium is a dielectric (ϵ_m is positive; if we further assume for the sake of simplicity that it is nearly nonabsorptive, then ϵ_m is real), the resonance condition is then satisfied at frequencies where the real part of the dielectric permittivity ϵ' of the scattering particle is negative. We know that the real part of the dielectric permittivity for metals assumes negative values at optical frequencies (e.g., gold and silver have negative values for ϵ' in the entire visible wavelength range). Thus sub-wavelength metallic particles can exhibit this kind of resonance behavior, known as *localized plasmon resonance.* In the limit of vanishingly small size parameters (as we have considered so far in this section), where only the first-order ($n = 1$) TM scattering (a_1) mode contributes, the corresponding resonance is interpreted as the *dipolar plasmon resonance*, as the condition of resonance directly follows from the condition of resonant enhancement of the dipolar polarizability. Note that the same condition of resonance could also be directly obtained from the vanishing of the denominator of the Mie coefficients a_n in Eq. (11.9), under the approximation of vanishing size parameter $x \rightarrow 0$ (and finite $|m|$). Once again, using series expansion of the spherical Bessel functions in Eq. (11.9) and with a bit of algebra, the resonance condition can be obtained as $m^2 = -\frac{n+1}{n}$. For the lowest-order mode ($n = 1$), this is identical to the condition for resonance of the dipolar polarizability $\epsilon_1 = -2\epsilon_m$. It may be worth noting that this lowest-order (dipolar) TM mode has almost uniform field distribution throughout the sphere and is sometimes referred to as the *mode of uniform polarization.*

The condition for resonance discussed above is strictly valid in the limit of vanishingly small-size parameter x. Here, we shall briefly address the effect of the finite size of particles on the localized plasmon modes. If the size parameter x increases to an extent that we are still within the quasi-static limit $x < 1$, the condition for resonance of the a_1-mode changes from the ideal condition of the resonance of the dipolar polarizability. In fact, the condition for resonance can still be obtained by retaining a few more terms in the expansions of the spherical Bessel functions in the Mie coefficient a_1 (Eqs. (11.7) and (11.21)). An approximate condition for the resonance of the a_1-mode (correct to terms of order x^2) can be obtained as

$$\epsilon_1 = -\left(2 + \frac{12}{5}x^2\right)\epsilon_m. \tag{11.28}$$

It is known that for most metals, the real part of the dielectric permittivity assumes more negative values as we approach higher wavelengths (ϵ is an increasing function of frequency). Therefore, the wavelength at which the Fröhlich condition is satisfied shifts toward longer wavelengths (as compared to the ideal resonance condition of the dipolar mode $\epsilon_1 = -2\epsilon_m$) as we incorporate finite size effects. Thus the first observable effect of increasing size is the shift of the dipolar plasmon resonance to higher wavelengths (lower frequencies). As noted before with Eq. (11.21), with increasing value of x the other lower-order modes (TM a_2-mode and TE b_1-mode, according to the

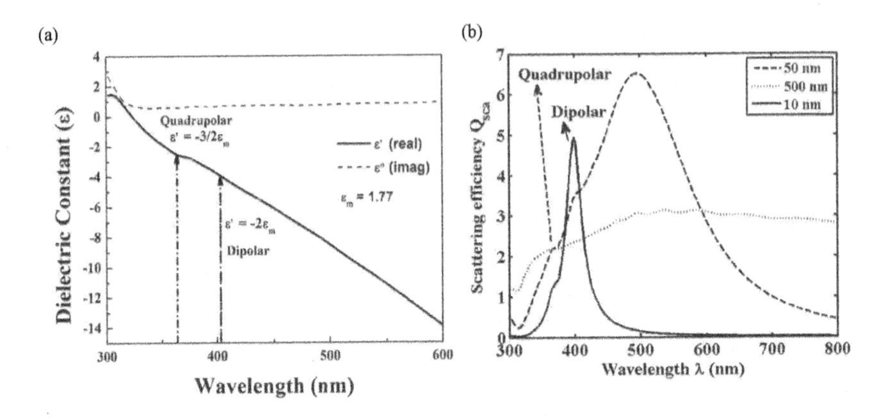

FIGURE 11.3: (a) The wavelength variation of the real and imaginary parts (ϵ' and ϵ'') of the dielectric permittivity of silver (Ag). The wavelengths corresponding to ideal dipolar and quadrupolar plasmon resonances are marked (the surrounding medium is taken to be water). (b) The Mie theory computed wavelength variation of scattering efficiencies(Q_{sca}) for Ag nanospheres of three different radii ($a = 10$ nm, 50 nm and 500 nm). The scattering efficiency for the 10 nm sphere is scaled-up fifteen times ($\times 15$) in order to show in the same scale. The wavelengths corresponding to the dipolar plasmon resonance (shown for the 10 nm spheres) and quadrupolar plasmon resonance (for the 50 nm spheres) are marked. With increasing size, the dipolar resonance broadens and shifts toward longer wavelengths. For the largest-sized sphere shown here ($a = 500$ nm), QSA is no longer valid and it is associated with significant broadening; contributions from higher-order plasmon modes and considerable radiative damping. For this figure, the surrounding medium is taken to be water and the values for ϵ' and ϵ'' are used from (a).

hierarchy) also start contributing to the scattering process. Specifically, the contribution of the a_2-mode becomes more prominent with increasing x. We note that the condition of resonance for the a_2-mode can be obtained from Eq. (11.21) as $\epsilon_1 = -\frac{3}{2}\epsilon_m$. This condition is identical to the resonance condition of electrostatic *quadrupolar polarizability*.

Thus, in the quasi-static limit, plasmon resonance associated with the a_2-mode can be identified as the quadrupolar plasmon resonance. Clearly, the resonance condition for the quadrupolar plasmon mode is satisfied at lower wavelengths (higher frequencies) than the corresponding ideal dipolar plasmon modes. These are illustrated in Fig. 11.3. Here, as an example, we have shown the wavelength variation of the real and imaginary parts (ϵ' and ϵ'') of the dielectric permittivity of silver (Ag) (Fig. 11.3(a)). The wavelengths corresponding to the resonances of ideal dipolar and quadrupolar plasmon modes are marked (the surrounding medium is taken to be water). Fig. 11.3(b) illustrates the finite size effect on the resulting plasmon resonance behavior, where the wavelength dependence of scattering efficiency Q_{sca} is shown for Ag nanospheres of three different radii (note, similar behavior is also observed for absorption and extinction efficiencies, which are not shown here). As discussed above, increasing size of the metal nanosphere is associated with the shift of the dipolar resonance toward longer wavelengths and subsequent broadening of the resonance peak. The broadening of the resonance peak is a manifestation of stronger radiative damping with increasing size of the scatterer. As expected, this is also associated with the appearance of a quadrupolar resonance peak. These behaviors may still be interpreted within the framework of the quasi-static approximation. Finally, when the size becomes too large ($x \gg 1$), the resonance behavior can no longer be described and interpreted using the quasi-static approximation. For such large sizes, the quasi-static approximation breaks down because (i) different volume elements within the sphere do not respond in phase (the retardation effect sets in) and (ii) it can no longer be assumed that each region within the sphere is only exposed to the incident field alone (the depolarization field becomes significant). Accurate description of the scattering process and the resonance behavior would necessitate the use of the Lorenz-Mie theory. Nevertheless, the resulting effects are contribution of higher-order normal scattering modes, significant broadening and radiative damping of the plasmon modes.

So far, we have restricted our discussion to sub-wavelength metallic spheres. It appears that within the small particle limit, the basic physics of localized surface plasmon resonance can be described adequately using the quasi-static approximation, and the resonance conditions can be determined using the resonance conditions of dipolar and quadrupolar polarizabilities. For conceptual and practical reasons, extension of the quasi-static approximation for other nonspherical particles, such as cylinders, disks, rods, ellipsoids, etc., is warranted. The resonance conditions can accordingly be determined from the condition of resonance of the polarizabilities along the different axes, within the quasi-static approximation. For example, if we consider an axially symmetric particle, such as an ellipsoid, it exhibits three different polarizabilities along the three major (minor) axes. Accordingly, an ellipsoidal metal particle would exhibit three different plasmon resonances at different wavelengths. In Fig. 11.4 we provide an illustrative example of plasmon resonances of spheroidal silver nanoparticles. A spheroid is an important special class of ellipsoids. For prolate (oblate) spheroids, the two major (minor) axes are of equal size, leading to two different polarizabilities along the major and the minor axes, respectively. The examples of oblate spheroids shown in Fig. 11.4 thus exhibit two plasmon resonance peaks. These two peaks are due to the plasmon resonances corresponding to the longitudinal and the transverse dipolar polarizabilities. The weaker peak at around 365 nm and the stronger peak at around 440 nm (e.g., for an aspect ratio $e = 1.5$) can be identified due to the surface plasmon resonance along the short axis (transverse) and the long axis (longitudinal) of the oblate spheroid, respectively. The relative intensities

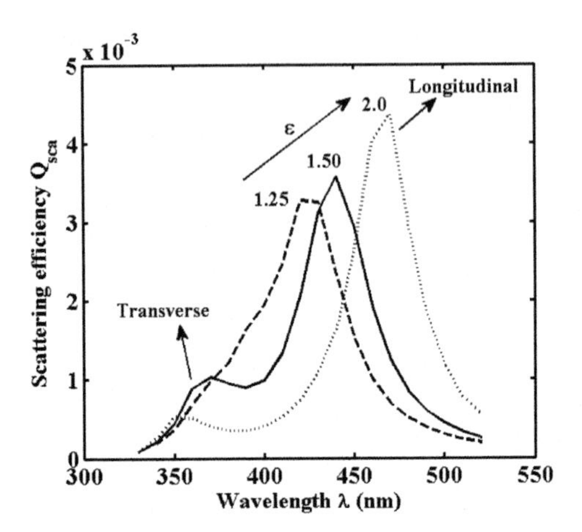

FIGURE 11.4: The scattering efficiency (Q_{sca}) as a function of wavelength for spheroidal (oblate) Ag nanoparticles of varying aspect ratios ($e = 1.25 - 2.0$). (Note that values of $e > 1$ correspond to oblate spheroids as shown here. Aspect ratios $e < 1$, on the other hand, correspond to prolate spheroids, not shown here.) The radius of equal surface area sphere is $a = 20$ nm (fixed). The surrounding medium is once again taken to be water and the values used for ϵ' and ϵ'' are from Fig. 11.3(a).

of the two bands and their spectral positions are controlled by the relative strength of the two orthogonal dipolar plasmon polarizabilities. Since this in turn is decided by the aspect ratio, with its increasing value, the two resonance peaks start moving apart (the transverse and the longitudinal we get blue- and red-shifted, respectively, as is apparent from the figure).

To conclude this section, we have briefly addressed the normal scattering modes of sub-wavelength particles under the quasi-static approximation. In this context, we have discussed the localized surface plasmon resonances of metal nanoparticles, with illustrative examples of the simplest form of nanoparticles (e.g., spheres and spheroids). A number of interesting and intricate optical effects associated with such localized plasmon resonances (in more complex nanostructures such as symmetric and symmetry-breaking nanoparticle arrays, clusters and so forth) have been observed in the recent past and are being actively pursued for both fundamental interests and numerous practical applications. Interested readers are referred to the monograph by S. A. Maier's monograph [99] for a more detailed account of these.

Chapter 12

Spin-orbit interaction of light

The term *spin-orbit interaction* refers to the interaction and coupling of the spin and orbital degree of freedom of spinning particles such as electrons or other quantum particles. The universal nature of such interaction has led to its manifestation in diverse fields of physics, ranging from atomic, condensed matter to optical systems. In optics, the angular momentum (AM) of light is related to the circular (elliptical) polarization of light waves or the helical phase fronts (vortex) of optical beams. In classical electromagnetic description, the former is associated with the rotation of the electric field vector around the propagation axis and is referred to as the spin angular momentum (SAM); the latter is associated with rotation of the phase structure of a light beam and is known as orbital angular momentum (OAM). From a fundamental point of view, coupling and interconversion between the spin and orbital AM degrees of freedom of light are thus expected under certain circumstances, and accordingly the evolution of polarized light in a trajectory should mimic the SOI effect of a massless spin 1 particle (photon) (that is the SOI effects exhibited by spin $\frac{1}{2}$ electrons while evolving under an external field). In this chapter, we address this issue and show that this effect can indeed be observed in a variety of light-matter interactions, e.g., by the tight focusing of fundamental or higher-order Gaussian beams, reflection/refraction of finite beams at dielectric interfaces, high numerical aperture imaging, scattering from micro-/nanosystems, propagation in gradient index media and so forth. We shall briefly address the various interesting manifestations of SOI, discuss corresponding mathematical frameworks (based on wave optics treatment) for describing SOI and provide illustrative examples of the resulting effects and discuss their potential implications. In this regard, we shall define SAM and OAM of light through classical electromagnetic (EM) wave optics formalism and introduce the concept of geometrical phase of light (the spin redirection Berry phase and Pancharatnam Berry phase), which is intimately related to the SOI of light.

12.1 Spin and orbital angular momentum of light

It is well known from Maxwell's theory that electromagnetic radiation carries both energy and momentum. The momentum may have both linear and angular contributions. The fact that light carries a linear momentum equivalent to $\hbar k$ per photon ($k = \frac{2\pi}{\lambda}$ is the wave vector) and that it can exert radiation pressure on atoms and matter was experimentally demonstrated long ago. Bethe, on the other hand, made the first experimental observation of the angular momentum of light and demonstrated that circularly polarized light could exert a mechanical torque on a birefringent plate by transfer of angular momentum [118, 119, 120]. In his celebrated experiment, a circularly polarized light beam was generated using a combination of a linear polarizer and a quarter-waveplate and was then made incident on a half-waveplate suspended by a fine quartz fiber. The half-waveplate transformed left circular polarization to right circular polarization and consequently transferred $2\hbar$ angular momentum per photon to the birefringent plate. The mechanical torque exerted on the half-waveplate by a circularly polarized light beam thus balances the flip of the angular momentum of light. The measured torque indeed agreed in sign and magnitude with that predicted by both the wave and quantum theories of light, in which the angular momentum associated with left/right circular polarization is described as the $\pm\hbar$ spin of individual photons and is accordingly termed the spin angular momentum (SAM) of light. For an idealized circularly polarized wave (with frequency ω), SAM is given by $J_z = N\hbar$ and the energy by $W = N\hbar\omega$, where N is the total number of photons. Thus the angular momentum to energy ratio is $\frac{J_z}{W} = \frac{1}{\omega}$. This definition can be made more generalized to include elliptically polarized light characterized by $-1 \leq \sigma \leq +1$ ($\sigma = \pm 1$ for left/right circularly polarized light, respectively, and $\sigma = 0$ for linearly polarized light), yielding an angular momentum-to-energy ratio $\frac{J_z}{W} = \frac{\sigma}{\omega}$.

It was not recognized until the early 1990s that in addition to SAM, light beams can also carry orbital angular momentum (OAM). In 1992, Allen et al., by a straightforward calculation, theoretically demonstrated that light beams with helical phase fronts characterized by a phase dependence in the transverse plane as $\sim \exp(il\phi)$ (where ϕ is the azimuthal angle and l can be any integer value, positive or negative) carry an OAM of $l\hbar$ per photon [121]. This type of beam has a phase dislocation on the beam axis and is often referred to as an *optical vortex*. In general, any beam with inclined phase fronts may carry OAM about the beam propagation axis, which is perfectly discernible and profoundly different in nature from SAM, being related to helical wavefronts rather than to polarization. The Laguerre-Gaussian (LG) laser mode is an example of an OAM-carrying beam and these so-called vortex beams can nowadays be readily generated in a standard optics laboratory. From an analogy of quantum mechanics and paraxial wave optics, it is easily recognizable that LG modes having an azimuthal angular dependence of amplitude as $\sim \exp(il\phi)$ are eigenmodes of the angular momentum operator and accordingly carry $l\hbar$ OAM per photon. In what follows, we show that the definitions of SAM and OAM of light also follow from purely classical (EM) wave optics treatments, and we subsequently discuss some of the salient features of such angular momentum-carrying light beams.

Let us consider a monochromatic (with frequency ω) EM wave propagating along a direction z, where x and y are the other two axes in Cartesian coordinates. The local densities of the linear momentum ($\boldsymbol{p}$) and angular momentum ($\boldsymbol{j}$) for such an EM wave can be calculated from the electric field $\mathbf{E}$ and the magnetic field $\mathbf{B}$ of the wave as [119]

$$\mathbf{p} = \epsilon_0 \mathbf{E} \times \mathbf{B} \text{ and } \mathbf{j} = \mathbf{r} \times \mathbf{p} = \epsilon_0 \mathbf{r} \times (\mathbf{E} \times \mathbf{B}). \tag{12.1}$$

The total linear ($\mathcal{P}$) and angular ($\boldsymbol{J}$) momentum of the field is then given by

$$\mathcal{P} = \int \epsilon_0 \, (\mathbf{E} \times \mathbf{B}) dr \text{ and } \mathbf{J} = \int \epsilon_0 \, \mathbf{r} \times (\mathbf{E} \times \mathbf{B}) dr. \tag{12.2}$$

In atomic physics, we expect that $\mathbf{J} = \mathbf{L} + \mathbf{S}$, where the first term is identified with the orbital angular momentum $\mathbf{L}$ and the second with the spin $\mathbf{S}$. But the important questions that arise here are (i) whether such a physically unambiguous separation of SAM and OAM is possible for any arbitrary optical vector (EM) field and (ii) whether they are separately physically observable. In other words, is it possible to single out some general prescriptions for the evaluation (and observation) of the angular momentum of a field from the phase and amplitude structure of the vector field? To answer these questions, a deeper insight into the definition of the angular momentum is necessary. If we consider a purely transverse plane EM wave (as we often tend to do while describing polarization and other properties of light) propagating in the z direction with the vibrations of the electric and magnetic fields contained in the xy plane, we end up in a seemingly paradoxical situation. For such an idealized transverse plane EM wave, the linear momentum (and Poynting vector $\mathbf{E} \times \mathbf{B}$) is along the propagation (z) direction and there cannot be a component of angular momentum $\mathbf{r} \times (\mathbf{E} \times \mathbf{B})$ in the same direction. At the most fundamental level, in order to have a component of angular momentum (of any form) along the propagation direction z, we need to have components of electric and/or magnetic fields in the z direction. Thus even if circularly polarized a plane wave cannot carry any angular momentum. This contradicts the results of Bethe's celebrated experiment and what we stated at the beginning of this section. The resolution of this seemingly paradoxical situation lies in the fact that a plane wave is purely an idealization that cannot be applied in the real world. Real optical beams are limited in spatial extent either by the beams themselves or by the finite extent of the measurement system used to detect them, and this leads to a nonzero longitudinal component of the field (an EM field is not purely transverse in such a situation). This longitudinal component of the field arises from the radial gradient of the field that occurs at the edge of the beam or the measurement system, which eventually leads to a value of angular momentum of $\pm\hbar$ per photon (for left/right circular polarization) when integrated over the cross-section of the entire beam. The origin of OAM for a finite optical beam having an azimuthal angular dependence of amplitude in the transverse plane as $\sim \exp(il\phi)$ can also be understood in an analogous fashion. Such beams have helical phase fronts, with the number of intertwined helices and the handedness depending on the magnitude and the sign of l. An EM field transverse to these phase fronts has longitudinal components. The Poynting vector, which is parallel to the surface normal of these phase fronts, has an azimuthal component around the beam and accordingly an angular momentum along the beam propagation direction. The arguments above can be put forward in a formal manner using the wave equation in paraxial approximation, which is quite conveniently used to model the distribution of the field amplitudes of laser modes and its propagation.

Paraxial approximation of wave equations describes the propagation of light beams whose transverse dimensions are much smaller than the characteristic longitudinal distance over which the field changes its magnitude. In this approximation, the beam waist w_0 (transverse dimension of the beam) is assumed to be much smaller than the diffraction length $l_d = kw_0^2$. The general form of an electric field in Cartesian coordinates for the simplest paraxial waves (propagating along the z direction) can be written as

$$\mathbf{E}(x, y, z) = \mathbf{F}(x, y, z)\exp(ikz). \tag{12.3}$$

Here, $\mathbf{F}(\mathbf{x}, \mathbf{y}, \mathbf{z})$ is the slowly varying spatial envelope. The small parameter (w_0/l_d) can be used as an expansion parameter, and under paraxial approximation, the derivative of $\mathbf{F}$ with respect to z is negligible when compared to the transverse derivatives. Accordingly, the field $\mathbf{F}$ satisfies the paraxial wave equation

$$2ik\frac{\partial}{\partial z}\mathbf{F} = -\left(\frac{\partial^2}{\partial x^2} + \frac{\partial^2}{\partial y^2}\right)\mathbf{F}. \tag{12.4}$$

This equation describes the propagation of the wave in the z direction for a given input field distribution. Both the polarization (linear, circular or elliptical) and the phase structure of the beam are encoded in $\mathbf{F}$. We also note that the longitudinal (z) component of the field is smaller than the transverse component by a factor w_0/l_d.

The z component of the angular momentum density j_z (defined in Eq. (12.1)) can now be determined for the paraxial EM wave defined in Eqs.

(12.3) and (12.4). By eliminating the magnetic field $\mathbf{B}$ from Eq. (12.1) using Maxwell's equation $i\omega\mathbf{B} = \nabla \times \mathbf{E}, \nabla \cdot \mathbf{E} = 0$ and by performing simple algebraic manipulations, the expression for the z component of the angular momentum density can be obtained as [121, 122]

$$\begin{aligned} j_z(x,y,z) &= \left[\frac{\omega\epsilon_0}{2i}\mathbf{E}^*(\hat{r} \times \nabla)\mathbf{E}\right]_z + \left[\frac{\omega\epsilon_0}{2i}\mathbf{E}^* \times \mathbf{E}\right]_z \\ &= \frac{\omega\epsilon_0}{2i}\left[F_k^*\left(x\frac{\partial}{\partial y} - y\frac{\partial}{\partial x}\right)F_k\right]_{k=x,y} + \frac{\omega\epsilon_0}{2i}(F_x^* F_y - F_y^* F_x). \end{aligned} \tag{12.5}$$

This expression for the angular momentum density answers the first question that we posed, 'whether such a physically unambiguous separation of SAM and OAM is possible for any arbitrary optical vector (EM) field.' A careful look at the equation reveals that the first term of the equation is related to the transverse distribution of the field (amplitude and phase) and the second term to the polarization of the field. The first term may therefore be identified as the OAM density. The second term, on the other hand, is directly proportional to the circularly polarized field component of the wave (the fourth Stokes vector element V as defined in Eq. (6.12) in Chapter 6) and vanishes for a linearly polarized wave. This therefore reflects the SAM density of the EM wave. The angular momentum density j_z can easily be evaluated for circularly polarized paraxial beams with helical phase fronts (the so-called vortex beams like the circularly polarized LG modes), whose transverse field can be written in cylindrically symmetric form as

$$\mathbf{F}(r,\phi) = u(r)\exp(il\phi)\hat{\mathbf{F}}. \tag{12.6}$$

In order to do so, we recognize $\left(x\frac{\partial}{\partial y} - y\frac{\partial}{\partial x} = \frac{\partial}{\partial \phi}\right)$ and represent the normalized form of the quantity $i(F_y^* F_x - F_x^* F_y)$ by wave helicity $\sigma(-1 \leq \sigma \leq 1)$,

$$j_z(r,\phi) = (\sigma + l)\omega\epsilon_0|u(r)|^2 = (\sigma + l)\epsilon_0\omega^2|u(r)|^2. \tag{12.7}$$

The energy density of the paraxial beam above can be written as

$$w = c\epsilon_0(\mathbf{E} \times \mathbf{B})_z = c\omega k\epsilon_0|u(r)|^2 = \epsilon_0\omega^2|u(r)|^2.$$

Thus, the ratio of angular momentum density to energy density of the field becomes

$$\frac{j_z}{w} = \frac{\sigma + l}{\omega}. \tag{12.8}$$

The ratio of the total angular momentum carried by the beam to the energy per unit length (J_z/W) can also be worked out by integrating over the xy plane, which would also yield the same value as above. These results are in agreement to what was stated at the beginning that circularly (or elliptically) polarized vortex beams carry $\pm\sigma\hbar$ SAM per photon ($\sigma = \pm 1$ for left/right circular and $\sigma < 1$ for elliptical polarization) and $l\hbar$ OAM per photon.

It is pertinent to note here that for the cylindrically symmetric paraxial beam treated above, the symmetry of the radial and azimuthal components about the axis ensures that integration over the beam profile leaves only the z component of the angular momentum (the transverse component of the angular momentum vanishes). Caution must be exercised however, in generalizing this treatment (based on paraxial approximation) for unambiguous separation of SAM and OAM for any arbitrary optical vector (EM) field. This seemingly clear separation of SAM and OAM is particularly complicated in the presence of tight focusing (or similar highly nonparaxial fields), and we need to use a more sophisticated treatment incorporating the three-dimensional nature of the vector field. We refer the reader to the relevant literature on this topic.

Both the types of angular momentum described above (SAM arising from circular polarization and OAM arising from helical phase front) do not depend upon the lateral position of the axis (choice of the calculation axis), and they are accordingly 'intrinsic' per se. They may also have an extrinsic contribution of the angular momentum, which depends upon the choice of the origin and the calculation axis. This extrinsic angular momentum cannot have a contribution from SAM (which is purely intrinsic in nature), but it can have a contribution of extrinsic OAM. As defined above, when integrated over the beam profile, the total angular momentum in the z direction is given by [123]

$$\mathbf{J}_z = \int \epsilon_0 \mathbf{r} \times (\mathbf{E} \times \mathbf{B}) dxdy. \tag{12.9}$$

If the axis is laterally displaced by $\mathbf{r_0} = (r_{0x}, r_{0y})$, the change in the z component of the angular momentum would be

$$\Delta J_z = r_{0x}\epsilon_0 \int (\mathbf{E} \times \mathbf{B})_y dxdy + r_{0y}\epsilon_0 \int (\mathbf{E} \times \mathbf{B})_x dxdy. \tag{12.10}$$

The angular momentum is said to be intrinsic if ΔJ_z vanishes for all values of r_{0x} and r_{0y}. This can be satisfied if z is chosen in such a way that the total transverse momenta vanish:

$$\int (\mathbf{E} \times \mathbf{B})_y dxdy = 0 \text{ and } \int (\mathbf{E} \times \mathbf{B})_x dxdy = 0. \tag{12.11}$$

Apparently, for the cylindrically symmetric paraxial beams (e.g., the LG beams having any value of l), the total transverse momenta are exactly zero, and accordingly the OAM associated with such beams is intrinsic. Breaking the symmetry of the beam (e.g., by truncating using apertures), on the other hand, would lead to finite values of the transverse momenta and nonzero ΔJ_z values, and the corresponding OAM can be termed extrinsic.

Note that even for such truncated beams, the SAM remains $\pm\sigma\hbar$ per photon, irrespective of the choice of the calculation axis, and so it can be treated as intrinsic. We shall stop here with this brief definition of the intrinsic and extrinsic angular momenta and their physical origins. As we shall see later,

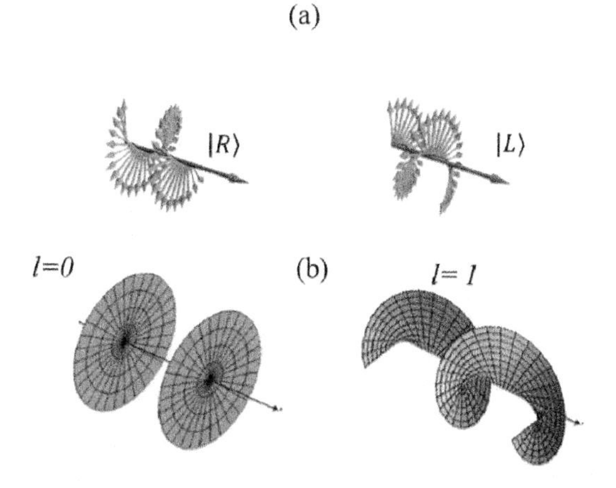

FIGURE 12.1: (a) Rotation of the electric field vector around the propagation axis for light with left (L) and right (R) circular polarization. This is connected to the SAM of light. (b) Rotation of the phase front for a phase front having $l = 0$ (no OAM-carrying) and for $l = 1$ (OAM-carrying) light beam.

the generation of the extrinsic OAM has an important role in the spin-orbit interaction of light and the resulting polarization-dependent shift of the beam trajectory or the beam's center of gravity (an effect known as the Spin Hall effect of light).

Now we turn to the other important question that we posed in the beginning, 'whether the angular momenta are separately physically observable.' To answer this question, we take a step back and look at the distinctive origin of the two different forms of angular momentum (SAM and OAM) based on simple geometrical arguments. SAM is related to the circular polarization, i.e., to the rotation of the electric field vector about the axis of propagation (Fig. 12.1(a)). OAM, on the other hand, is related to the rotation of the phase structure (inclination of the phase front) of the light beam (Fig. 12.1(b)) [124]. The fact that circularly polarized light should carry SAM of $\pm\hbar$ per photon follows from the following geometric argument. A circular path of circumference λ has a radius of $\lambda/2\pi$ and a linear momentum of $\hbar k$ directed around this circle yields an angular momentum of $\hbar$. Similarly, the quantization of OAM in units of $\hbar$ can also be understood by noting that the inclination of the phase front and hence the Poynting vector at radius r, with respect to the beam propagation axis, is $l\lambda/2\pi r$. This results in an azimuthal component of the linear momentum of light as $\frac{\hbar k l \lambda}{2\pi r}$ per photon. This quantity, when multiplied with the radius vector, yields an angular momentum of $l\hbar$ per photon. With this geometrical concept discerning the nature of the two different angular momenta of light, we now briefly look at their mechanical

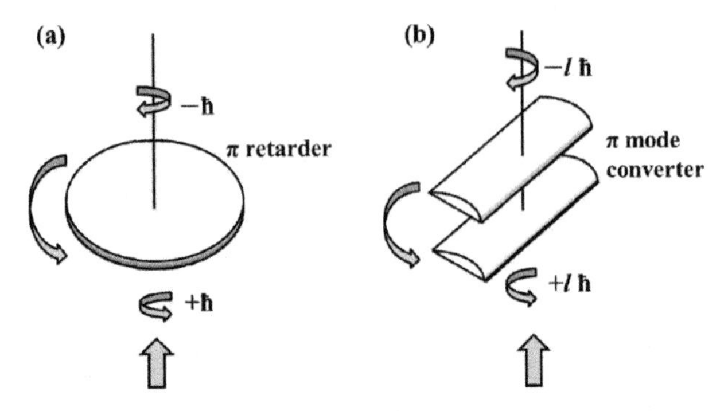

FIGURE 12.2: (a) Torque applied to a birefringent half-waveplate as input left circularly polarized light is converted into outgoing right circularly polarized light. The angular momentum balance demands a transfer of SAM of $2\hbar$ per photon to the waveplate, leading to rotation of the plate. (b) In an analogous experiment proposed by Allen et al., the transfer of OAM to a π mode converter that consists of two cylindrical lenses. The π mode converter converts $l = +1$ LG mode to $l = -1$ LG mode, thereby transferring OAM of $2\hbar$ per photon to the mode converter.

equivalence, i.e., how or whether the mechanical effects they produce are different and separately observable when an AM-carrying light beam interacts with any physical object. As previously discussed, Bethe demonstrated the mechanical equivalence of SAM by observing the transfer of SAM from light to a birefringent crystal waveplate, leading to a torque exerted on the waveplate. This mechanical torque exerted on the half-waveplate (which converts left circular polarization to right circular polarization) by the incident circularly polarized light beam was shown to balance the flip of the SAM of light, resulting from the conservation of the angular momentum at the face of the birefringent plate (see Fig. 12.2(a)). An analogous experiment was also proposed by Allen et al. [121], where the OAM present within the Laguerre-Gaussian (LG) laser mode would cause the rotation of a π mode converter (see Fig. 12.2(b)). Note that this π mode converter in the OAM basis resembles the role of a half-waveplate in the SAM basis, in that this converts $l = +1$ mode to $l = -1$ mode (just like half-waveplate converts $\sigma = +1 \to -1$). Such a π mode converter can be made by using a pair of cylindrical lenses kept at $2f$ distance away (f is the focal length), or alternatively a Dove prism may also act as a π mode converter (transforming any mode into its mirror image) [122, 124]. Even though such an experiment has actually not been performed (since this has proved to be too technically demanding), from a conceptual point of view, transfer of OAM to the π mode converter in the form of mechanical torque (thereby balancing the flip of OAM) is warranted. The first experimental observation of the existence of OAM as a mechanical property was made in optical tweezers [125]. Optical tweezers uses the gradient force of tightly focused laser beams to trap microscopic dielectric particles in three dimensions within a surrounding fluid (described in detail in Chapter 13). In this experiment, microscopic absorptive particles trapped by optical tweezers were transferred to the OAM carried by a LG beam, which resulted in the rotation of the particle at several hertz. This rotation was correctly attributed to the transfer of OAM from light to matter. In fact, for an on-axis trapped particle, it was demonstrated that the rotation speed of the particle can be increased or reduced by combining SAM with OAM (circularly polarized LG beam with $\sigma = \pm 1, l = \pm 1$ or more). In other words, the SAM can be added or subtracted from the OAM component consistent with the statement that the total AM component of light beam is $(l + \sigma)\hbar$. In an off-axis trap, on the other hand, the particle behaves differently. It responds to OAM by orbiting about the axis and also spins about its own axis because of SAM. Finally, the AM corresponding to these distinctive properties seems to exhibit different mechanical features. SAM plays important roles in birefringent materials, driving the rotation of the local optical axis, while OAM can set in motion, along trajectories circulating around the beam axis, the molecule center of mass or elemental volumes in inhomogeneous macroscopic bodies. A detailed account of the transfer of AM to yield mechanical motion (by exerting torque) of microscopic particles in optical tweezers is provided in Chapter 13.

Finally, we end this section with a brief mention of the various experimental methods for generating angular momentum-carrying light beams. As discussed in some detail in Chapter 6 (on polarization), the SAM-carrying beam can be conveniently generated or converted using appropriate combinations of linear polarizers and birefringent wave plates (e.g., a linear polarizer and a quarter-wave plate with its axis making an angle 45° with respect to the axis of the polarizer can be used to generate left or right circularly polarized light). In recent years several methods have been developed to generate an OAM-carrying beam or to convert the OAM of any light beam. A detailed discussion on these various techniques is not within the scope of this book. Here, we just briefly identify the adopted methods and leave the reader with appropriate references for further reading on this subject. The helical phase front, having an azimuthal angular dependence of amplitude in the transverse plane as $\sim \exp(il\phi)$, can be directly imparted on a laser beam by using a spiral phase plate. A spiral phase plate has an optical thickness d, given by $d = \frac{\lambda l \phi}{2\pi}$. Upon transmission, a plane wave input beam is transformed into a helically phased beam characterized by an azimuthal phase structure $\exp(il\phi)$. Alternatively, this can be achieved using diffractive optical elements, such as 'forked gratings' (which gives rise to first-order diffracted spots with annular intensity cross-sections, a natural consequence of the $\exp(il\phi)$ phase structure) or computer-generated holograms. The astigmatic mode converters consisting of a pair of cylindrical lenses (as previously mentioned) are also frequently used to generate OAM-carrying beams or for OAM conversion. When a pair of cylindrical lenses is kept at $2f$ distance away, they act as π mode converters; when they are kept at $\sqrt{2}f$ distance away, they act as $\pi/2$ mode converters. The $\pi/2$ mode converter in the OAM basis plays the role of a quarter-waveplate in the SAM basis, and accordingly it can convert Hermite-Gaussian (HG) laser modes into LG modes (carrying OAM). The Pancharatnam-Berry optical elements (such as the q-plate) exploits the spin-orbit interaction of light to generate the desired OAM state (this is discussed subsequently in the context of spin-orbit interactions of light in inhomogeneous anisotropic medium). It may be pertinent to note here, despite the various sophisticated approaches that have been developed to generate helically phased beams, they are not unique features in optical settings and are in fact quite ubiquitous. This follows from the fact that while interference of two plane waves yields sinusoidal fringes, interference between three or more plane waves leads to points with perfect destructive interference, around which the phase advances or retards by 2π. These are the phase singularities characteristic of OAM and can easily be observed in optical speckles. Of course, over the extent of the speckle pattern, there are equal numbers of clockwise and counterclockwise singularities, leading overall OAM to be zero.

Having introduced the concept of spin and orbital angular momentum of light, we now turn to the interaction and interconversion (known as SOI) between them in various light-matter interactions, which is of course the focus of this chapter.

12.2 Spin-orbit interaction (SOI) of light

Spin-orbit interaction (SOI) is a well-known phenomenon in quantum physics. It describes a weak coupling between the spin and orbital angular momentum of quantum particles, such as electrons. This is usually interpreted as an electromagnetic interaction of the moving magnetic moment of the electron with an external electric field. In the quantum mechanical description, SOI of electrons is described (and included) as a relativistic correction to the Schrödinger equation in the presence of an external field, which eventually leads to a spin-dependent term in the Hamiltonian. Without going into the quantum mechanical details of the relativistic corrections to the Schrödinger equation here, we briefly illustrate the effect in the framework of classical electrodynamics, where the vectors of the electromagnetic field depend on the reference system. In a reference system that moves with velocity $\mathbf{v}$ relative to an external electric field $\mathbf{E}$, we find a magnetic field

$$\mathbf{B} = -\frac{1}{c}\mathbf{v} \times \mathbf{E} = \frac{1}{mc}\mathbf{E} \times \mathcal{P}, \tag{12.12}$$

where m is the mass of the electron and $\mathcal{P}$ is the momentum. The moving electron thus experiences a magnetic field in its rest frame that arises from the Lorentz transformation of the static (external) electric field. This field in turn interacts with the electron spin. The energy of the electron in this field, due to its magnetic moment μ, is

$$-\mu \cdot \mathbf{B} = -\frac{e}{mc}\mathbf{S} \cdot \mathbf{B} = -\frac{e}{m^2c^2}\mathbf{S} \cdot (\mathbf{E} \times \mathcal{P}), \tag{12.13}$$

where e is the electron charge and its spin is $\mathbf{S} = \frac{\hbar}{2}\sigma$.

In case of centrally symmetric electrical fields, such as the one for orbital motion of an electron in an electric field of an atomic nucleus, we have

$$\mathbf{E} = -\frac{1}{e}\frac{\mathbf{r}}{r}\frac{dV}{dr}, \tag{12.14}$$

where V is the potential given by $V = \frac{Ze^2}{4\pi\epsilon_0 r}$ and Ze is the charge of the atomic nucleus.

The additional energy of the electron in Eq. (12.13) can therefore be written as

$$-\frac{e}{m^2c^2}\mathbf{S} \cdot (\mathbf{E} \times \mathcal{P}) = \frac{1}{m^2c^2}\frac{1}{r}\frac{dV}{dr}(\mathbf{S} \cdot \mathbf{L}), \tag{12.15}$$

where we have used the orbital angular momentum $\mathbf{L} = \mathbf{r} \times \mathcal{P}$. This additional spin-dependent energy term is referred to as the spin-orbit energy, as it results from an interaction of the spin with the magnetic field that is experienced by the moving electron. This spin-dependent correction term in the Hamiltonian (which also follows from a more rigorous quantum mechanical treatment on relativistic generalization of Schrödinger's equation and Dirac's equation for electronic systems dealing with electron spin and its relativistic behavior) results in the splitting of the energy levels of doubly degenerated bands (for spin-up and spin-down electrons).

The universal nature of this effect dealing with coupling of spin and orbital degrees of freedom of spinning particles leads to its manifestation in diverse fields of physics and at different scales, ranging from stellar objects to fundamental particles. The SOI effects observed in various other fields of physics (atomic, condensed matter physics, etc.) are beyond the scope of this book, and we are only interested in its manifestation in the optical domain. As discussed in the preceding section, both classical light and quantum photons possess SAM (intrinsic) and OAM (intrinsic + extrinsic); a coupling between the spin and orbital AM degrees of freedom of light should therefore be expected under certain circumstances. The evolution of polarized light in trajectory (which may be set by a number of processes involving light-matter interaction, e.g., propagation through inhomogeneous isotropic/anisotropic media) may thus mimic the evolution of a massless spin 1 particle (photon) in an external scalar field. We clarify here that by 'inhomogeneous isotropic medium' we mean media having spatially varying refractive indices (continuously or discretely varying) but exhibiting no birefringence effects; 'inhomogeneous anisotropic medium' on the other hand, refers to media exhibiting spatially varying birefringence effects (spatial variation of birefringence or the orientation axis of anisotropy). As we will describe, the SOI effect can indeed be produced by a variety of light-matter interactions, e.g., by a tight focusing of light beams, reflection/refraction at dielectric interfaces, high numerical aperture imaging, scattering from micro-/nanosystems, propagation through inhomogeneous anisotropic media and so forth. This has led to the observation of a number of tiny (typically sub-wavelength), interesting and intricate effects associated with the SAM and OAM of optical fields. Studies on the SOI of light and its various interesting manifestations are thus currently attracting growing attention owing to both fundamental interests and potential nano-optical applications [126, 127, 128]. In the following, we shall briefly address the various interesting manifestations of SOI, discuss mathematical frameworks (based on wave optics treatment) for treating SOI, provide illustrative examples of the resulting effects and discuss their potential implications. We note here that the SOI of light has an inherent geometrical origin and is thus intimately related to the generation of geometric phase of light. It is thus imperative that we define the geometrical phase of light and discuss its role in SOI.

According to the division of SAM and OAM in terms of their intrinsic and extrinsic nature (as described in the preceding section, while SAM is purely intrinsic, OAM can either be intrinsic or extrinsic), the resulting interactions may also be broadly classified in the following three categories [129]:

- *Interaction between SAM and intrinsic OAM of light*, dealing with interconversion between SAM and intrinsic OAM of light beam. This phenomenon leads to the generation of spin-induced optical vortices (characterized by helical phase fronts) and usually occurs in cylindrically or spherically symmetric systems. Examples of such systems are propagation of light through inhomogeneous anisotropic media (space-variant anisotropies), the tight focusing of fundamental or higher-order Gaussian beams, and scattering from micro-/nanosystems. The associated effect can be described as the effect of a trajectory of light on the polarization state, and the generation of spin-dependent vortices in this case can be related to the generation of azimuthal geometric phase.

- *Interaction between SAM and extrinsic OAM of light*, dealing with the reverse effect of polarization (spin) on the trajectory of light (extrinsic OAM). This effect is manifested as a spin-dependent shift of the trajectory of the light beam and is usually associated with the breaking of the symmetry (observed, e.g., in reflection/refraction in inhomogeneous media, in asymmetric focusing or scattering). The resulting effect is known as the Spin Hall effect of light.

- *Interaction between intrinsic OAM and extrinsic OAM*, dealing with intrinsic OAM-dependent shift of the beam trajectory. This phenomenon is similar to (b) and is observed in similar systems but for higher-order beams (carrying intrinsic OAM), accordingly known as the Orbital Hall effect of light. This effect may also be termed the Orbit-orbit interaction (OOI) of light.

As we discuss below, generation of geometric phases and subsequent conservation of total angular momentum of light is inherent to all the optical SOI phenomena. Before we proceed further on defining the geometric phase of light, we note two important features of geometric phases in the context of SOI of light. First, generation of the azimuthal geometric phase (for spherically or cylindrically symmetric systems) is the origin of the spin-to-orbital angular momentum conversion (and the subsequent generation of spin-induced vortices). When we deal with finite beams (such as the fundamental or higher-order Gaussian beams, which have a spread in k vector space), the different k vectors of the beam acquire slightly different geometrical phases. The resulting k gradient of the geometric phase eventually leads to the polarization (the intrinsic SAM) or the intrinsic OAM-dependent shift in the trajectory of the beam (or the center of gravity of the beam). Next, it is this second effect that is analogous to the splitting of the energy levels of doubly degenerated bands for spin-up and spin-down electrons, as a consequence of SOI (the spin-orbit energy in Eq. (12.15)). Here, the SOI (or OOI) of light increases the degeneracy in the spatial modes (spatial distribution) between the opposite circular polarization (intrinsic SAM) or the optical vortex (intrinsic OAM) states. This can also be treated as the dynamical manifestation of geometric phases.

12.3 Geometric phase of light

It is well known that the propagation of light is associated with a phase factor that depends upon the optical path length ($\Theta_d = \frac{2\pi}{\lambda} \times$ *optical path*). This phase factor is termed the 'dynamical phase,' and it is responsible for most of the observable interference effects. The 'geometric phase,' on the other hand (as its name suggests), is independent of the optical pathlength and is determined solely by the geometry, or more specifically by the topology of the evolution of the electromagnetic wave. This phase is intimately connected to the change in the polarization state of the EM wave when it undergoes evolution in an inhomogeneous isotropic/anisotropic medium. There are two types of geometric phase:

- *The spin redirection Berry phase*: This arises from a parallel transport of the wave field under continuous variation of the direction of propagation of the wave. In this case, the wave vector **k** (representing the direction of propagation of the wave) changes smoothly (adiabatically) so that in the local reference frame attached to the wave, the state of polarization of the wave does not change. In this so-called 'adiabatic evolution' of the **k** vector imparted by slowly varying changes in the local environment (e.g., the refractive index gradient in a smoothly inhomogeneous medium), when the wave completes a full cyclic evolution (in the k-space), it acquires an additional phase factor independent of the pathlength. This is manifested as a change in the direction of the polarization (rotation of the polarization vector or the polarization ellipse) of the wave when observed from a global reference frame. This topological phase factor was originally discovered by Berry in the 1980s in the context of the quantum interference effect [130, 131]. It was subsequently shown by a series of papers that this phase factor is universal and can be observed for classical polarized light also in its evolution in curved trajectory (set by refractive index variation).

- *Pancharatnam-Berry phase*: This arises for a wave propagating in a fixed direction (fixed k vector) but undergoing a continuous change in the state of polarization while propagating through an anisotropic (birefringent) medium [131]. When the wave completes a full cyclic evolution in the polarization state space (closed loop in the Poincaré sphere, defined in Section 6.1.5 of Chapter 6), it acquires an additional geometrical phase factor. This geometric phase related to the continuous and cyclic evolution of the polarization state of light was discovered by Pancharatnam in the 1950s.

In the following section, we briefly address the origin of the geometric phase factors and their manifestations, with selected examples.

12.3.1 Spin redirection Berry phase

As noted before, this type of the geometric phase is associated with the adiabatic evolution of the wave-vector **k**, when the wave propagates in a curved trajectory set by the spatial variation of the local optical parameters (refractive index) [126, 128]. A convenient example of this is the propagation of the polarized wave in a helically wound circular wave guide (optical fiber, shown in Fig. 12.3(a)). The condition for perfect adiabatic evolution requires (i) that there be no sharp kink (in the scale of wavelength) in the fiber so that the helicity of the wave (handedness of the circular/elliptical polarization) does not change locally as it propagates, and (ii) that the medium have no birefringence (anisotropy) effect that can cause local changes in the state of polarization of light. In the example shown in Fig. 12.3(a), the laboratory coordinate frame is represented by Cartesian coordinates (X, Y, Z) and the local coordinate frame attached to the ray is shown by (v, w, t). The corresponding direction of the polarization vector is denoted by the vector **e**. In the general case of elliptically polarized light, the direction of **e** corresponds to the orientation of the major axis of the ellipse. As is apparent from Fig. 12.3(a), when the wave propagates

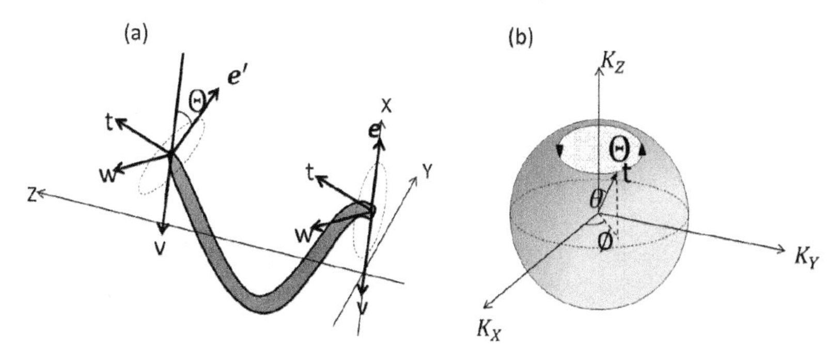

FIGURE 12.3: (a) Propagation of polarized light in a helical wound circular waveguide (optical fiber). The laboratory reference frame is represented by (X, Y, Z) Cartesian coordinates, the local reference frame attached to the ray is shown by (v, w, t), and the corresponding direction of the polarization vector (or orientation of the polarization ellipse) is denoted by the vector $\mathbf{e}$. One full cyclic evolution of the polarized wave leads to a rotation of the polarization vector by an angle Θ. (b) Representation of one full cyclic evolution in momentum space (wave-vector k space). The spherical angles of the k sphere are represented as (θ, ϕ); θ is the angle between the local waveguide axis and the axis of the helix (z-axis) (the pitch angle of the helix) and ϕ is the azimuthal angle. The closed loop at the k-sphere corresponding to one full cyclic evolution subtends a solid angle Θ at the center of the sphere.

through the helical waveguide, the local coordinate frame (v, w, t) undergoes continuous rotation; so does the polarization vector $\mathbf{e}$. Therefore, in the local coordinate frame attached with the ray, the polarization does not change. It is convenient to represent such adiabatic evolution processes of polarized waves in the momentum space (k space), where the direction of propagation of the wave is represented by the three Cartesian components of the wave vector (k_x, k_y, k_z) in the k sphere (which is the parameter space here with k_x, k_y, k_z as the three axes, shown in Fig. 12.3(b)). The spherical angles of the k sphere are represented as (θ, ϕ), and accordingly the direction of the wave momentum with respect to the laboratory coordinate frame (X, Y, Z) can be represented as

$$(k_x, k_y, k_z) = k(\sin\theta\cos\phi, \sin\theta\sin\phi, \cos\theta). \tag{12.16}$$

Note that θ is the angle between the local waveguide axis and the axis of the helix (z-axis) (i.e., the pitch angle of the helix) and ϕ is the azimuthal angle related to the winding of the helix (one full winding period corresponds to $\phi = 2\pi$). One full cyclic evolution of the k vector (closed loop at the k sphere) corresponds to one period of the trajectory of the helically wound optical fiber. As apparent from Fig. 12.3(a), in one full cyclic evolution, although the polarization vector $\mathbf{e}$ (orientation of the polarization ellipse) never changes its direction in the local frame (attached to the ray), in the global laboratory frame (X, Y, Z), it is rotated by an angle Θ. The corresponding closed loop at the k sphere is shown in Fig. 12.3(b). The surface spanned over the closed loop is observed to subtend a solid angle (area enclosed by the projection of the loop onto the k sphere) at the center of the sphere, which is exactly equal to Θ. This solid angle can be calculated as

$$\Theta = \int_0^{\theta} \sin\theta d\theta \int_0^{2\pi} d\phi = 2\pi(1 - \cos\theta). \tag{12.17}$$

Let's now try to interpret the observed rotation of the polarization in terms of the geometric phase of light. For simplicity, we assume that the input light is linearly polarized, in principle; this can also be generalized for any arbitrary elliptically polarized wave. The rotation Θ of the linear polarization vector due to one full cyclic evolution of the wave in the helical waveguide can be interpreted as an 'optical rotation' effect. As we have seen in Chapter 6 (Section 6.2.3), that optical rotation is a manifestation of the circular birefringence effect (which arises from a phase difference between orthogonal circular polarization states). However, the waveguide has no local intrinsic anisotropies or circular birefringence. The observed circular birefringence thus appears to be a geometrical effect. This can be understood in a simple way if we consider linearly polarized light to be superposition of equal amplitude of left (L) and right (R) circular polarization. We may write the input linear polarization state (say, oriented along the x-axis of the laboratory coordinate) as

$$|x\rangle = \frac{1}{\sqrt{2}}\left(|L\rangle + |R\rangle\right). \tag{12.18}$$

The state of linear polarization of the wave after one fully cyclic evolution (rotated by an angle Θ) is given by

$$|x_{out}\rangle = \frac{1}{\sqrt{2}}\left(\exp(i\Theta)|L\rangle + \exp(-i\Theta)|R\rangle\right). \tag{12.19}$$

It appears from Eq. (12.19) that while propagating through the helical wave guide, the constituent left and right circular polarization modes of the linearly polarized wave have acquired equal and opposite phases ($\pm\Theta$), which do not originate from any intrinsic anisotropies and are thus purely geometric in nature. This is the so-called spin redirection Berry phase. As is apparent from Eq. (12.17), for one full cyclic evolution (due to continuous change in the trajectory of the polarized light wave), the spin redirection Berry phase is determined by the solid angle subtended by the closed loop in the k sphere (in the momentum domain representation).

The spin redirection Berry phase may also be directly measured in the following interference experiment. A circularly polarized (either left or right) laser beam is coupled to an optical fiber, which in turn couples equal amounts of light into two helically wound optical fibers. Each of these fibers has N number of turns but in the opposite sense (right and left helix), which are adjusted to have equal optical pathlengths. These two oppositely wound helical fibers constitute the two arms of the interferometer. The fibers are then once again brought together and coupled to another single output optical fiber. The interference is observed at the output optical fiber using a detector. As apparent from Eq. (12.17), circularly polarized light while propagating through the two arms acquire equal and opposite amounts of geometric phases ($\Theta = \pm 2\pi N(1 - \cos\theta)$) (the factor N arises due to N number of cyclic evolutions). The resultant intensity pattern would therefore be

$$I = I_0 \cos^2[2\pi N(1 - \cos\theta)]. \tag{12.20}$$

The resultant output intensity after the interference would thus be determined by the pitch angle of the helix (θ) and the number of turns N.

12.3.2 Pancharatnam-Berry phase

The Pancharatnam-Berry phase arises when a polarized wave undergoes continuous change in the state of polarization, keeping the direction of propagation fixed (fixed k vector). An additional geometrical phase factor is introduced when the wave completes a full cyclic evolution in the polarization state space (in the Poincaré sphere) [131, 132, 133]. Clearly, such a situation may arise when polarized light propagates through homogeneous/inhomogeneous anisotropic (birefringent) medium. In order to illustrate this, in Fig. 12.4(a) we show a Michelson interferometer arrangement for observing the Pancharatnam phase [133].

Consider linearly polarized (say, a polarization axis oriented along the x-axis of the laboratory coordinate) light from a laser is divided into two equal beams by a 45°, 50: 50 beam-splitter. Beam 1 travels to a perpendicular mirror M_1 and is reflected back. Beam 2 first travels through a quarter-wave plate $QP1$, whose optical axis is fixed at ($\theta_1 = \pi/4$) relative to the incident x-polarization. This converts the linearly polarized light into right circularly polarized, which then passes through another quarter waveplate, $QP2$, whose optical axis makes and angle $\theta_2 = \left(\frac{3\pi}{4} + \beta\right)$ with the x-axis. The light emerging from $QP2$ strikes the perpendicular mirror $M2$, from which it is reflected and made to retrace its path. The beam-splitter BS combines portions of the returning light from arms 1 and 2 and gives rise to an interference pattern in arm 3 that allows the phase difference between the two return beams to be determined. In this system, the presence of geometrical phase can be inferred by keeping the lengths of the two arms fixed (thus keeping the dynamical phase difference fixed between two arms) and by changing the orientation angle θ_2 of the second quarter-wave plate $QP2$. Changes in geometrical phase associated with polarization evolution in arm 2 of the interferometer will be manifested as shift of the interference phase (fringe shifts). In order to understand the Pancharatnam-Berry phase resulting from polarization transformations in arm 2 of the interferometer, we model each of the polarizing interactions using Jones matrices (following Section 6.2.3 of Chapter 6). The first quarter-wave

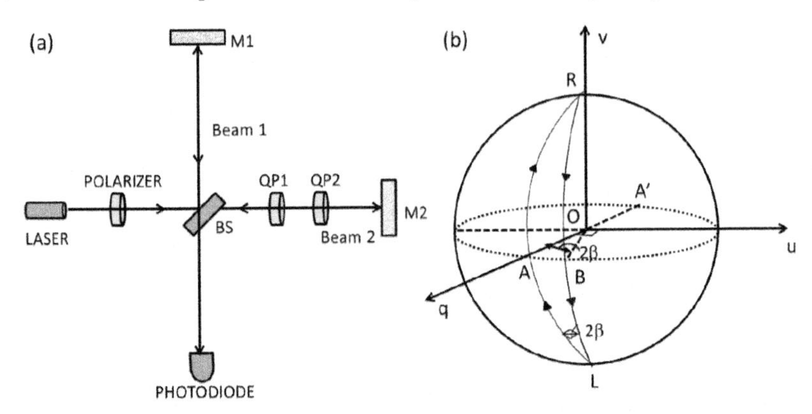

FIGURE 12.4: (a) Schematic of the Michelson interference experiment for observation of Pancharatnam-Berry phase. (b) The evolution of the state of polarization of light in arm 2 of the Michelson interferometer (Fig. 12.4(a)) is shown in the Poincaré sphere (shown for incident x-polarized light). Equatorial points A and A′ represent x and y linear polarization, respectively, the polar points R and L represent right and left circular polarization respectively. The action of QP1 and QP2 are shown by trajectories AR and RB. After reflection, the polarization state evolves in similar manner but in the opposite sense, represented by trajectories BL and LA. The closed loop corresponding to full cyclic evolution of polarization state (ARBLA) subtends a solid angle 4β, and the corresponding geometric phase is half of this solid angle (2β).

plate (oriented at $\pi/4$ with respect to the x-axis) transforms the incident x-polarized light (represented by Jones vector $\begin{bmatrix} 1 & 0 \end{bmatrix}^T$) into right circularly polarized (RCP) light:

$$\begin{bmatrix} 1+i & 1-i \\ 1-i & 1+i \end{bmatrix} \begin{bmatrix} 1 \\ 0 \end{bmatrix} = \begin{bmatrix} 1 \\ -i \end{bmatrix} \exp(i\phi_1). \tag{12.21}$$

Here, we have used the Jones matrix of a quarter-wave plate oriented at an angle $\pi/4$ (orientation angle $\theta_1 = \pi/4$, retardance $\delta = \pi/2$ in Eq. (6.37). As is apparent from Eq. (6.37), the phase factor ϕ_1 here is related to the thickness and refractive index of the birefringent waveplate and is therefore a dynamic one. The resulting RCP light passes through the second quarter-wave plate (QWP2 oriented at $\theta_2 = (\frac{3\pi}{4} + \beta)))$ to yield linearly polarized light:

$$\begin{bmatrix} \cos^2\theta_2 + i\sin^2\theta_2 & (i-1)\sin\theta_2\cos\theta_2 \\ (i-1)\sin\theta_2\cos\theta_2 & \sin^2\theta_2 + i\cos^2\theta_2 \end{bmatrix} \begin{bmatrix} 1 \\ -i \end{bmatrix} = \begin{bmatrix} \cos\beta \\ \sin\beta \end{bmatrix} e^{(i\phi_2)} \times e^{(-i\beta)}. \tag{12.22}$$

It is important to note that the polarized light acquires an additional phase factor $\exp(-i\beta)$ depending upon the orientation angle (θ_2 is related to β with a constant factor) of the wave plate in addition to the dynamical phase factors

(ϕ_1 and ϕ_2). The linearly polarized light then gets reflected from the mirror and undergoes similar evolution through the two quarter-wave plates. The action of QP2 (which is now oriented at an angle $\pi/4$ with respect to the direction of linear polarization of the reflected wave) will convert the linearly polarized light into a left circularly polarized light (LCP, when represented in the same reference frame) and will introduce another phase factor $\exp(-i\beta)$ in addition to the dynamical phase factor. Finally, when the LCP light passes through QP1, it will be converted back to x-polarized light $\begin{bmatrix} 1 & 0 \end{bmatrix}^T$. Thus, in this process, the input x-polarized wave completes a full cyclic evolution in the polarization state space. During this evolution, it acquires a total dynamic phase

$$\phi_d = 4 \times \frac{2\pi}{\lambda} \times \frac{(n_e + n_o)}{2} d, \tag{12.23}$$

where n_e and n_o are the refractive indices of the slow and fast components in the quarter-wave plates and d is the thickness of the wave plate so that the retardance $\delta = \frac{2\pi}{\lambda}(n_e - n_o)d = \frac{\pi}{2}$. Importantly, in addition to this dynamical phase, the polarized wave acquires a geometric phase factor $\exp(-2i\beta)$ (geometric phase of 2β), which is independent of the optical pathlength and is only determined by the orientation angle of the second quarter-wave plate (thus it depends upon how the polarization state has evolved during the cyclic process).

The evolution of the geometric phase can be conveniently represented by the polarization transformations in the Poincaré sphere (defined in Section 6.1.5 of Chapter 6). The trajectory of the polarization evolution for this particular experimental configuration is shown in Fig. 12.4(b). The four basic interactions in arm 2 of the interferometer can be represented by the following transformations in the Poincaré sphere:

1. The incident x-polarized light (noted by position A in the equator region of the sphere) transforms to RCP light (represented by point R in the north pole) by passing through QP1 (trajectory shown by AR).

2. The action of QP2 brings back the state to linear polarization, represented by point B in the equator region (trajectory RB). After reflecting back from the mirror, the polarization state evolves in a similar manner but in the opposite sense.

3. The action of QP2 on the polarization state of the reflected light takes it to L in the south pole (LCP).

4. QP1 then brings the state of polarization to the initial linear polarization (A, representing x-polarization).

Thus the state of polarization performs a closed loop in the Poincaré sphere. As shown by the geometry of the Poincaré sphere in Fig. 12.4(b) (and corresponding analogy of the k-sphere shown in Fig. 12.3(b) for the case of spin redirection Berry phase), the solid angle subtended by this closed loop (ARBLA) at the center of the sphere is given by 4β. The corresponding geometric phase will be half of this solid angle and will thus be equal to 2β. The origin of this $\frac{1}{2}$ factor in the case of Poincaré sphere representation (as compared to the k sphere) can be understood by noting that the rotation of the polarization ellipse (or polarization vector) by an angle ϕ in the real space corresponds to 2ϕ rotation in the Poincaré sphere (see Eq. (6.46) and subsequent discussion in Chapter 6). We note here that the closed loop shown in Fig. 12.4(b) is for input x-polarized light. In the case of input y-polarized light, the evolution would also be cyclic but given by the reflection of the loop through the origin. The two loops are traversed in opposite senses, and hence they subtend at equal and opposite solid angles at the origin. Thus, input x- and y-polarizations would acquire the Pancharatnam-Berry phase of $\mp 2\beta$.

We now briefly touch upon an interesting effect associated with dynamical manifestation of the Pancharatnam-Berry phase in context with the polarization evolution in anisotropic medium. We now consider the same experiment as above, however, in our experiment the second quarter-wave plate (QP2) is uniformly rotated with an angular velocity $\Omega = \frac{d\theta_2}{dt}$. The segment RAL (of the closed loop ARBLA shown in Fig. 12.4(b)) remains unchanged under this rotation, whereas the segment RBL continuously rotates about the axis RL with an angular velocity of 2Ω. This will make the geometric phase evolve with time as $\beta(t) = \beta(0) \mp 2\Omega t$, where the $\mp$ signs are for input x- and y-polarizations, respectively. This linear time variation of the geometric phase therefore contributes to a shift in the frequency. If the input light has a frequency ω, the frequency of the output light (after making the round trip) would become $\omega' = \omega \pm 2\Omega$, for input x- (plus) and y- (minus) polarizations, respectively. Note that the plus and minus signs in the frequency shift are relative and depend upon the convention (how the positive or negative phases are defined). It is also important to note that this effect may also be observed with a single rotating half-wave plate (whose retardation is given by $\delta = \frac{2\pi}{\lambda}(n_e - n_o)d = \pi$). The half-wave plate transforms input RCP light to LCP light ($\sigma = -1 \rightarrow +1$), and the corresponding evolution of the polarization state in the half-wave plate can be represented in the Poincaré sphere as shown in Fig. 12.5(a). Note, we may reach to LCP (south pole) from RCP (north pole) using different paths on the sphere, which is determined by the orientation angle of the wave plate (θ) with respect to the laboratory polarization axis (say, the x-axis). The difference in geometric phases between two such paths is half of the solid angle enclosed by the loop formed by these two trajectories (shown by the shaded region in Fig. 12.5(a)). If the wave plate is rotated with uniform angular velocity $\Omega = \frac{d\theta}{dt}$, akin to the previous case, the geometric phase would evolve with time as $\approx \mp 2\Omega t$. However, unlike the previous case, here $\mp$ signs are for input right and left circular polarizations, respectively. This would accordingly lead to a shift of the frequency of the input wave $\omega' = \omega \pm 2\Omega$, for input right (plus) and left (minus) polarizations, respectively. This is the so-called rotational Doppler shift for the SAM-carrying light beam (displayed in Fig. 12.5(b)). As

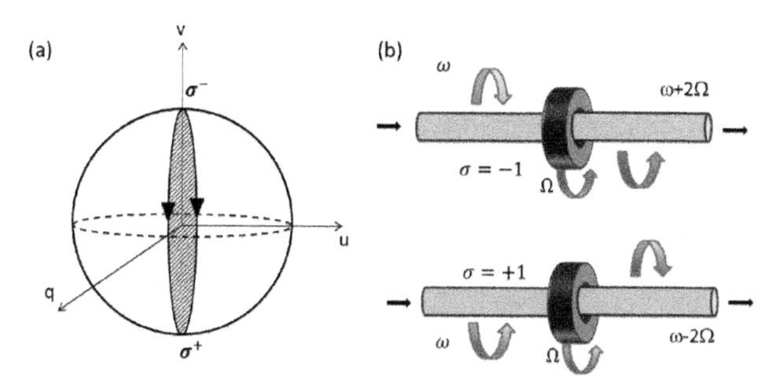

FIGURE 12.5: Dynamical manifestation of the Pancharatnam-Berry phase for a rotating half-wave plate. (a) The half-wave plate converts incident RCP light to LCP light ($\sigma = -1 \rightarrow +1$). The trajectory depends upon the orientation angle of the optical axis of the wave plate with respect to the laboratory axis (say, the x-axis). Two different trajectories corresponding to two orientations (at two different instances) of the rotating wave plate are shown. The difference in geometric phases between two such paths is half of the solid angle enclosed by the loop formed by these two trajectories, shown by the shaded region. (b) The evolving geometrical phase leads to equal and opposite shifts in the frequency for input right and left circularly polarized light. Here, Ω is the angular velocity of the rotating wave plate.

we shall discuss later (in context with the spin-orbit interaction of light in inhomogeneous anisotropic medium), this effect is closely related to SOI and the polarization-dependent shift in the trajectory of a beam (or the center of gravity of a beam).

12.3.3 Geometric phase associated with mode transformation

The geometric phases discussed earlier (spin redirection Berry phase and Pancharatnam-Berry phase) are associated with the polarization state of light (or SAM). From a conceptual point of view, it is expected that analogous geometric phases should also be observed for beam-carrying intrinsic OAM (where the role of SAM is played by intrinsic OAM) [128]. As we now know that the intrinsic OAM of light is associated with the mode structure of a light beam (transverse distribution of the field amplitude and phase), mode transformation should also lead to the generation of both the variants of the geometric phase. One deals with the geometry of the path in the configuration space (continuous variation of the direction of propagation of the wave and the wave-vector k, like the helical trajectory described earlier), leading to a change in the global orientation of the mode structure of the beam. Waves propagate in a fixed direction (fixed k vector) but undergo a continuous mode transformation (the geometry of the path can be represented in the mode space as was done in the polarization state space for SAM-carrying beams). This can be understood from the following analogy of SAM and OAM [134].

As we have previously discussed, the Laguerre-Gaussian (LG) laser modes have a helical phase front characterized by transverse field distribution $\approx \exp(il\phi)$ (where ϕ is the azimuthal angle) and accordingly carry OAM. In fact, the LG modes follow from the solutions to the paraxial wave equation in cylindrical coordinates, and a general LG mode (LG_p^l) is characterized by radial and azimuthal indices p and l, respectively, carrying $l\hbar$ OAM per photon. The Hermite-Gauss (HG) modes, on the other hand, are solutions to the paraxial wave equation in rectangular coordinates (represented by HG_{nm}) and do not carry OAM as such. The order of these modes is generally given by $N = 2p + |l| = n + m$. We note that, for modes with order $N = 1$, in phase superposition of left-handed ($l = +1, p = 0$) and right-handed ($l = -1, p = 0$) helical LG modes form an HG mode with indices $m = 1, n = 0 (HG_{10})$. Similarly, the HG_{01} mode can be obtained by superposition of left and right LG modes with a phase difference of π between them:

$$HG_{10} = LG_0^{+1} + LG_0^{-1} \quad HG_{01} = LG_0^{+1} - LG_0^{-1}. \tag{12.24}$$

Conversely, the first-order LG modes may also be obtained by superposition of orthogonal HG modes:

$$LG_0^{+1} = HG_{10} + iHG_{01} \quad LG_0^{-1} = HG_{10} - iHG_{01}. \tag{12.25}$$

This set of equations provides the basis for an analogy between Jones vector representation of linear and circular polarization states and those of the first-order HG and LG modes. If we represent the states of the HG_{10} and HG_{01} modes as $[1,0]^T$ and $[0,1]^T$ (the equivalent representation of horizontal and vertical linear polarization states), the corresponding orthogonal LG modes (with $l = \pm 1$) can be represented as $\frac{1}{\sqrt{2}}[1, \pm i]^T$ (equivalent representation of left and right circular polarization). In fact, we can decompose any arbitrarily oriented HG mode (determined by the orientation of the phase structure and intensity distribution) and LG modes using the basis of the HG_{10} and HG_{01} modes, just like any arbitrarily oriented linear polarization state and circular (elliptical) polarization states can be decomposed using horizontal and vertical linear polarization basis. This yields a one-to-one correspondence between the first-order modes with the polarization state of light, as shown in Fig. 12.6(a). The corresponding analogy between the Poincaré sphere representation of the polarization states and the mode states in the sphere of the first-order modes (the orbital Poincaré sphere) is shown in Fig. 12.6(b). In the orbital Poincaré sphere of the first-order modes, the poles correspond to right- and left-handed LG modes ($l = \mp 1$, north and south poles), and the equator region corresponds to HG modes oriented at different angles.

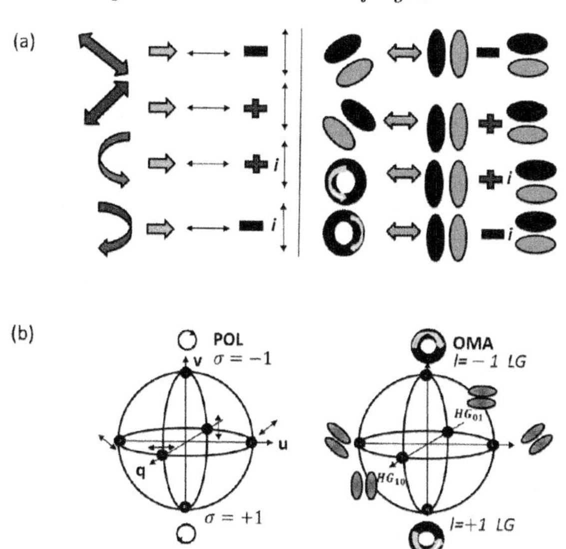

FIGURE 12.6: (a) Decomposition of the polarization states (±45° linear and left/right circular polarizations) using the horizontal and vertical linear polarization basis states (left panel). The corresponding decomposition of ±45° oriented HG modes and the left- and right-handed LG modes ($l = \pm 1$) using the HG_{10} and HG_{01} modes. Note that the orientation of the lobes in an HG mode corresponds to the direction of the linear polarization vector (e.g., HG_{10} horizontal orientation of the lobes → horizontal linear polarization). (b) Analogy between the polarization Poincaré sphere (left panel) and orbital Poincaré sphere of first-order modes (right panel). Linear and circular polarization states are shown in the polarization Poincaré sphere, while their analogous HG and LG modes are shown in the orbital Poincaré sphere.

From this analogy it is apparent that both the variants of the geometric phase (spin redirection Berry phase and Pancharatnam-Berry phase) associated with the polarization state of light should be manifested in the case of the evolution of mode structure in curved trajectory of a light beam or for continuous mode transformation (a change of the OAM state of the beam). As in the case of the spin redirection Berry phase (Fig. 12.3(a)), let us now consider the propagation of an HG laser mode through a helically wound circular waveguide (shown in Fig. 12.7). Here also, for one full cyclic evolution of the k vector (corresponding to one period of the trajectory of the helically wound optical fiber), although mode structure does not change in the local coordinate frame attached with the beam, in the global laboratory frame it is rotated by an angle Θ. The corresponding closed loop at the k sphere and the solid angle subtended by this at the center of the k sphere can also be shown to be equal to Θ (similar to Eq. (12.17)). If we consider the evolution

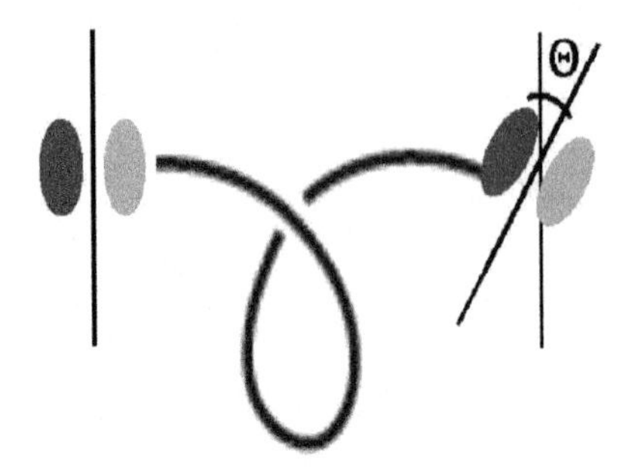

FIGURE 12.7: Propagation of an HG laser mode through a helically wound circular waveguide. One full cyclic evolution of the k vector (corresponding to one period of the trajectory of the helically wound optical fiber) leads to a rotation of the mode structure by an angle Θ.

of the HG_{10} mode (as an example), the output rotated (by an angle Θ) mode structure after one full cyclic evolution may also be interpreted as

$$|HG_{out}\rangle = \frac{1}{\sqrt{2}}\left(\exp(i\Theta)|LG_0^{+1}\rangle + \exp(-i\Theta)|LG_0^{-1}\rangle\right). \tag{12.26}$$

Apparently, while propagating through the helical waveguide, the constituent left- ($l = +1$) and right-handed ($l = -1$) LG modes of the HG_{10} mode acquire equal and opposite phases ($\pm\Theta$), which is the geometric phase. In an analogy with polarization, this may be termed the *OAM redirection Berry phase.* The other variant of the geometric phase arises when the laser beam propagates in a fixed direction (fixed k vector) but undergoes a continuous mode transformation (the Pancharatnam-Berry equivalent geometric phase for mode transformation). This type of geometric phase can be observed using a Michelson interferometer arrangement similar to that of the Pancharatnam phase associated with polarization transformation (Fig. 12.4(a)). The polarization states of light in this experiment have to be replaced by the corresponding mode states (according to Fig. 12.6(a)). We may start with an HG_{10} mode (in place of the horizontal linear polarization), which is then split into the two arms of the interferometer using a beam-splitter. Beam 1 once again travels to a perpendicular mirror M_1 and is reflected back. Beam 2 undergoes continuous mode transformations in the second arm of the interferometer. As we have previously discussed, when a pair of cylindrical lenses is kept at $2f$ distance away with their focal lines parallel, they act as $\pi/2$ mode converter. The direction of the focal lines of the cylindrical lenses is referred to as the principal axis of the converter. The $\pi/2$ mode converter in the OAM basis plays the role of a quarter-wave plate in the SAM basis, and accordingly it can convert HG laser modes into LG modes (carrying OAM) when the principal axis of the converter is kept at an angle $\pm 45°$ with respect to the axis of the HG mode. Thus in this experiment, the two quarter-wave plates (QP1 and QP2 of Fig. 12.4(a)) may be replaced by two $\pi/2$ mode converters. The orientation of the principal axis of the first mode converter can be fixed at 45°, whereas the orientation of the second mode converter may be changed to observe the resulting geometric phase associated with the mode transformation (which would be manifested in the interferogram). The trajectory of the mode transformation and the evolution of the OAM state here can be represented in the orbital Poincaré sphere of first-order modes (Fig. 12.6(b)) in a similar fashion to what was done for polarization evolution in the polarization Poincaré sphere (Fig. 12.4(b)) (in order to avoid repetition, the corresponding representation in the orbital Poincaré sphere is not shown here). The corresponding geometric phase associated with mode transformation would also be half of the solid angle enclosed by the closed loop in the center of the orbital Poincaré sphere (the solid angle $= 4\beta$ and the geometric phase $= 2\beta$, where β is the orientation of the principal axis of the second $\pi/2$ mode converter).

Having described the various types of geometric phases, we are now in a position to discuss spin-orbit interaction of light. In what follows, we describe three specific cases of SOI with illustrative examples: SOI in inhomogeneous anisotropic media, in scattering and in tight focusing of a fundamental Gaussian beam. We shall outline the mathematical framework for describing SOI and discuss the role of geometric phases in the resulting SOI.

12.4 Spin-orbit interaction of light in inhomogeneous anisotropic medium

Thus far, we have seen that when a polarized light beam propagates through an anisotropic medium (either birefringent or dichroic), its state of polarization changes. This aspect was dealt with in some detail in Chapter 6, where it was discussed that the two of the important anisotropic properties of the medium, linear birefringence (retardance) and linear dichroism (diattenuation) arise from the differences in the real and imaginary parts of the refractive indices for different polarization states, respectively (described in terms of ordinary and extraordinary axes and indices). Propagation of polarized light through such a medium generally leads to change in the SAM of light. For example, when circularly polarized light propagates through a half-wave plate (linear retardance $\delta = \pi$), it results in the flipping of the SAM ($\sigma = +1 \rightarrow -1$), which is balanced by the transfer of angular momentum to the birefringent plate (see Fig. 12.2). On the other hand, continuous evolution of the state of the polarization of light in such an anisotropic medium leads to the generation of geometric phase (Pancharatnam-Berry phase, discussed in Section 12.2). We note here that the kind of anisotropic media discussed so far are assumed to be homogenous per se. By the term *homogeneous anisotropic medium*, we mean that the anisotropic media (it may be either birefringent or dichroic) are associated with an axis of anisotropy and that the orientation of the anisotropy axis is the same across the transverse dimension of a light beam (examples are a homogeneous wave plate and a polarizer). In this section, we address the interaction of polarized light with the *inhomogeneous anisotropic medium*, i.e., a medium having a spatially varying axis of anisotropy [127]. Recall the Michelson interferometer experiment shown in Fig. 12.4, where it was demonstrated that the evolution of Pancharatnam-Berry phase is determined by the orientation of the anisotropy axis of the waveplate (the trajectory in the Poincaré sphere is decided by this). Conceptually, it may thus be reasonable to anticipate spatially varying anisotropy (across the transverse dimension of a light beam) should lead to a space-variant geometric phase, which should eventually lead to the generation of OAM. We shall illustrate here that indeed, in rotationally symmetric geometries such space-varying anisotropy leads to the generation of input spin-dependent OAM of light and that under certain circumstances, the SAM variation of light may be entirely converted into intrinsic OAM. Under the circumstances, there would be no net angular momentum transfer to the material and the total angular momentum conservation is maintained by 'spin-to-orbital angular momentum conversion.' Such interconversion between SAM and intrinsic OAM of light beams in inhomogeneous anisotropic media can be conveniently dealt with using Jones and Stokes-Mueller polarization algebra. However, before we do so, we shall address this issue using elementary treatment on evolving geometric phases in such inhomogeneous anisotropic medium.

Recall the discussion of a dynamical manifestation of Pancharatnam-Berry phase in the context of polarization evolution in anisotropic media (Fig. 12.5). It was shown that a rotating half-wave plate (a homogenous retarder with a magnitude of linear retardance $\delta = \pi$) leads to a time-varying geometric phase. If the wave plate is rotating with angular velocity $\Omega = \frac{d\theta}{dt}$ (θ is the orientation of the anisotropy axis of the wave plate), then the time-varying geometric phase is given by $\Phi_g(t) = 2\Omega t$. Note that a half-wave plate converts input SAM $\sigma = +1 \rightarrow -1$. In the general case, for an arbitrary anisotropic medium (any retarder/waveplate with magnitude of retardance δ or even a diattenuator/polarizer with magnitude of diattenuation D), the time varying geometric phase can be written as [128]

$$\begin{aligned} \Phi_g(t) &= (S - S_0)\Omega t, \\ \frac{d}{dt}\Phi_g(t) &= (S - S_0)\Omega, \\ \Phi_g(t) &= \int (S - S_0)\Omega dt, \end{aligned} \tag{12.27}$$

where S_0 and S are the input and the output SAM of the light beam, respectively (for a given SAM input, the output state is determined by the magnitude of the anisotropy of the medium). For example, $S_0 = +1,\ S = -1$ for a half-wave plate ($\delta = \pi$); and $S_0 = +1,\ S = 0$ for a quarter-wave plate

($\delta = \pi/2$) or for a polarizer ($D = 1$). It was also shown that such evolving geometric phases dynamically manifest as a shift of the frequency (ω) of light as

$$\Delta\omega = (S - S_0)\Omega. \tag{12.28}$$

In this case, we considered temporal evolution of the coordinate frame (the rate of rotation Ω of the orientation of the anisotropy axis of the medium, in rad/s). Similar evolution and dynamical manifestation of the geometric phase are also warranted if we consider spatial rotation, with respect to a chosen spatial coordinate ξ, in a plane transverse to the propagation direction of the beam:

$$dt \to d\xi;\ \Omega \to \Omega_\xi,$$
$$\Omega_\xi = \frac{d\theta}{d\xi}. \tag{12.29}$$

Here, Ω_ξ (in units of rad/length) is the rate of spatial rotation of the orientation axis (for an inhomogeneous anisotropic medium) in a plane transverse to the propagation direction of the beam. In such a case, the geometric phase can be written in an analogous fashion as

$$\Phi_g(\xi) = \int (S - S_0)\Omega_\xi d\xi. \tag{12.30}$$

For simplicity, let us now consider cylindrically symmetric geometry, with the radius vector and azimuthal angle in a plane transverse to the propagation direction of the beam (z) given by (r, ϕ). The spatial rotation rate $\Omega_\xi(\xi \to \phi)$ can thus be written as

$$\Omega_\xi = \frac{d\theta}{d\xi} = \frac{d}{d\phi}\theta(\phi). \tag{12.31}$$

Here, we have considered that the orientation axis of the anisotropic medium changes uniformly as a function of the azimuthal angle ϕ. Uniform spatial rotation rate can be achieved by setting the azimuthal variation of the orientation axis as $\theta(\phi) = q\phi + \theta_0$ (where q and θ_0 are constants), so that

$$\Omega_\xi = \frac{d}{d\phi}\theta(\phi) = q. \tag{12.32}$$

It immediately follows from Eq. (12.30) and (12.32) that

$$\Phi_g(\phi) = q\phi(S - S_0). \tag{12.33}$$

The generation of such an *azimuthal geometric phase* would imprint helical phase front $\exp(iq\phi(S - S_0))$ on an input non-OAM-carrying beam and would lead to the generation of OAM. The magnitude of the generated OAM (l) would depend upon the parameter q and change in the SAM ($S - S_0$). This is therefore a manifestation of spin-orbit interaction (SOI): *The interaction between SAM and intrinsic OAM of light, dealing with interconversion between SAM and intrinsic OAM.*

If we consider the case of a half-wave plate with its anisotropy axis changing direction in the transverse plane as $\theta(\phi) = q\phi + \theta_0$, the azimuthal geometric phase is $\Phi_g(\phi) = \pm 2q\phi$ ($\pm$ corresponds to input circular polarization state, positive for input left circular and negative for input right circular polarization state). For $q = 1$, it leads to perfect conversion and angular momentum balance between SAM and OAM so that no net angular momentum is transferred to the wave plate:

$$(\sigma = \pm 1, l = 0) \Rightarrow (\sigma = \mp 1, l = \pm 2). \tag{12.34}$$

In the case of an inhomogeneous quarter-wave plate (linear retardance $\delta = \pi/2$) or inhomogeneous linear polarizer (linear diattenuator with diattenuation $D = 1$) with varying anisotropy axis $\theta(\phi) = q\phi + \theta_0$, the azimuthal geometric phase would be $\Phi_g(\phi) = q\phi$, because $|S - S_0| = 1$ (a quarter-wave plate converts linearly polarized light to circular polarization or vice versa; a linear polarizer, on the other hand, can convert input circular polarization to linear polarization). For a value of $q = 1$, the resulting spin-to-orbital angular momentum conversion can be written as

$$(\sigma = \pm 1, l = 0) \Rightarrow (\sigma = 0, l = \pm 1). \tag{12.35}$$

Note that finite light beams (such as the fundamental or higher-order Gaussian beams) are associated with the transverse component of momentum (having a spread in the wave-vector k space); in addition to a component of the wave vector along the propagation direction (k_z), they also possess a component of k in the transverse (xy) plane ($k_\perp = k_x$ and k_y). The geometric phase given by Eq. (12.33) acquired by the beam while propagating through the inhomogeneous anisotropic medium may thus be assigned to the central wave-vector of the beam. The different constituent k vectors of the beam in fact acquire slightly different geometrical phases. This should therefore manifest as an input SAM-dependent splitting of the wave-vector distribution in the transverse plane ($\Delta k_\perp$) of the beam. This may also be put forward using the analogy between the temporal and the spatial rotation of the wave plate. As we discussed, the temporal rotation of the homogeneous wave plate leads to an evolving geometric phase, which eventually manifest is as a shift of the frequency ($\Delta\omega$) of the light as per Eq. (12.28). In the case of the inhomogeneous anisotropic medium having a spatial rotation rate Ω_ξ of the anisotropy axis, this would be manifested as a shift in the transverse spatial frequency distribution of the beam ($\Delta k_\perp$) (spatial frequency replacing the temporal frequency);

$$\begin{gathered} dt \to d\xi;\ \Omega \to \Omega_\xi;\ \Delta\omega \to \Delta k, \\ \Delta k_\perp = (S - S_0)\Omega_\xi = q(S - S_0). \end{gathered} \tag{12.36}$$

Thus, the spin-orbit interaction (spin-to-orbit AM interconversion) is accompanied by a fine polarization-dependent splitting of the transverse momentum distribution. Note that in this particular cylindrically symmetric case, the splitting is radial, which does not cause a shift of the overall center of gravity (intensity barycenter) of the beam. However, this polarization-dependent splitting of the beam's spatial frequency distribution may be enhanced significantly by breaking the symmetry of the system, which eventually leads to a polarization-dependent shift of the center of gravity of the beam. This effect is analogous to the fine structure energy level splitting of the spin-up and spin-down electron as a consequence of SOI (the spin-orbit energy of Eq. (12.15)). Here, the SOI of light lifts the degeneracy in the spatial modes (intensity distribution) of opposite SAM (circular polarization) states.

We mentioned earlier that the spin-to-orbit AM conversion in inhomogeneous anisotropic medium may also be dealt with adequately using conventional Jones matrix algebra, which we address now. As an example, we consider the case of the inhomogeneous half-wave plate with a magnitude of linear retardance $\delta = \pi$ and a spatially varying orientation angle given by $\theta(\phi) = q\phi + \theta_0$. Using Eq. (6.37) of Chapter 6, the Jones matrix for this birefringent medium can be written as

$$J = \begin{bmatrix} \cos\theta & -\sin\theta \\ \sin\theta & \cos\theta \end{bmatrix} \begin{bmatrix} 1 & 0 \\ 0 & -1 \end{bmatrix} \begin{bmatrix} \cos\theta & \sin\theta \\ -\sin\theta & \cos\theta \end{bmatrix} = \begin{bmatrix} \cos 2\theta & \sin 2\theta \\ \sin 2\theta & -\cos 2\theta \end{bmatrix}. \tag{12.37}$$

Let us now consider a left circularly polarized plane wave (the plane wave is an idealization corresponding to the central wave-vector of a Gaussian beam) that is represented by Jones vector $E_in = [1, i]^T$ and is incident on the inhomogeneous half-wave plate. The output field emerging from the medium can be written as

$$\mathbf{E_{out}} = J\mathbf{E_{in}} = \exp(2i\theta) \begin{bmatrix} 1 \\ -i \end{bmatrix} = \exp(2iq\phi) \times \exp(2i\theta_0) \begin{bmatrix} 1 \\ -i \end{bmatrix}. \tag{12.38}$$

The emerging wave is thus uniformly right circularly polarized (as we would expect for a half-waveplate), but in addition to that, it acquires an azimuthal phase factor $\exp(2iq\phi)$. This conforms with the azimuthal geometric phase derived in Eq. (12.33). For $q = 1$, it leads to the same spin-to-orbit AM conversion as described in Eq. (12.34).

Similarly, spin-to-orbit AM conversion for an inhomogeneous quarter-wave plate ($\delta = \pi/2$) or for an inhomogeneous linear polarizer with varying orientation of axis $\theta(\phi) = q\phi + \theta_0$ can be obtained using their corresponding Jones matrices as

$$\begin{aligned} \mathbf{E_{out}} &= J_{qwp} \begin{bmatrix} 1 \\ i \end{bmatrix} = e^{(iq\phi)} e^{(i\theta_0)} \begin{bmatrix} \cos(q\phi + \theta_0 - \frac{\pi}{4}) \\ \sin(q\phi + \theta_0 - \frac{\pi}{4}) \end{bmatrix}, \\ \mathbf{E_{out}} &= J \begin{bmatrix} 1 \\ i \end{bmatrix} = e^{(iq\phi)} \times e^{(i\theta_0)} \begin{bmatrix} \cos(q\phi + \theta_0) \\ \sin(q\phi + \theta_0) \end{bmatrix}. \end{aligned} \tag{12.39}$$

In both cases, for input left circular polarization state, the output state becomes inhomogeneously linearly polarized (orientation of the polarization vector depends upon the azimuthal angle ϕ) and acquires azimuthal phase factor

$\exp(iq\phi)$. In the specific case when $(q = 1, \theta_0 = 0)$, the output beam acquires a phase vortex (OAM) with $l = +1$. The output polarization state for the inhomogeneous polarizer becomes radially polarized (the linear polarization vector is directed along the radial direction at all points of space, given by the Jones vector $[\cos\phi, \sin\phi]^T$). The corresponding state for the inhomogeneous quarter-wave plate is neither radial nor azimuthal polarization (Jones vector $[\cos(\phi - \pi/4, \sin(\phi - \frac{\pi}{4}))]^T$ (it becomes azimuthally polarized, with the polarization vector directed perpendicular to the radial direction, given by the Jones vector $[\sin\phi, \cos\phi]^T$ when θ_0 is $-\pi/4$). Nevertheless, none of the inhomogeneous linear polarization states above (either radial or azimuthal) carries any SAM. Accordingly, the spin-to-orbit AM conversion is in agreement with Eq. (12.35).

The inhomogeneous anisotropic media (inhomogeneous retarder and polarizer) described earlier achieves spin-to-orbit AM conversion by generating space (azimuthal) varying Pancharatnam-Berry geometric phase. These are accordingly known as *space variant Pancharatnam-Berry optical elements* (sometimes termed vortex retarders/q-plate and vortex polarizers) [127]. These types of optical elements are therefore used for generating input spin (circular polarization) dependent optical vortices, and for generating radially and azimuthally polarized light beams. The resulting OAM state (the vortex charge l) and the radial/azimuthal polarization states can be controlled by tuning the two parameters (q and θ_0). For practical purposes, these are fabricated either by using patterned liquid crystal cells or by using nanostructured sub-wavelength gratings. Figure 12.8(a) shows illustrative examples of such structured inhomogeneous anisotropic optical elements with varying orientation of the anisotropy axis (determined by the q and θ_0 parameters). Fig. 12.8(b) illustrates of spin-induced vortex generation for an inhomogeneous half-wave plate having a q-parameter value of unity.

12.5 Spin-orbit interaction of light in scattering

The spin-to-orbit AM conversion described in the preceding section deals with SOI due to interaction of paraxial light with inhomogeneous anisotropic media having certain azimuthal symmetries. The SOI effect may also be produced by nonparaxial optical fields in locally isotropic media, e.g., in scattering from micro-/nanosystems or by tight focusing of fundamental or higher-order Gaussian beams.

While SOI in anisotropic paraxial systems is produced by the azimuthal Pancharatnam-Berry phase, which is an extrinsic phenomenon that, produced in nonparaxial fields (scattering and focusing), owes its origin primarily to the intrinsic properties of light, the geometrical transformations of the field and the resulting geometric Berry phase. Nevertheless, as we have previously

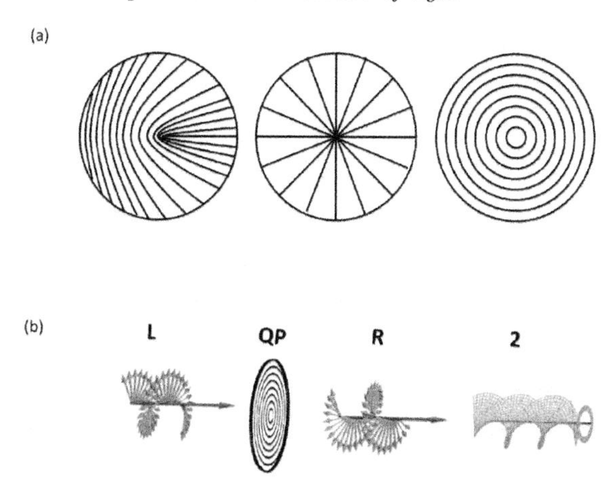

FIGURE 12.8: (a) Examples of inhomogeneous anisotropic optical elements with varying orientation of the anisotropy axis given by $\theta(\phi) = q\phi + \theta_0$). The (q, θ_0) parameters for these are $(q = 1/2, \theta_0 = 0)$, $(q = 1, \theta_0 = 0)$ and $(q = 1, \theta_0 = \pi/2)$, respectively. If we use half-wave plates (retardance $\delta = \pi$) with these inhomogeneity parameters, we can generate OAM with charge $l = \pm 2$. In the case of the second and the third wave plates, there would be perfect spin-to-orbit AM conversion with no net AM transferred to the wave plate. (b) Pictorial illustration of the corresponding spin-to-orbit AM conversion.

noted, SOI in either of these systems (paraxial and nonparaxial) occurs in cylindrically or spherically symmetric systems. Here, we shall address the SOI effect produced by scattering [128, 135].

In order to study the SOI of light mediated by the scattering process, we begin with the simplest case, that of scattering of incident plane wave by a spherical scatterer. Generalization of this approach for other regular-shaped scattering particles (rotationally symmetric particle shapes such as spheroids, rods, etc.) and for scattering of finite light beams (where many constituent plane waves with different wave vectors k need to be considered) is warranted. Consider the Cartesian coordinate system with the incident plane wave propagating in the z direction, the two orthogonal axes x and y representing the polarization axes in the laboratory reference frame (schematics shown in Fig. 12.9). The scattered electric field ($\mathbf{E^s}$) can be related to the incident field ($\mathbf{E^i}$) in the laboratory frame by the transfer function (J) as

$$\mathbf{E^s} \approx T_z(-\phi)T_y(-\theta)S(\theta)T_z(\phi)\mathbf{E^i} = J\mathbf{E^i},$$
$$J = \begin{pmatrix} E_\alpha + E_\beta \cos 2\phi & E_\beta \sin 2\phi & E_\gamma \cos\phi \\ E_\beta \sin 2\phi & E_\alpha - E_\beta \cos 2\phi & E_\gamma \sin\phi \\ -E_\gamma \cos\phi & -E_\gamma \sin\phi & E_\alpha + E_\beta \end{pmatrix}, \tag{12.40}$$

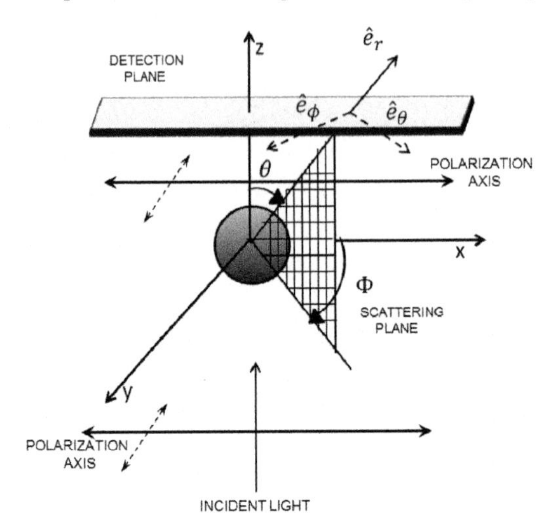

FIGURE 12.9: The scattering geometry showing the laboratory frame (x, y, z) and the scattering frame (r, θ, ϕ).

where θ is the scattering angle and ϕ is the azimuthal angle. The transformation matrices $T_a(\psi)$ represent geometrical rotational transformation about the a-axis by an angle ψ. The transformation matrix $T_z(\phi)$ transforms the laboratory frame field vector of the incident wave to the scattering plane. The inverse transformation matrices $T_z^{-1}(\phi)$ and $T_y^{-1}(\theta)$ transform the field vector of the scattered wave from the scattering coordinate (r, θ, ϕ) to the laboratory coordinate (see Fig. 12.9). The matrix $S(\theta)$ (defined in the scattering plane) includes the effect of the scattering in its elements; $S_2(\theta)$ and $S_1(\theta)$ are scattered field polarized parallel and perpendicular to the scattering plane $(\hat{\theta}, \hat{\phi})$, respectively.

The scattered field descriptors, $E_\alpha(\theta), E_\beta(\theta)$ and $E_\gamma(\theta)$, are related to the elements of $S(\theta)$ as

$$E_\alpha = S_2 \cos\theta + S_1; \; E_\beta = S_2 \cos\theta - S_1; \; E_\gamma = -S_2 \sin\theta. \tag{12.41}$$

Here, the expressions for the amplitude scattering matrix elements $S_2(\theta)$ and $S_1(\theta)$ for a spherical scatterer can be obtained from the Mie theory solution (discussed in Chapter 11) as

$$S_1 = \sum_{n=1}^{\infty} \frac{(2n+1)}{n(n+1)}(a_n \pi_n + b_n \tau_n), \; S_2 = \sum_{n=1}^{\infty} \frac{(2n+1)}{n(n+1)}(a_n \tau_n + b_n \pi_n), \tag{12.42}$$

where a_n and b_n are the coefficients of the normal scattering modes (*TM* (electric) and *TE* (magnetic) modes, respectively) and τ_n, π_n are the corresponding angle (θ) dependent functions (see Chapter 11). For small scatterers (with a dimension much smaller than the wavelength of light, radius

$a << \lambda$), scattering is primarily dominated by the lowest-order TM mode (electric dipolar a_1 mode). The corresponding expressions for $S_2(\theta)$ and $S_1(\theta)$ can be given as $S_2 = \frac{3}{2}a_1 \cos\theta$ and $S_1 = \frac{3}{2}a_1$ with $a_1 = -i\frac{2x^3}{3}\left(\frac{m^2-1}{m^2+1}\right)$. Here, x is known as the size parameter of scatterer ($x = \frac{2\pi}{\lambda}an_{medium}$) and m is the ratio of the refractive index of the scatterer to that of the surrounding medium ($m = \frac{n_s}{n_{medium}}$).

The phenomenon of SOI in scattering can immediately be understood by applying the Jones vector $\begin{bmatrix} 1 & i & 0 \end{bmatrix}^T$ of incident left circularly polarized (LCP) light on the system Jones matrix J. The resulting field can be decomposed into three uniform polarization components as

$$E_\alpha \begin{bmatrix} 1 \\ i \\ 0 \end{bmatrix} + E_\beta \exp(i2\phi) \begin{bmatrix} 1 \\ -i \\ 0 \end{bmatrix} - E_\gamma \begin{bmatrix} 0 \\ 0 \\ 1 \end{bmatrix}. \tag{12.43}$$

As is obvious, the second (transverse) component of Eq. (12.43) represents flipping of helicity (reversal of spin $\sigma = +1 \rightarrow -1$) and subsequent generation of the phase vortex-carrying orbital angular momentum $l = +2$. Similarly, the third (longitudinal) component represents conversion of the circular to the linear polarization state ($\sigma = +1 \rightarrow 0$) and is accordingly associated with the generation of orbital angular momentum $l = +1$. It appears from Eq. (12.43) that the amplitude and relative phases of the scattered field descriptors $E_\alpha, E_\beta, E_\gamma$ (and consequently the scattering matrix elements $S_2(\theta)$ and $S_1(\theta)$) play important roles in the resulting spin-orbit interactions.

Since in a typical experiment (e.g., Fig. 12.9), we detect the scattered far field picked up by the analyzing optics and detectors usually in the xy plane, it suffices to consider only the transverse field components henceforth. The corresponding transfer function (see the first two rows and columns of Eq. (12.40)) is the conventional 2×2 Jones matrix. In order to interpret the SOI via the conventional polarization parameters (namely, the *diattenuation* and *retardance* parameters defined in Chapter 6), we now derive the Mueller matrix corresponding to this Jones matrix. Note that diattenuation and retardance are conventionally defined as differential attenuation of orthogonal polarizations and the phase shift between orthogonal polarization states, respectively. These two polarization effects thus deal with the amplitude and the phase parts of the scattered field, respectively. The Mueller matrix corresponding to the 2×2 Jones matrix (see the first two rows and columns of Eq. (12.40)) can be derived using the standard relationship connecting the two (see Eq. 6.34 of Chapter 6). The resulting matrix is a diattenuating retarder Mueller matrix characterized by diattenuation $D(\theta)$, retardance $\delta(\theta)$ and orientation angle of the axes of the diattenuating retarder ϕ (azimuthal angle), the elements of which are M_{ij} (i-row, j-column):

$$\begin{aligned}
&M_{11} = 1; M_{12} = M_{21} = D\cos 2\phi; M_{13} = M_{31} = D\sin 2\phi;\\
&M_{14} = M_{41} = 0; M_{22} = \cos^2 2\phi + x\cos\delta\sin^2 2\phi;\\
&M_{23} = M_{32} = \sin 2\phi\cos 2\phi - x\cos\delta\sin 2\phi\cos 2\phi;\\
&M_{24} = -M_{42} = -x\sin\delta\sin 2\phi; M_{33} = \sin^2 2\phi + x\cos\delta\cos^2 2\phi;\\
&M_{34} = -M_{43} = x\sin\delta\cos 2\phi; M_{44} = x\cos\delta;\\
&\quad x = |\sqrt{1-D^2}|.
\end{aligned} \tag{12.44}$$

Here, D and δ are related to the scattering matrix elements as

$$D(\theta) = \left\{\frac{|S_2(\theta)|^2\cos^2\theta - |S_1(\theta)|^2}{|S_2(\theta)|^2\cos^2\theta + |S_1(\theta)|^2}\right\}, \quad \delta(\theta) = \tan^{-1}\left[\frac{Im(S_2^*(\theta)S_1(\theta))}{Re(S_2^*(\theta)S_1(\theta))}\right]. \tag{12.45}$$

Eq. (12.44) is a Mueller matrix of an azimuthal diattenuating retarder, and thus in principle it can exhibit all the different types of spin-to-orbit AM conversion as exhibited by the vortex retarder (inhomogeneous quarter- and half-wave plate) and/or the vortex polarizer (diattenuator) discussed in the previous section (inhomogeneous anisotropic medium). The nature of SOI in scattering (whether it would act like a vortex retarder or a vortex polarizer or a mixture of both) is crucially determined by the scattering matrix elements $S_2(\theta)$ and $S_1(\theta)$ because these parameters determine the magnitude of the scattering-induced diattenuation D and retardance δ parameters. Since $S_2(\theta)$ and $S_1(\theta)$) of a scattering particle can be tuned by selecting a suitable wavelength, size and shape, various interesting regimes of SOI can be realized by clever manipulation of the diattenuation and retardance polarimetry parameters of scattering. In the following, we provide examples of three pure cases of scattering-mediated SOI effects depending upon the different natures of SAM-to-OAM conversion and the resulting evolution of the geometric phases [135].

Case 1: $D = 0, \delta = \pi, \leftrightarrow E_\alpha = 0$ ***or*** $S_2\cos\theta = -S_1$; ***Geometric phase vortex formation by SAM flipping***

The Mueller matrix for this case represents generation of geometric phase vortex with topological charge $l = \pm 2$ (similar to the case of the vortex half-wave plate described in the previous section). For any incident circular polarization state (e.g., LCP), the resulting SOI is $(\sigma = +1, l = 0) \Rightarrow (\sigma = -1, l = 2)$. Subsequent flipping of SAM is evident from the matrix element $M_{44} = -1$. For incident horizontal linear polarization (Stokes vector $S_i = \begin{bmatrix} 1 & 1 & 0 & 0 \end{bmatrix}^T$), on the other hand, the output state becomes $S_0 = MS_i = \begin{bmatrix} 1 & \cos 4\phi & \sin 4\phi & 0 \end{bmatrix}^T$, which does not contain any spin per se but implies rotation of the initial linear state acquiring phase vortices with opposite topological charges ($l = \pm 2$).

Note, however, the condition ($S_2 \cos\theta = -S_1$) cannot be fulfilled for single scattering from dielectric Rayleigh scatterers (radius $a << \lambda$) either for forward or backscattering angles. This can be recognized by noting that for such scatterers (where scattering is contributed solely by the lowest-order TM electric dipolar a_1 mode), the amplitude-scattering matrix elements are ($S_2 \sim a_1 \cos\theta$ and $S_1 \sim a_1$). Although in the single backscattering ($\theta = \pi$) from Rayleigh scatterers, the overall helicity flips, there is no reversal of the z component of the spin angular momentum (helicity flipping is associated with reversal in the direction, $+z \rightarrow -z$), and thus it is not associated with acquisition of the phase vortex (or generation of OAM). In contrast, this can be observed in backscattering from a random medium. In this case, the incident polarized light suffers a series of forward scattering events to eventually emerge through the backward direction of the random medium. The helicity is preserved throughout the multiple scattering trajectory (adiabatic evolution of polarization through helicity preserving forward scattering paths); thus while emerging through the backscattering direction ($+z \rightarrow -z$), there is a complete reversal of the z component of the spin angular momentum ($\sigma = +1 \rightarrow -1$, or vice versa). The resulting scattered fields accordingly acquire phase vortices ($l = +2$).

Case 2: $D = \pm 1, \delta = 0, \leftrightarrow E_\alpha = \pm E_\beta$ ***or*** $S_2 = 0/S_1 = 0$; ***SAM to OAM conversion by pure diattenuation effect of scattering***

The corresponding Mueller matrix is a pure azimuthal diattenuator matrix signifying complete conversion of SAM to OAM ($\sigma = \pm 1 \rightarrow 0, l = \pm 1$). As evident, for the input RCP state ($S_i = \begin{bmatrix} 1 & 0 & 0 & 1 \end{bmatrix}^T$), the output is an azimuthal linear polarization state: $S_o = \begin{bmatrix} 1 & \cos 2\phi & \sin 2\phi & 0 \end{bmatrix}^T$, carrying no SAM. The angular momentum of the scattered light is thus entirely OAM.

This effect can be observed even for Rayleigh scattering from dielectric particles. The required condition is fulfilled for a Rayleigh scatterer at a scattering angle $\theta = 90°$, where the amplitude scattering matrix element S_2 vanishes. For incident circularly polarized light, the scattered light at $\theta = 90°$ becomes completely linearly polarized, which carries no SAM. Accordingly, at this angle, the angular momentum is entirely carried by the OAM. In the case of larger-sized Mie scatterers ($a \geq \lambda$), on the other hand, such a complete conversion of SAM to OAM may occur at several narrow ranges of scattering angle θ, depending upon the size parameter of the scatterer.

Case 3: $D = 0, \delta = \pi/2, \leftrightarrow E_\alpha = iE_\beta$ ***or*** $S_2 \cos\theta = -iS_1$; ***SAM to OAM conversion by pure retardance effect of scattering***

The resulting Mueller matrix assumes the form of a pure retarder matrix (similar to the azimuthal quarter-wave plate discussed in the previous section). For the input horizontal linear polarization state, the output state is $S_o = \begin{bmatrix} 1 & 0.5 \times (1 + \cos 4\phi & 0.5 \times \sin 4\phi & \sin 2\phi \end{bmatrix}^T$, implying generation of the azimuthal angle $\phi-$separated lobes of opposite circular polarization states

($\sigma = \pm 1$). For the input RCP state, on the other hand, the SAM-to-OAM conversion is similar to the previous effect: $S_o = \begin{bmatrix} 1 & -\sin 2\phi & \cos 2\phi & 0 \end{bmatrix}^T$, implying complete SAM-to-OAM conversion and subsequent generation of phase vortex ($\sigma = -1 \rightarrow \sigma = 0, l = -1$).

The condition for this type of SOI can be fulfilled for scatterers where the scattering process is contributed by more than one scattering mode. For example, this arises for larger-sized dielectric scatterers ($a \geq \lambda$), where in addition to the lowest-order *TM* mode (the a_1 electric dipole mode, which is the dominant mode for Rayleigh scatterers with $a << \lambda$), the other higher-order modes (e.g., the a_2 electric quadrupolar *TM* mode, b_1 magnetic dipole *TE* mode and so forth) contribute to the scattering process. Finally, It has been demonstrated that all the three pure cases of SOI can be realized exploiting the resonance effect in scattering, for example, by clever manipulation of the localized plasmon resonance modes in metal nanoparticles/nanostructures. A detailed discussion of this aspect is beyond the scope of this book.

Before we conclude this section, the nature of the geometric phase produced by the scattering process is worth a brief mention. Note that the geometric phases associated with the SOI by diattenuation and retardance effects of scattering can neither be treated as the *pure* spin redirection Berry phase nor the *pure* Pancharatnam-Berry phase. This follows because, in this case, both the polarization state and the propagation direction of the wave changes (due to scattering-induced diattenuation and retardance effects and subsequent evolution in different scattering trajectories). Nevertheless, the resulting geometric phase for such space-varying polarization can be determined from the general definition of geometric phase. It should also be noted that although the three pure cases of SOI have been discussed, in the general case ($D \neq 0, \delta \neq 0$), all three effects may manifest, the strength of each one being determined by the magnitudes of D and δ. As evident from the general scattering Mueller matrix of Eq. (12.44), the Case 1 type SOI will take place for $\delta > \pi/2$ (the negative value of M_{44}), and the other two effects may take place for any values of D and δ.

12.6 Spin-orbit interaction of light in tight focusing of Gaussian beam

Spin-orbit interaction of light by tight focusing of fundamental or higher-order Gaussian beams may also be described in a way similar to that of scattering, where the geometric transformations of the field vector (as a result of focusing) is responsible for the observed effects. The SOI effects in tight focusing can be described using the so-called Debye-Wolf theory of focusing (also referred to as the angular spectrum method) with a spherical lens. The

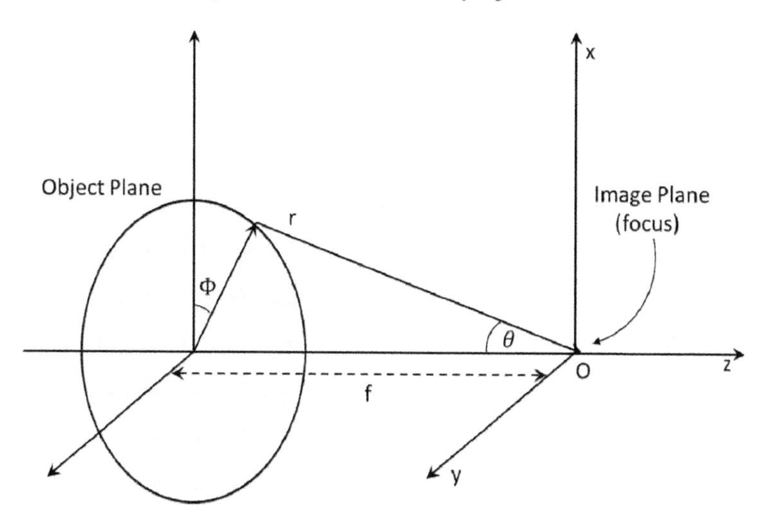

FIGURE 12.10: The geometry of tight focusing showing the Cartesian coordinates (x, y, z) at a point P in the object space. The corresponding spherical polar coordinates (r, θ, ϕ) are also shown.

geometry of tight focusing is shown in Fig. 12.10 [128, 136]. With the incident field at the entrance pupil of the figure, $\mathbf{E_{in}}(r)$ is assumed to be paraxial and propagating along the z direction, so that we can neglect the z component of the incident field (the input polarization is conveniently described by two component Jones vectors corresponding to the transverse fields).

It is convenient to introduce spherical polar coordinates $(r, \theta, \phi)(r > 0, 0 \leq \theta \leq \pi, 0 \leq \phi \leq 2\pi)$, with the polar axis $\theta = 0$ along the z direction, and with azimuth $\phi = 0$ containing the electric field vector in the object space. After refraction, the partial rays converge at the focal point (P) and have the nonparaxial k vectors. The lens performs a sort of Fourier transform, translating the initial real-space distribution $\mathbf{E_{in}}(r)$ into the momentum distribution $\mathbf{E_p}(k)$ in the image space. Thus the expressions for the real-space field distribution at any point $P(x, y, z)$ in the image region can be written in the angular spectrum form (see Chapter 14) as

$$E_p(x, y, z) \approx \int\int E_p(k_x, k_y) e^{i(k_x x + k_y y + k_z z)} dk_x dk_y. \tag{12.46}$$

Here, the angular spectrum integral is based on the fact that the real-space electric field near the focal point is determined by the interference of the constituent partial plane waves having different k vectors. Here, the components k_x, k_y, k_z along a ray in the image space and the coordinates (x, y, z) of the point P (shown in Figure 12.10) in the image region can be expressed as

$$\begin{aligned} k_x &= k\sin\theta\cos\phi, \quad k_y = k\sin\theta\sin\phi, \quad k_z = k\cos\theta, \\ x &= r_p\sin\theta_p\cos\phi_p, \quad y = r_p\sin\theta_p\sin\phi_p, \quad z = r_p\cos\theta_p, \end{aligned} \tag{12.47}$$

so that the term in the exponent of the integral (12.46) can be written as

$$\begin{aligned} k_x x + k_y y + k_z z &= k r_p \cos\nu, \\ \cos\nu &= \cos\theta\cos\theta_p + \sin\theta\sin\theta_p\cos(\phi - \phi_p). \end{aligned} \tag{12.48}$$

Here, the set of spherical polar coordinates (r_p, θ_p, ϕ_p) is defined with respect to the origin at the focal point. Equipped with this set of equations relating the real-space field $\mathbf{E_p}(x, y, z)$ and the momentum space field $\mathbf{E_p}(k)$ at the focal region (point P), we now consider the evolution of the polarization associated with each partial wave (characterized by a constituent k due to the focusing transformation). In the case of pure focusing (with no other effects, e.g., refraction/reflection due to the presence of any interface in between), the partial waves do not change their polarization state in the local basis attached to the ray, and the electric fields experience pure meridional rotations by the refraction angle θ together with the k vector. As previously discussed, such an evolution of the polarization state is referred to as an adiabatic evolution. Accordingly, the geometric phase associated with this polarization evolution can be treated as a pure spin redirection Berry phase. The resulting focused field spectrum for a partial wave $\mathbf{E_p}(\theta, \phi)$ can be related to the input field $(\mathbf{E_{in}}(\theta, \phi))$ (when represented in terms of the angles θ, ϕ) using the purely geometrical rotational transformation

$$\begin{aligned} E_p =& A E_{in}, \\ A =& \begin{bmatrix} \cos\phi & -\sin\phi & 0 \\ \sin\phi & \cos\phi & 0 \\ 0 & 0 & 1 \end{bmatrix} \begin{bmatrix} \cos\theta & 0 & -\sin\theta \\ 0 & 1 & 0 \\ \sin\theta & 0 & \cos\theta \end{bmatrix} \begin{bmatrix} \cos\phi & \sin\phi & 0 \\ -sin\phi & \cos\phi & 0 \\ 0 & 0 & 1 \end{bmatrix} \\ =& \begin{bmatrix} a - b\cos 2\phi & -b\sin 2\phi & c\cos\phi \\ -b\sin 2\phi & a + b\cos 2\phi & c\sin\phi \\ -c\cos\phi & -c\sin\phi & a - b \end{bmatrix}. \end{aligned} \tag{12.49}$$

Here, the coefficients a, b and c for the case of pure focusing are given by

$$a = \frac{1}{2}(1 + \cos\theta), \quad b = \frac{1}{2}(1 - \cos\theta), \quad c = \sin\theta.$$

The polarization distribution of the real-space electric field near the focal point can then be related to the input field by considering interference of all the constituent partial plane waves (using Eq. (12.46) as

$$E_p(x, y, z) \approx \int_0^{\theta_m} \int_0^{2\pi} E_p(\theta, \phi)\, e^{ikr_p\cos\nu} d\theta d\phi, \tag{12.50}$$

where the limit of the θ integral (θ_m) is set by the numerical aperture of the focusing lens. This integral can be evaluated using Eq. (12.48) and Eq. (12.50) and by performing a few simple algebraic steps. The resulting Cartesian components (x, y, z-polarizations) field at the focal point P (with corresponding representation in spherical polar coordinates r_p, θ_p, ϕ_p) can be related to the input polarization of the field as

$$\begin{bmatrix} E_x \\ E_y \\ E_z \end{bmatrix}_p = \begin{pmatrix} I_0 + I_2\cos 2\phi_p & I_2 \sin 2\phi_p & 2iI_1\cos\phi_p \\ I_2\sin 2\phi_p & I_0 - I_2\cos 2\phi_p & 2iI_1\sin\phi_p \\ -2iI_1\cos\phi_p & -2iI_1\sin\phi_p & I_0 + I_2 \end{pmatrix} \begin{bmatrix} E_x \\ E_y \\ E_z \end{bmatrix}_{in}, \tag{12.51}$$

where the coefficients I_0, I_2 and I_1 are the diffraction integrals originating from the integration of Eq. (12.50) and are given as

$$\begin{aligned} I_0 &\approx \int_\theta^{\theta_m} E_{in}\sqrt{\cos\theta}(1+\cos\theta)J_0(kr_p\sin\theta\sin\theta_p)e^{ikr_p\cos\theta\cos\theta_p}\sin\theta\, d\theta, \\ I_1 &\approx \int_\theta^{\theta m} E_{in}\sqrt{\cos\theta}J_1(kr_p\sin\theta\sin\theta_p)e^{ikr_p\cos\theta\cos\theta_p}\sin^2\theta\, d\theta, \\ I_2 &\approx \int_\theta^{\theta m} E_{in}\sqrt{\cos\theta}(1-\cos\theta)J_2(kr_p\sin\theta\sin\theta_p)e^{ikr_p\cos\theta\cos\theta_p}\sin\theta\, d\theta. \end{aligned} \tag{12.52}$$

Here, J_n is the Bessel function of the first kind and order n.

Apparently, the polarization transformation matrix of Eq. (12.51) resembles the corresponding transformation matrix for scattering (Eq. (12.40). Thus the focusing process would also mediate similar types of spin-to-orbit AM conversion. For example, for input left circularly polarized light, the resulting field at the focal region may also be decomposed into three components: the first component has the same helicity as that of the incident circular polarization, the second component has opposite helicity and is associated with an orbital angular momentum component of $l = +2$ and the third component (the z-polarization component) is linearly polarized, carrying an orbital angular momentum of $l = 1$. The associated coefficients I_2 and I_1 of the transverse (second term) and the longitudinal (third term) field components thus determine the strength of the spin-orbit angular momentum conversion. We once again note here that the spin-to-orbit AM conversion described above for the case of pure focusing arises due to adiabatic evolution of polarization during focusing and generation of the spin redirection geometric Berry phase. The presence of any interface near the focal point can also be dealt with in a similar way, but we then have to incorporate the local polarization changes in the local basis attached to the ray (due to refraction/reflection). The corresponding effect may be incorporated by introducing the Fresnel transmission (reflection) matrix within the polarization transformation given in Eq. (12.49):

$$\begin{aligned} E_p(\theta,\phi) &= AE_{in}(\theta,\phi), \\ A^{\frac{t}{r}} &= T_z(-\phi)T_y(-\theta)FT_z(\phi). \end{aligned} \tag{12.53}$$

As previously defined, $T_a(\psi)$ represents geometrical rotational transformation about the a-axis by an angle ψ. The matrix F is the Fresnel polarization transformation matrix, which describes the amplitude transmission (reflection) coefficients for p- (parallel to the plane of incidence) and s-(perpendicular to the plane of incidence) polarization states, respectively, T_p (R_p) and T_s (R_s). For example, in the case of transmission,

$$F = \begin{pmatrix} T_p & 0 & 0 \\ 0 & T_s & 0 \\ 0 & 0 & T_p \end{pmatrix}. \tag{12.54}$$

The final polarization transfer function (Eq. (12.51)) will of course remain similar but the coefficients I_0, I_1 and I_2 will now be modified to include the transmission (reflection) coefficients:

$$\begin{aligned} I_0 &\approx \int_\theta^{\theta_m} E_{in}\sqrt{\cos\theta}(T_s + T_p\cos\theta)J_0(kr_p\sin\theta\sin\theta_p)e^{ikr_p\cos\theta\cos\theta_p}\sin\theta\, d\theta, \\ I_1 &\approx \int_\theta^{\theta_m} E_{in}\sqrt{\cos\theta}J_1(kr_p\sin\theta\sin\theta_p)e^{ikr_p\cos\theta\cos\theta_p}T_p\sin^2\theta\, d\theta, \\ I_2 &\approx \int_\theta^{\theta_m} E_{in}\sqrt{\cos\theta}(T_s - T_p\cos\theta)J_2(kr_p\sin\theta\sin\theta_p)e^{ikr_p\cos\theta\cos\theta_p}\sin\theta\, d\theta. \end{aligned} \tag{12.55}$$

A comparison of Eq. (12.40) for SOI in scattering indicates that in the case of focusing the role of the scattering matrix elements $S_2(\theta)$ and $S_1(\theta)$ (with the scattered field polarized parallel and perpendicular to the scattering plane) is played here by the Fresnel transmission (reflection) coefficients T_p (R_p) and T_s (R_s), respectively. In other words, the role scattering plane is played by the plane of incidence (for planar interface) in the case of focusing through reflecting/refracting planar interfaces. We note here that in the case of a multilayered (stratified) medium, the Fresnel transmission (reflection) coefficients are generally complex (as in the case of scattering where $S_2(\theta)$ and $S_1(\theta)$ may be complex, depending upon the nature and the number of contributing scattering modes). Thus with suitable choice of the multilayered or stratified medium, we may in principle realize all the various interesting regimes of SOI in the case of focusing of light through such a medium.

To summarize, in this chapter we have introduced the concept of angular momentum of light and defined both its variants, namely, the spin and orbital angular momentum via classical wave optics treatment. It has been shown that under certain circumstances, the spin and orbital degrees of freedom of a light beam may get coupled. The resulting spin-orbit interaction of light is manifested as (i) the interconversion between SAM and the intrinsic OAM of a light beam, which occurs due to the evolution of azimuthal geometric phase, which this eventually leads to the generation of spin-induced optical vortices (characterized by helical phase fronts), and (ii) the reverse effect of polarization (spin) on the trajectory of a light beam, which deals with the interaction between SAM and the extrinsic OAM of light. This effect is manifested as a spin-dependent shift of the trajectory of the light beam [137, 138]. It has been shown that the former effect arises usually in cylindrically or spherically symmetric systems, such as in scattering, in tight focusing and for propagation through inhomogeneous (having azimuthal symmetry) anisotropic media. In contrast, the latter effect is associated with the breaking of the symmetry of the system and the generation of extrinsic OAM. Each of the aforementioned effects have been dealt with using the appropriate mathematical framework and illustrated with selected examples. The various kinds of geometric phases and their roles in the SOI process have also been discussed. Finally, as discussed in this chapter, classical light captures all of the basic features of interaction and coupling between spin and orbital degrees of freedom of relativistic spinning particles in external fields. The relative purity and simplicity of optical systems (as compared to condensed matter and high-energy physics) enables us to observe the fundamental effects associated with SOI with relative ease and extrapolate results to a range of physical systems where such observations are impossible. In addition to the fundamental interests, the SOI of light are also finding potential nano-optical applications, in developing novel sensors, nanoprobes and so forth, exploiting the tiny optical effects associated with SOI.

Chapter 13

Optical tweezers

13.1 Basic theory

Light possesses momentum. When treated as a wave, the momentum is often manifested as a radiation pressure; that was first asserted by Maxwell in 1862. In the particle treatment that is introduced by the de Broglie relation, we calculate the momentum of a single photon to be $\hbar k$. In this chapter, we will study how light can be used to exert forces on microparticles so as to provide a means of confining and manipulating them. These effects can be analyzed by considering the ray or wave picture of light, and the quantum picture need not be invoked.

13.2 Force and torque on a dipole

The first thing to understand is how light can interact with matter. Light is essentially an electric field (of course the magnetic field is also there, but it is much weaker in magnitude), and from the multipole expansion of charges, we know that it is the dipole component that the electric field interacts with to exchange energy. Thus, the initial step is to understand the force that an electric field can exert on a dipole.

13.2.1 Force and torque on a dipole in a uniform electric field

Figure 13.1 shows a dipole having dipole moment $\mathbf{p} = q\mathbf{L}$, where $\mathbf{L}$ is the separation between charges $+q$ an $-q$ constituting the dipole. If the dipole is placed in a uniform electric field E, the net force on the dipole about the center point O is $\mathbf{F}_{+} + F_{-} = q\mathbf{E} - q\mathbf{E} = 0$. However, since the direction of the forces F_{+} and F_{-} is opposite, it is clear that there would be a nonzero moment of the forces about O. This moment, or torque is given by

$$
\begin{aligned}
\mathbf{T} &= \frac{\mathbf{L}}{2} \times \mathbf{F}_{+} + (\frac{-\mathbf{L}}{2}) \times \mathbf{F}_{-}, \\
&= \frac{\mathbf{L}}{2} \times q\mathbf{E} + (\frac{-\mathbf{L}}{2}) \times (-q)\mathbf{E}, \\
&= q\mathbf{L} \times \mathbf{E}, \\
&= \mathbf{p} \times \mathbf{E}.
\end{aligned} \tag{13.1}
$$

The torque T results in the dipole being aligned in the direction of the electric field E.

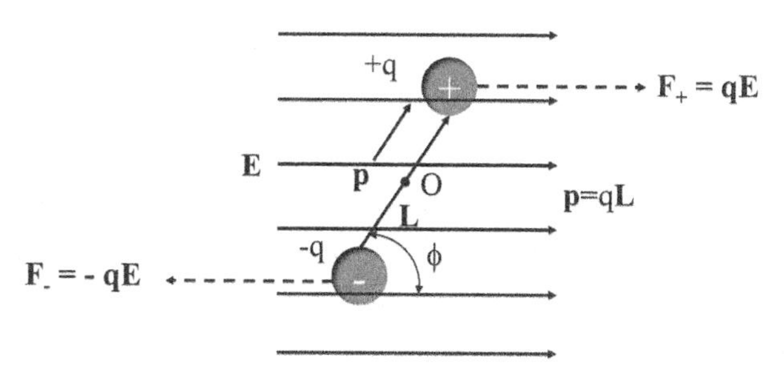

FIGURE 13.1: Forces on a dipole in a uniform electric field. The net force is zero, but there exists a torque around the center point O.

13.2.2 Force and torque on a dipole in a nonuniform electric field

In this case, where the electric field is dependent on r and is not the same everywhere in space, the situation is very different. The net force on the dipole, where the two charges are separated by distance $\mathbf{L}$, is then

$$\begin{aligned}\mathbf{F} &= \mathbf{F}_+ + \mathbf{F}_-,\\ &= q\mathbf{E}(r+L) - q\mathbf{E}(r),\end{aligned} \tag{13.2}$$

so that

$$\begin{aligned}F_x =& q\left[E(x+L_x, y+L_y, z+L_z) - E_x(x,y,z)\right],\\ =& q\left[E_x + L_x\frac{\partial E_x}{\partial x} + L_y\frac{\partial E_x}{\partial y} + L_z\frac{\partial E_x}{\partial z} + ... - E_x\right],\\ =& (\mathbf{p}.\nabla)E_x.\end{aligned} \tag{13.3}$$

Adding all components, we finally have

$$\mathbf{F} = (\mathbf{p}.\nabla)\mathbf{E}. \tag{13.4}$$

In order to calculate the torque, we use again

$$\begin{aligned}\mathbf{T} &= \frac{\mathbf{L}}{2} \times \mathbf{F}_+ + (\frac{-\mathbf{L}}{2}) \times \mathbf{F}_-,\\ &= q\frac{\mathbf{L}}{2} \times \left[\mathbf{E}(r+L) + \mathbf{E}(r)\right],\\ &= q\frac{\mathbf{L}}{2} \times \left[\mathbf{E}(r+L) + \mathbf{E}(r) + \mathbf{E}(r) - \mathbf{E}(r)\right],\\ &= q\frac{\mathbf{L}}{2} \times \left[\mathbf{E}(r+L) - \mathbf{E}(r) + 2\mathbf{E}(r)\right],\\ &= \frac{\mathbf{L}}{2} \times (\mathbf{p}.\nabla)\mathbf{E}(r) + \mathbf{p} \times \mathbf{E}(r),\end{aligned} \tag{13.5}$$

where we have used Eq.(13.4) in the last step. Thus, generally, the torque

$$\mathbf{T} = \frac{\mathbf{r}}{2} \times (\mathbf{p}.\nabla)\mathbf{E} + \mathbf{p} \times \mathbf{E}. \tag{13.6}$$

13.2.3 Potential energy of a dipole placed in an electric field

For electrical charges, the potential energy may be understood as the work required to move the charges from infinity to a particular position. For a dipole, this translates to no work when the two equal and opposite charges are placed at the center of the dipole (point O in Fig. 13.1), or when they are moved normal to the field, since the force is perpendicular to the direction of the field. Work is done, however, when the dipole is rotated with respect to the field. Thus, if the dipole oriented at an angle ϕ with respect to the field $\mathbf{E}$ is rotated by an angle $d\phi$, the work done is obviously $dW = T\ d\phi$, where T is the torque induced on the dipole due to the field. From Eq. (13.1), $T = pE\ \sin\phi$, so that the potential energy changed as the dipole is rotated from an initial angle ϕ_i to a final angle ϕ_f is given by

$$\begin{aligned} U_f - U_i &= \int_{\phi_i}^{\phi_f} T\ d\phi, \\ &= \int_{\phi_i}^{\phi_f} pE\ \sin\phi\ d\phi, \\ &= pE\ (\cos\phi_f - \cos\phi_i). \end{aligned} \tag{13.7}$$

We choose the initial angle ϕ_i to be 90° and the initial energy U_i to be zero. This implies that Eq. (13.7) can be written as

$$\begin{aligned} U_f = U &= -pE\ \cos\phi, \\ \text{or}\ \ U &= -\mathbf{p}.E. \end{aligned} \tag{13.8}$$

13.3 Exerting controlled forces on particles using light: Optical trapping

As we saw earlier, light can exert forces and torques on dipoles. For macroscopic matter this is manifested in the form of radiation pressure, which for the case of the sun is around 1361 W/m^2 on the earth. The effects of the force emanating from this radiation pressure are nontrivial, and must be taken into account while setting the trajectory of spacecrafts. However, quite understandably, the momentum of light would be of maximum consequence to small particles. A simple calculation shows that if light having 1 watt of power is incident on a particle of radius around a wavelength of light (say 600 nm), the particle would feel a recoil force of around 1,000 dynes, assuming that it is a perfect mirror and reflects all the light incident upon it. More important, the acceleration of the particle is extremely high—close to 10^5 g, where g is the acceleration due to gravity. In 1969, Arthur Ashkin realized that this force could actually be used directly to manipulate particles and even confine them using focused laser beams. This led to the first series of experiments in optical micromanipulation, where Ashkin demonstrated that even with a mildly focused laser beam, transparent latex microspheres dispersed in water could be made to assemble near the beam axis where the intensity was highest [139]. The microparticles could also be translated by moving the laser beam. In 1986, Askin and co-workers once again demonstrated that making a tighter focus could actually trap single microparticles, and a new tool in studying light matter interactions was born [140]. Ashkin proceeded to name this tool 'optical tweezers'—where the focused laser beam (Gaussian) performed the optical equivalent of mechanical tweezers to achieve similar manipulation of micro-objects.

13.3.1 Gradient and scattering forces

13.3.1.1 Ray optics picture

Fig. 13.2 describes the forces exerted by a focused Gaussian beam on a single dielectric spherical object whose diameter is much greater than the wavelength of light (λ) so that a ray optics treatment of the process is permissible. The particle is shown to be slightly off-axis in the transverse direction with respect to the beam. We consider a pair of rays symmetrically distributed about the center of the particle. Since a Gaussian beam (the intensity distribution of which we have dealt with in detail in the previous chapters) has the highest intensity at the center with a rapidly falling intensity radially outward (remember that $I(r) = I_0 \ \exp(-\frac{r^2}{2w^2})$ where I_0 and w are the peak intensity and beam waist radius, respectively), a ray emanating from near the center of the beam has a higher intensity associated with it (ray 1, in bold), while that from the low-intensity fringe of the beam is much weaker in intensity (ray 2). The rays get refracted by the dielectric sphere in a direction toward the center of the sphere. This is because we assume that the sphere is placed in a medium that has a lower refractive index compared to it. The sphere thus changes the momentum of light so that by Newton's third law, the light exerts an equal and opposite momentum on the sphere. If we consider all pairs of such symmetric rays, we see that the net force can be resolved in terms of a scattering force $\mathrm{F}_{\mathrm{scatt}}$ in the longitudinal direction (in the direction of the light beam), and a gradient force $\mathrm{F}_{\mathrm{grad}}$ in the transverse direction toward the region

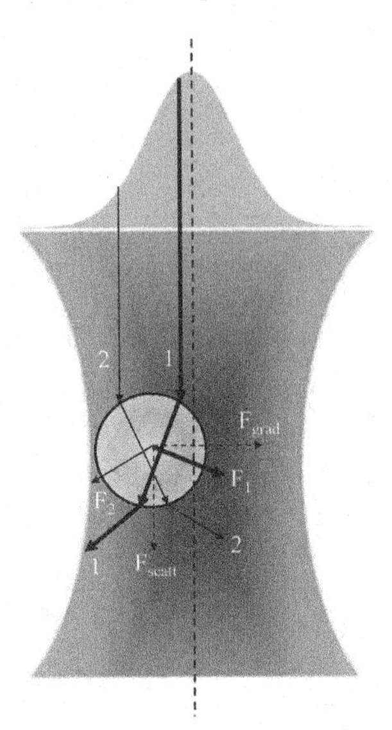

FIGURE 13.2: Ray optics analysis of the forces exerted by a focused Gaussian beam on a transparent particle of diameter >> λ, where λ is the wavelength of light. The net gradient force F_{grad} is directed toward the center of the beam, while the scattering force F_{scatt} is in the direction of the beam. The refractive index of the sphere is higher than that of the medium.

of high intensity of the beam. For particles on the beam axis, $F_1 = F_2$ and the particle feels no net force. Now, the tighter the beam focus, the higher is the gradient force—and the particle feels a strong restoring force toward the center of the beam. The scattering force is dependent on the difference of refractive index of the microparticle from its environment, and for most dielectric objects, it is weaker than the gradient force. The interplay of the two force components, however, results in the equilibrium position of the particle to be shifted slightly from the center of the trap in the direction of the beam. A ray optics–based calculation elaborated in Ref. [141] gives analytical expressions for the longitudinal component F_{scatt} and the transverse F_{grad} as

$$F_{scatt} = \frac{n_m P}{c}\left(1 + R\cos(2\alpha) - \frac{T^2[\cos(2\alpha - 2\beta) + R\cos(2\alpha)]}{1 + R^2 + 2R\cos(2\beta)}\right), \tag{13.9}$$

and

$$F_{grad} = \frac{n_m P}{c}(R\sin(2\alpha) - \frac{T^2[\sin(2\alpha - 2\beta) + R\sin(2\alpha)]}{1 + R^2 + 2R\cos(2\beta)}), \tag{13.10}$$

where R is the Fresnel reflection and T the Fresnel transmission coefficient, α the angle of incidence and n_m the refractive index of the surrounding medium, while β is the angle of refraction.

13.3.1.2 Rayleigh regime: Dipole picture

For particles that are much smaller than the wavelength of light, the theory is very different, and we are not allowed to invoke the ray optics picture. However, in this regime, we can consider the particle to be a point dipole and then use the results of Section 13.2 to calculate the forces applied on it due to the electric and magnetic fields of light. Thus, let us consider the force $\mathbf{F}$ on a point charge q placed at x_1 and moving with a velocity v_1 in an electric field E and magnetic field B, which is nothing other than the well-known Lorentz force written as

$$\mathbf{F} = q\left(\mathbf{E}_1 + \frac{d\mathbf{x}_1}{dt} \times \mathbf{B}\right). \tag{13.11}$$

For a dipole, this translates to

$$\mathbf{F} = (\mathbf{p} \cdot \nabla)\mathbf{E} + \frac{d\mathbf{p}}{dt} \times \mathbf{B}. \tag{13.12}$$

This equation can be generalized for a particle having macroscopic polarization P so that we have

$$\mathbf{F} = (\mathbf{P} \cdot \nabla)\mathbf{E} + \frac{d\mathbf{P}}{dt} \times \mathbf{B}. \tag{13.13}$$

Here we make an important assumption that the particle has no permanent electric dipole moment but one induced by light due to its polarizability $\alpha(\omega)$ (which is of course a function of the light frequency ω and is complex). The polarization induced can then be related to the electric field as $\mathbf{P} = \alpha(\omega)\mathbf{E}$. Using this in Eq. (13.13), we get

$$\mathbf{F} = \alpha\left((\mathbf{E} \cdot \nabla)\mathbf{E} + \frac{d\mathbf{E}}{dt} \times \mathbf{B}\right). \tag{13.14}$$

We now use a well-known vector identity

$$(\mathbf{E} \cdot \nabla)\mathbf{E} = \nabla((E)^2/2) - \mathbf{E} \times \nabla \times \mathbf{E}, \tag{13.15}$$

so that, using the Maxwell's relation,

$$\nabla \times \mathbf{E} = -\frac{d\mathbf{B}}{dt}, \tag{13.16}$$

we obtain

$$\mathbf{F} = \alpha\left(\frac{1}{2}\nabla(\mathbf{E}^2) + \frac{d(\mathbf{E} \times \mathbf{B})}{dt}\right). \tag{13.17}$$

The last term is essentially the time derivative of the Poynting vector or energy flow. Now, since the sampling frequencies used to study particle motion in optical tweezers are typically $\sim 10^4$ Hz, the time average of the time derivative of the Poynting vector becomes zero, considering that the frequency of light is $\sim 10^{14}$ Hz. Thus, just the first term in Eq. (13.17) remains, and we have

$$\langle\mathbf{F}\rangle = Re\left\langle\frac{\alpha}{2}\nabla(\mathbf{E}^2)\right\rangle. \tag{13.18}$$

Now we can write the complex electric field as $\mathbf{E} = \mathbf{E}_0 e^{-i\phi}$, where $\mathbf{E}_0$ is the real amplitude and ϕ a phase factor. Then $(\mathbf{E})^2 = \mathbf{E}_0^2 e^{-2i\phi}$, so that representing the complex polarizability $\alpha = \alpha' + i\alpha''$ and taking the time average, we obtain

$$\langle \mathbf{F} \rangle = \frac{\alpha'}{4} \nabla \mathbf{E}_0^2 + \frac{\alpha''}{2} \mathbf{E}_0^2 \nabla \, \phi. \tag{13.19}$$

This is the mathematical form of the time-averaged total mechanical force exerted by the trapping laser on a particle. The first term, also called the gradient force, is proportional to the gradient of intensity and the real or dispersive component of the polarizability of the particle, while the scattering force depends on the phase gradient of the field and imaginary or dissipative component of the polarizability. The latter is a consequence of the radiation pressure and is a measure of the momentum transfer between the field and the particle. On the other hand, the gradient force exerts a pulling force on a particle toward the region of light where the intensity is maximum. In fact, for a fundamental Gaussian (TEM_{00}) laser beam, an evaluation of the derivative term of Eq. (13.18) reveals that $F = -kr$, with k being a constant (that we will later identify as the trap stiffness). The force is therefore linear and directed toward the region of intensity maximum, at which point it is zero (since the derivative is zero at the maxima). The negative sign implies that the force is restoring in nature, so that any particle will reach a position of equilibrium at the beam center (intensity maximum) and will experience a force pulling it back if it moves away from the center. Thus we have an 'optical trap' at the center of the beam. It is also clear that the magnitude of the restoring force will be increased if the intensity gradient is high. This can be achieved in two ways: (i) by increasing the laser power, which increases the intensity gradient linearly since $I \sim P/r^2$, where P is the laser power and r the beam waist radius, and (ii) by tightly focusing the laser beam, which increases the gradient quadratically, as apparent from the expression of I. It is for this reason that optical tweezers are typically developed around microscope objective lenses having high numerical aperture, so that very small laser spot sizes (less than the wavelength of the trapping laser) can be achieved, and large intensity gradients can be created to exert commensurately high restoring forces on a trapped particle.

In order to understand the physical origin of the scattering force, we write the electric field as $\mathbf{E}(\mathbf{r}) = \mathbf{E_0}(\mathbf{r}) \cos(\omega t - \phi(r))$, so that with $\dfrac{\partial \mathbf{B}}{\partial t} = -\nabla \times \mathbf{E}$, we have

$$\mathrm{E}_0^2 \, \nabla\phi = 2\omega \left\langle \mathbf{E} \times \mathbf{B} \right\rangle, \quad \mathrm{E}_0^2 = 2 \left\langle |\mathbf{E}|^2 \right\rangle. \tag{13.20}$$

On substitution in Eq. (13.19), we obtain the following expression for the total force:

$$\langle F \rangle = \frac{\alpha'}{2} \nabla |\mathbf{E}^2| + 2\omega\alpha'' \left\langle \mathbf{E} \times \mathbf{B} \right\rangle. \tag{13.21}$$

Thus it is apparent that the scattering force is proportional to the time average of the Poynting vector or the field momentum.

This treatment is for point dipoles. In reality, we actually deal with particles that have finite radii (less than the wavelength of light for nanoparticles), and the treatment has to be thus modified. We need to consider higher-order multipoles, and a treatment can be developed similar to Mie scattering (see Chapter 11). In that case, the gradient force can be written as

$$\langle F \rangle = Q(r) \frac{\epsilon_m^2 \langle P \rangle}{c}, \tag{13.22}$$

where $\langle P \rangle$ is the incident power and $Q(r)$ is the trapping efficiency that is closely related to the polarizability α which for a sphere is identical within the quasi-static limit to the electrostatic field of a dipole placed in the center of the sphere. Thus, α is given by

$$\alpha = 4\pi n_m^2 \epsilon_0 a^3 \frac{m^2 - 1}{m^2 + 2}, \quad m = \frac{n_1}{n_m}, \tag{13.23}$$

where n_m (n_1) is the refractive index of the medium (particle), ϵ_0 is the dielectric constant of vacuum and a the radius of the particle. Eq. (13.23) shows that the gradient force increases as the cube of the particle radius, which implies that smaller particles are more difficult to trap.

We now consider the scattering force, the magnitude of which for a sphere having $a \ll \lambda$ would obviously depend on the scattering cross-section (discussed in Chapter 11), so that we can write

$$F_{scatt} = n_0 \langle P \rangle C_{scatt}/c, \tag{13.24}$$

where n_0 is the refractive index of the medium, and C_{scatt} is the scattering cross-section that depends on the real component of the complex polarizability α given as $C_{scatt}{=}k^4|\alpha|^2/4\pi$, k being the wave-vector of the incident light. The scattering force is thus proportional to $|\alpha|^2$, implying a dependence of a^6 from Eq. (13.23).

Finally, there is a third force that needs to be considered, especially when dealing with nontransparent objects such as metallic micro/nanoparticles. This is the absorption force that arises from the complex component of α that was identified in Eq. (13.19) with the scattering or dissipative force for point dipoles. For finite-sized particles, we associate this force with the absorptive force that can be represented as (see Chapter 11)

$$F_{abs} = n_1 \langle P \rangle C_{abs}/c, \tag{13.25}$$

where C_{abs} is the absorption cross-section given as $C_{abs}{=}k\ \alpha''$ with α'' being the complex component of the polarizability. This force is of particular significance in the trapping of metals near the plasmon resonance where absorption is very high, leading to large heating in the vicinity of the particle such that trapping becomes increasingly difficult due to increased kinetic energy of the particle. It is also obvious that the recoil the particle experiences when absorbing from the trapping laser is in the direction of beam propagation, and therefore it attempts to kick the particle out of the trap akin to scattering forces.

13.4 Dynamics of trapped particles

In the previous section, we saw that a focused Gaussian beam behaves like an optical trap for microparticles since it exerts a finite restoring force on them when they attempt to leave the center or region of highest intensity of the beam. Since the force is conservative and linear in nature, it must be derived from a potential of the form $V(r) = \frac{1}{2}kr^2$, which is parabolic in nature. Thus, the potential is well shaped in nature, just as in the case of a harmonic oscillator. If a particle comes in the vicinity of the potential well having kinetic energy greater than the well depth, it will not fall into the well — or in other words, it will not be trapped. Thus it is unlikely that particles moving about in air in room temperature (calculate the kinetic energy for particles of mass around 10^{-12}, assuming the average velocity of the particles to be the velocity of sound) can be easily trapped just by a focused Gaussian beam. The situation would be much improved if the medium is a viscous fluid instead of air, so that the viscosity of the fluid acts as a frictional force and slows down the particles enough to lower their kinetic energy to be smaller than the potential depth of the trap. Thus most trapping experiments are performed in viscous fluids, of which water is the most common, owing to its low absorption at most visible and near-IR wavelengths, and the value of its viscous drag coefficient is just about perfect to allow particles to diffuse about slowly. Once trapped inside the potential well, the particle feels no restoring force due to the light at the trap center but would still execute Brownian motion (it's still in a fluid after all!). Let us now study the dynamics of Brownian motion more carefully.

13.4.1 Langevin equation

A small particle of density similar to the fluid it is immersed in executes random or Brownian motion inside the fluid. The size of the particle is of course much larger than the atoms and molecules that make up the fluid. While Robert Brown observed Brownian motion for the first time in pollen grains immersed in water, it was Einstein who related the macroscopic and microscopic properties of matter with his famous equation

$$D = \frac{k_B T}{6\pi\eta a}, \tag{13.26}$$

where D is the diffusion constant, a is the radius of the particle, η the coefficient of viscosity, k_B the Boltzmann constant and T the temperature. Note that $k_B = \frac{R}{N_A}$, where R is the universal gas constant, and N_A is Avogadro's number. Now, the motion of objects in such a particle-fluid system has three distinct time scales: τ_a, τ_b, and τ_v. Of these, τ_a has the shortest duration, being the time scale associated with the fluid atomic and molecular motion and the resulting collisions with the particle, and is typically of the order of picoseconds or even smaller. Next, τ_b is the time scale for the ballistic Brownian motion of the particle inside the fluid, which is the time scale at which the inertia of the particle is damped out due to collisions occurring as a result of rapid density fluctuations of the fluid. We have $\tau_b = \frac{m}{\gamma}$, with $\gamma = 6\pi a\eta$, where m is the particle mass. Typical orders of magnitude of τ_b vary from nanoseconds to tens of microseconds for particles in water. Finally, τ_v is the diffusion time for the particle to a distance equal to its own radius inside the viscous fluid, and its value is given by $\tau_v = \frac{a^2}{D}$. Thus, τ_v is limited by the friction (viscosity) of the fluid and can be really slow for highly viscous fluids (on the order of seconds or even minutes). Hence, in terms of duration, $\tau_a \ll \tau_b \ll \tau_v$.

We now consider in detail the motion of the particles. Even for random motion, the particles must obey Newton's equation (assuming motion in a single dimension)

$$m\frac{dv(t)}{dt} = F(t), \tag{13.27}$$

where m is the mass, $v(t)$ is the instantaneous velocity, and $F(t)$ the instantaneous force on the particle at time t, which arises due to interaction with the medium. Thus if the particle positions are exactly known as a function of time, the force would also be completely determined so that the motion would not really be random. However, in a fluid, while it can be expected that the dominant force would be the viscous force given by $F(t) = -6\pi a\eta v(t) = -\gamma v(t)$, there would also be a stochastic or random force $\zeta(t)$ whose origin would be the random density fluctuations of the fluid. Thus, the equations of motion would be

$$\begin{aligned} \frac{dx(t)}{dt} &= v(t), \\ \frac{dv(t)}{dt} &= -\frac{\gamma}{m}v(t) + \frac{\zeta(t)}{m}, \end{aligned} \tag{13.28}$$

Note that if we neglect the stochastic term in Eq. (13.28), the solution for velocity would simply be $v(t) = v(0)\exp(-t/\tau)$, where the time constant $\tau = \frac{m}{\gamma}$. This implies that at infinite time, $v(t)$ would actually decay to zero, i.e., the Brownian motion would stop altogether. This is unphysical, since we know that according to the equipartition theorem, at equilibrium,

$$\langle v^2(t)\rangle_{fin} = \frac{k_B T}{m}, \tag{13.29}$$

and is not zero—as Eq. (13.28) would suggest for the equilibrium velocity $\langle v^2(t)\rangle_{fin}$ without the stochastic term. Thus, it is clear that the stochastic term is indeed imperative to describe a physical origin of the Brownian motion. With a little intuition, we understand that the stochastic force originates from random collisions of the Brownian particle with the molecules of the surrounding medium, so that the average of the force would be zero. The average is what in statistical mechanics is known as the ensemble average, where we have several realizations of the force in different independent manifestations of the system. Note that the ensemble average of the second moment would be nonzero. However, for pure Brownian motion, there cannot be any correlation between collisions at time intervals dt_1 and dt_2. This is because the atomic time scale is of the order of ps, so that even by assuming dt to be of the order of μs, there would be 10^6 collisions of the particle with the atoms during this time interval, so that any correlation or memory effect of forces at different time scales vanish due to the large number of collisions. Thus, we may finally represent the effects of the force as

$$\langle\zeta(t)\rangle_\zeta = 0, \text{ and } \langle\zeta(t_1)\zeta(t_2)\rangle_\zeta = g\delta(t_1 - t_2). \tag{13.30}$$

Here, $\langle\ldots\rangle_\zeta$ denotes ensemble average, and g is the measure of the strength of the fluctuating force. However, this does not imply that there exists a unique solution for $\frac{dv}{dt}$, or even that $\frac{dv}{dt}$ exists at all points. We thus need to delve into the problem more deeply and impose conditions on $\zeta(t)$ such that at least local solutions of $\frac{dv}{dt}$ exist. We thus write the solution of Eq. (13.28) as

$$v(t) = \exp\left(\frac{-t}{\tau_a}\right) v(0) + \frac{1}{m}\int_0^t \exp\frac{-(t-s)}{\tau_a}\zeta(s)ds. \tag{13.31}$$

We now need to ensure that the integral in Eq. (13.31) exists. For this, we rewrite Eq. (13.28) as

$$dv(t) = -\frac{\gamma}{m}v(t)dt + \frac{1}{m}dU(t), \tag{13.32}$$

where $dU(t) = \zeta(t)dt$.

Integrating Eq. (13.32) between 0 and t, we obtain

$$v(t) - v(0) = \frac{\gamma}{m}\int_0^t v(s)ds + \frac{1}{m}\left[U(t) - U(0)\right], \tag{13.33}$$

$$= \frac{\gamma}{m}\left[x(t) - x(0)\right] + \frac{1}{m}\left[U(t) - U(0)\right]. \tag{13.34}$$

If we now discretize $U(t)$ into small time intervals that are very close to each other, we may write

$$U(t) - U(0) = \sum_{k=1}^{n}\left[U(t_k) - U(t_{k-1})\right]. \tag{13.35}$$

$U(t)$ is a continuous Markov process and the continuity follows from Eq. (13.32), so that

$$U(t) = U(0) + \int_0^t \zeta(s)ds, \tag{13.36}$$

where the integral should be a continuous function of its upper limit. The Markovian nature is also understandable since the particle undergoes $\sim 10^{12}$ collisions per second with the atoms and molecules of the fluid, so that even with the time intervals $t_k - t_{k-1}$ being very small (a few tens of microseconds), there is a very large number of collisions. Thus, the correlations in time only exist for the two intermediate steps (t_k, t_{k+1}), while those that occurred earlier than t_k are of no consequence—a typical property of Markovian processes. This implies that $U(t_k)$ depends only on $U(t_{k+1})$, and all the differential increments are independent of each other; they are also stationary, having zero mean. Thus, we can apply the central limit theorem to $U(t)$, which implies that $U(t)$ has a Gaussian nature with zero mean and a standard deviation of 1. Finally, we rewrite Eq. (13.32) as

$$v(t) = \exp\left(\frac{-t}{\tau_a}\right) v(0) + \frac{1}{m}\int_0^t \exp\frac{-(t-s)}{\tau_a} dU(s). \tag{13.37}$$

This is thus the solution of the Langevin equation for pure Brownian noise, also known as the Ornstein-Uhlenbeck process.

13.5 Brownian motion in a harmonic potential: The optical trap

The optical trap created by a focused Gaussian beam acts as a harmonic potential, which constrains the motion of Brownian particles that are introduced inside it. In such a case, Eq. (13.28) is modified to

$$\begin{aligned} \frac{dx(t)}{dt} &= v(t), \\ \frac{dv(t)}{dt} &= -\frac{\gamma}{m}v(t) - \omega^2 x(t) + \frac{\zeta(t)}{m}), \end{aligned} \tag{13.38}$$

where $\omega^2 = \frac{\kappa}{m}$, κ being the spring constant or stiffness of the trap (recall the discussion earlier), while $\zeta(t)$ is of course the Gaussian noise we discussed in the previous section, which has the properties $\langle\zeta(t) = 0\rangle$ and $\langle\zeta(t)\zeta(t')\rangle = \delta(t-t')$. Now, $t_{inert} \equiv m/\gamma_0$ is the characteristic time for loss of kinetic energy via friction. Since $t_{inert} \ll$ experimental time resolution (around tens of microseconds corresponding to typical sampling rates of tens of kHz), the inertial term can be dropped. This simplifies Eq. (13.38) considerably so that we now have

$$\dot{x}(t) + 2\pi f_c x(t) = (2D)^{1/2}\eta(t). \tag{13.39}$$

Here, we have introduced the diffusion constant D using Einstein's equation $D = \dfrac{k_B T}{\gamma_0}$ and the corner frequency $f_c = \dfrac{\kappa}{2\pi\gamma_0}$. As we shall see shortly, f_c

is one of the most important parameters used to characterize an optical trap. To solve Eq. (13.39), it makes sense to work in the Fourier domain. Now, in an experiment, we measure the particle displacement $x(t)$ for a time interval t_m, so that going into the Fourier domain, we have $x_k = \int_{-T_m}^{T_m} dt\, e^{i2\pi f_k}\, x(t)$, where $f_k = k/T_m$, $\eta(k)$ is the Fourier transform of $\eta(t)$, k being an integer. A solution for Eq. (13.39) can now be obtained in the Fourier domain as

$$\tilde{x}_k = \frac{(2D)^{1/2}\tilde{\zeta}_k}{2\pi(f_c - if_k)}. \tag{13.40}$$

Here, in Fourier transforming $\dot{x}_t$, we ignore the contributions from the ends of the integration limits (known as 'leakage' terms) in the partial integration, since the power spectral density that we will now consider is smoothly behaving without any discontinuities. We turn our attention to the real and imaginary components of $\tilde{\zeta}_k$, which would follow a Gaussian distribution of uncorrelated random variables since the process $\zeta(t)$ is Gaussian. Then $|\zeta_k|^2$ would form a series of uncorrelated exponentially distributed variables that are nonnegative. Thus, the experimentally measured power spectral density would be written as

$$P_k = \frac{|\tilde{x}_k|^2}{T_m} = \frac{D/(2\pi^2 T_m)\tilde{\zeta}_k^2}{f_c^2 + f_k^2}. \tag{13.41}$$

We finally have the time-averaged experimental power spectrum as

$$P_k^{exp} = \langle P_k \rangle = \frac{|\tilde{x}_k|^2}{T_m} = \frac{D/(2\pi^2 T_m)\tilde{\zeta}_k^2}{f_c^2 + f_k^2}. \tag{13.42}$$

As can be observed from Eq. (13.42), the power spectrum is clearly a Lorentzian. A Lorentzian fit would thus yield the coefficients f_c and D, which could be related to the trap stiffness and diffusion coefficient of the trapped particle, respectively. The former helps in characterizing the trap, while the latter serves as a consistency check and also helps in understanding local effects that can affect diffusion at the microscopic level.

The knowledge of corner frequency gives quantitative values for the stiffness of the optical trap for a particular bead diameter, which now enables us to set a limit for the minimum measurable displacement of a bead having a certain diameter for a particular averaging time [142, 143]. This is the so-called thermal limit, which is basically decided by the extent of the Brownian motion of the bead at a given trap stiffness over an averaging time t_{av}. From Ref. [142], this is given by

$$\Delta s_{min} = \frac{1}{\kappa}\sqrt{\frac{k_B T 6\pi\beta a}{t_{av}}}, \tag{13.43}$$

where Δs_{min} is the thermal resolution limit.

13.5.1 Generalized Langevin equation: The effect of hydrodynamic mass

While the viscous medium results in very fast damping of the inertial term so that it is dropped from Eq. (13.38), a careful look at the problem yields something more interesting. In fact, there does appear an inertial term if we consider the solution of the Navier-Stokes equation for a sphere in a harmonic potential inside an incompressible, viscous fluid that has a very low Reynold's number [144]:

$$F = -\gamma_0 \left(1 + \frac{a}{\delta}\right) \dot{x} - \left(3\pi\rho a^2\delta + \frac{2}{3}\pi\rho a^3\right) \ddot{x}. \tag{13.44}$$

In this equation, ρ is the density of the fluid, whereas δ is the penetration depth which is basically an estimate of the exponential decay of the fluid's velocity field with change in distance from the sphere executing harmonic oscillations. Thus, we observe that the total frictional force actually leads to an effect of inertia represented by the coefficient of $\ddot{x}$ in Eq. (13.44) — but this is basically due to the fluid entrained due to the sphere motion, while the coefficient of $\dot{x}$ acts as usual as the dissipating or damping term. Thus, this effect would be more pronounced if the volume of the fluid entrained would be larger, which implies that it would be more pronounced as the size of the particle (radius) increases. After some manipulation, we can then write down the generalized Langevin equation [145] in Fourier space for a sphere trapped in a harmonic potential well in an incompressible fluid as

$$m(-i2\pi f)^2 + [\gamma_s(f)(-i2\pi f) + \kappa]\,\tilde{x}(f) = [2k_B T\ Re(\gamma_s)]^{1/2}\,\tilde{\zeta}(f), \tag{13.45}$$

$$\gamma_s(f) = \frac{F}{-i2\pi f\tilde{x}(f)} = \gamma_s\left(1 + (1-i)\frac{a}{\delta} + -i\frac{2a^2}{9\delta^2}\right).$$

Note that the Ornstein-Uhlenbeck equation is recovered as $f \to 0$. However, the experimentally measured time-averaged power spectrum is modified to [146]

$$P_k^{hydro} = \frac{D/(2\pi^2)\,[1 + (f/f_\nu)^2]}{(f_c - f^{3/2}/f_\nu^{1/2} - f^2/f_m)^2 + (f + f^{3/2}/f_\nu^{1/2})^2}, \tag{13.46}$$

where

$$f_\nu = \nu/\pi a^2, \tag{13.47}$$

ν being the kinematic viscosity of the fluid, and

$$f_m = \gamma_0/2\pi m^* = 3f_\nu/2, \tag{13.48}$$

with $m^* = m + 2\pi\rho a^3/3 = 3m/2$, m^* being the effective hydrodynamic mass of the sphere. We can conceptually understand f_ν as the frequency at which the penetration depth of the flow pattern produced by the oscillatory motion

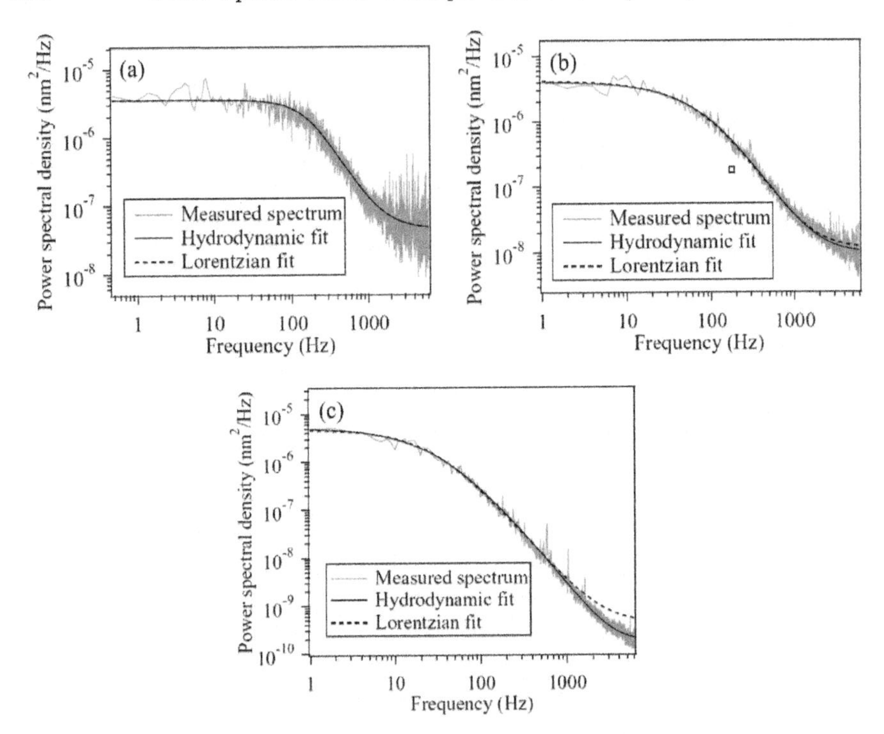

FIGURE 13.3: Typical power spectra of a trapped (a) 1 μm diameter bead, (b) 3 μm and (c) 16 μm bead. All spectra were fit to hydrodynamically correct power spectra (Eq. (13.46), black lines) as well as Lorentzians (dotted lines).

of the sphere equals the radius of the sphere. This term of course has no mass dependence, unlike f_m, which is the friction-mediated time constant of the sphere and depends entirely on the inertia of the sphere.

As mentioned earlier, the effects of the hydrodynamic corrections are clearly seen in the case of larger microparticles. Figure 13.3 demonstrates this for polystyrene beads having diameters 1, 3 and 16 μm. Clearly, as the bead size increases, the power spectrum deviates from the Lorentzian to that of the hydrodynamically corrected spectrum given in Eq. (13.46).

13.5.2 Experimental configurations

Optical tweezers require a large intensity gradient to trap and manipulate particles, which necessitates the use of a lens having small focal length (remember that $w = \dfrac{\lambda f}{\pi d}$, with w, f and d being the spot size radius, lens focal length and beam diameter, respectively) — something that is possible only with microscope objective lenses. It is for this reason that most optical tweezers are developed around microscopes. For a microscope objective, the minimum spot size $w = 1.22\frac{\lambda}{NA}$, where NA is the numerical aperture of the objective lens. Now, $NA = n \sin\theta$, with n being the refractive index of the medium in contact with the lens aperture, while θ is the angle that the refracted ray subtends with the lens axis. Thus, to obtain sub-wavelength focal spot sizes, we need $NA > 1$, so that typically oil-immersion objective lenses with oil of refractive index ~ 1.5 in contact with the lens are used. The highest NA available in objective lenses presently is around 1.5.

13.5.2.1 Typical setups: Upright versus inverted microscopes

Microscopes can be both upright and inverted, the difference being that in the former, the objective lens is placed above the microscope stage bearing the sample chamber, whereas in the latter it is placed below the chamber. For upright microscopes, detection is performed using forward scattered light after it passes through the objective, while for inverted microscopes, backscattered light is typically employed for detection. However, for forward scattering, one may face issues of saturating the detector, since a large portion of light incident on the detector comes directly from the laser source without being scattered from the source. The situation is not the same for backscattered detection, where the technique known as back focal plane interferometry is used, wherein the scattered light from the sample and unscattered light from other regions is superposed on the detector to study particle dynamics. Often, inverted microscopes are preferred since they facilitate backscattered detection, and they can also be used to detect the forward scattered light by appropriate optical arrangements.

13.5.2.2 Lasers and choice of wavelength

From the force equations it is clear that the gradient force that is responsible for trapping has dependence on the intensity of the laser beam only and very little dependence on wavelength. Thus light of any wavelength can be used to trap microparticles, but the most commonly used wavelength is at 1064 nm, since this is a bio-friendly wavelength and causes limited damage to cells due to the low energy associated with it.

13.5.2.3 Detectors: Cameras versus quadrant photo diodes

Video cameras are commonly used to track particle motion, but these are both expensive and require complex image-analysis software to track particle motion. In fact, most cameras are also limited in bandwidth, with their cost increasing almost exponentially with increasing bandwidth. Thus photodiodes, which detect light scattering, are also commonly used to track particles, and provide a much cheaper and user-friendly means of studying fast dynamics. Note that a special type of photodiode, called a quadrant photodiode (QPD), is employed since we are interested in the displacement of the bead. A schematic of a QPD is shown in Fig. 13.4. It has four quadrants, and each

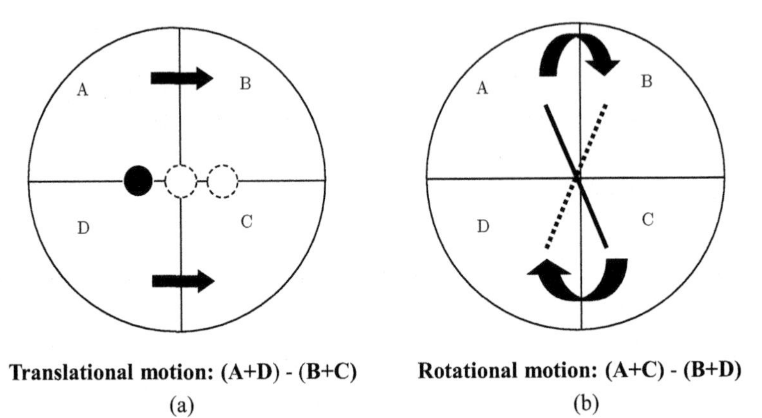

FIGURE 13.4: (a) Method to measure translation of a sphere; (b) Method to measure rotation of a rod-shaped particle.

quadrant gives an independent output, so that to measure horizontal displacement of a spherical particle, we perform an operation $(A+D)-(B+C)$, i.e., the difference of the sum of the vertical quadrants, while for vertical displacement we measure $(A+B)-(C+D)$, i.e., the difference of the sum of the horizontal components. It has been recently shown that QPDs can be used to measure rotation rates of asymmetric particles where the difference of the sum of the diagonal components $((A+C)-(B+D))$ yields the rate of rotation of such objects [147].

13.6 Trap calibration and measurements

One of the biggest advantages of optical tweezers is that they facilitate the application of controlled forces on particles. In fact, most of the applications in biology are based on this utility (which we shall discuss later), but the bottom line is that we need to measure the force applied by the focused beam, as well as the displacement of the trapped particle under the action of this force. Both the quantities are somewhat related as we saw in the previous sections, since the measurement of the displacement or Brownian motion of the particle leads to the measurement of the spring constant of the trap. However, there are several different procedures used to achieve this and calibrate optical traps, which we shall now study.

13.6.1 Power spectrum method

We discussed this method extensively in the previous section. As mentioned there, the Brownian noise of the particle is measured using detectors

(QPDs or cameras), and a power spectral analysis of the noise yields the trap stiffness by fitting a Lorentzian to the data and extracting the fit parameters (corner frequency f_c and diffusion coefficient D). However, while the parameters are measured, there is an issue about the consistency check of the parameter D, with known values of this constant for single particles existing only for spherical particles for which the Stokes drag is exactly known for a particular fluid. The measurement of D by this method becomes rather difficult to verify for nonspherical particles where the Stokes drag is difficult to calculate theoretically. The estimate of the corner frequency by this method, however, is rather reliable and widely performed by the optical tweezers community. A rather simple method to calibrate the trap is to vary the power of the trapping laser, so that the gradient force and thus the trap stiffness varies linearly as is seen from Eq. (13.22), and proceed to measure a linear response of the corner frequency. For measurement of the Brownian noise or displacement of a trapped particle, a separate detection laser is typically employed, since otherwise the calibration by varying trapping laser power would be a hindrance for detection, with the power on the detector also varying proportionately. However, care has to be taken that the intensity of the detection laser is low and that it itself does not apply a force on the trapped particle.

13.6.2 Calibration from time series of measured Brownian noise

While the power spectrum method works in the frequency domain, information about trap stiffness can be obtained by the measured time series of the Brownian noise itself. One of the ways is by equating the variance of the noise to the energy per degree of freedom, i.e., by using the equipartition theorem

$$\frac{1}{2}k_B T = \frac{1}{2}\kappa \left\langle x^2 \right\rangle, \tag{13.49}$$

where T is the absolute temperature, and x the displacement from equilibrium. The advantage of this equipartition theorem method is that it does not explicitly depend upon the viscous drag of the medium, so that the shape of the particle, its position with respect to nearby surfaces and the viscosity of the medium need not be known to measure the trap stiffness. The method is also susceptible to drifts in the particle position due to external perturbations (temperature, etc.), since any added noise and drift in position measurement serves only to increase the overall variance, effectively reducing the estimate of the trap stiffness.

The second approach for using the time series of the noise for calibration directly is by considering the fact that the probability distribution for the displacement of a particle trapped in a potential well is given as

$$P(x) = A\ exp(-\frac{U(x)}{k_B T}) = A\ exp(-\frac{\kappa x^2}{2k_B T}), \tag{13.50}$$

where $U(x)$ is the potential energy. When the potential is harmonic, this probability distribution is a Gaussian parametrized by the trap stiffness κ. Thus, a Gaussian fit to a time histogram of the measured variance yields the spring constant directly.

13.6.3 Viscous drag method

This method is very different from the previous methods, and uses the fluid properties to characterize the trapped particle. Thus, the particle is subjected to an external force, which may be achieved by perturbing the particle directly, or the fluid surrounding it. The former is achieved by applying a sinusoidal driving force to the particle, while for the latter the microscope stage holding the sample chamber with the particle may be subjected to a periodic force. When a sinusoidal driving force of amplitude A_0 and frequency ω is applied to the trapped particle, the response of the particle is given as

$$\begin{aligned} x(t) &= \frac{A_0\omega}{\sqrt{\omega_0^2+\omega^2}} exp(-i(\omega t-\phi)), \\ \phi &= -tan^{-1}(f_0/f), \end{aligned} \tag{13.51}$$

where f_0 is the characteristic roll-off frequency and ϕ the phase delay. The trap stiffness may be determined by measuring either amplitude or phase in Eq. (13.51). Drag force measurements are slower compared to the thermal motion of the particles so that the bandwidth limitations of the detector are not as stringent.

13.7 Some uses of optical tweezers

Optical tweezers have a wide spectra of applications in different areas of science, all of which cannot be given due justice in this chapter. We can only highlight the most important studies and applications of tweezers in brief. We start by categorizing the research using optical tweezers into three main areas, namely, in photonic force microscopy, the study of interactions between mesoscopic matter as well as between light and matter, and the induction and study of exotic dynamics in mesoscopic systems.

13.7.1 Photonic force microscopy

As has been extensively described previously, optical tweezers are all about applying controlled forces on microscopic matter using light. Just as a micro-cantilever measures very tiny forces in an atomic force microscope (AFM), a micron-sized optically trapped object can be used to probe forces at even femto-Newton levels. This can be understood from the fact that the displacement of the particle in the presence of the trap and other external perturbations can be measured very accurately using the detectors described earlier, and the information about the force applied onto a trapped bead is actually contained in its displacement. This is indeed the basis of photonic force microscopy (PFM) using optical tweezers, and it has been used in diverse applications, including imaging surface topographies to nanometer precision [149], colloidal physics [150] and especially biophysics [151].

To understand PFM, we revert to some classic studies in biology, which is the area that has the largest number of applications for optical tweezers. PFM has yielded a wealth of information in single molecule biophysics (for an excellent review consult Ref. [151]), especially in molecular motors (molecular motors are simply enzymes that move), such as kinesin and myosin, including translocation rates, pauses and step sizes. Moreover, the application of force via optical traps allows the motors' motion to be measured as a function of applied stress, which helps us better understand the different mechanisms that work in the movement of these molecular motors. In addition, diverse experiments have also been performed on single DNA and RNA, including information on structure and unfolding dynamics.

What is the magnitude of the displacements we are talking about? Typically, a single step size of a molecular motor is in the few nanometer regime, ranging from 8 nm for kinesin [152] and 5.5 nm for myosin [153]. This is, however, much higher than the atomic scale step sizes of proteins along DNA, which is around 0.34 nm. Optical tweezers have been employed to measure even at this level of precision; single base-pair stepping of RNA polymerase along DNA was measured to be 3.7 ± 0.6 Å [154]. The forces exerted by molecular motors also vary between 1 and 100 pN—a range that is perfectly within the range of forces that can be applied by optical tweezers.

We will now describe the methods by which such ultra-high precision measurements are performed. To do experiments on proteins, the first requirement is to develop a protein assay—that is, have an assembly of proteins on cover slips or even beads anchored on a cover slip. A typical single protein assay, and a schematic of the pioneering experiment in this direction [155], is shown in Fig. 13.5. In this experiment, a single molecule of RNA polymerase (RNAP), whose function is to copy information from DNA to RNA, is fixed or anchored to the surface, and it is then bound to a single DNA molecule as we can see in Fig. 13.5. The opposite end of the DNA is attached to a micron-sized polystyrene bead. Thus, the bead is 'tethered' to the surface by the DNA molecule. Many such tethered beads are prepared, and then one of them is trapped by the optical tweezers to subsequently carry out experiments. The main measurement is the displacement of the bead under tension due to the trap and the DNA tethered to it, and this displacement is measured by a detection laser. However, we need to carefully understand how the measurement of the displacement of the bead Δx_b would be related to the extension of the DNA x_{DNA}. Remember that the aim of the experiment was to measure the

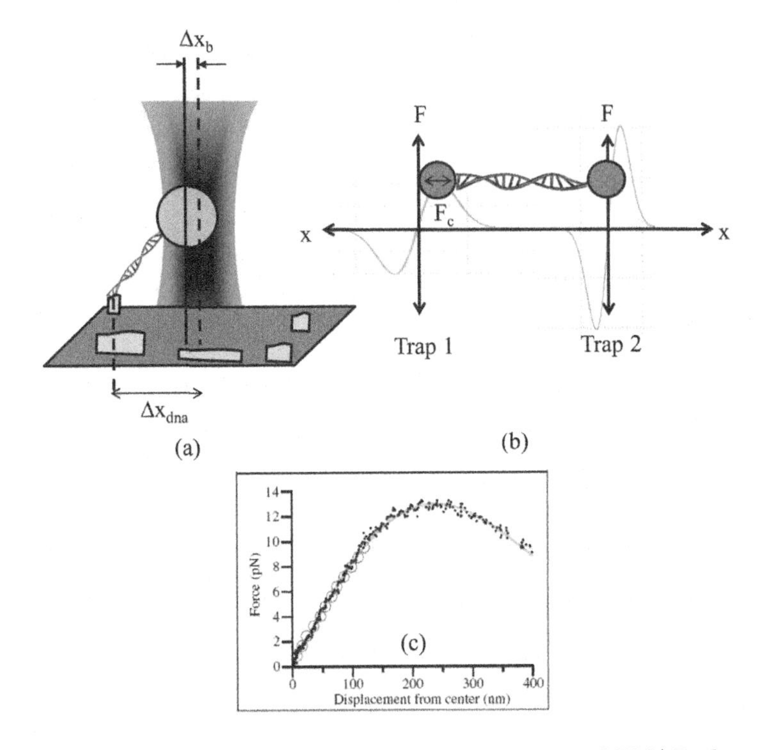

FIGURE 13.5: (a) Schematic to measure the motion of RNAP along DNA by a surface tethering experiment. Surface-anchored RNAPs are shown by gray pillars (not to scale). (b) A passive optical force clamp. Two traps next to each other have two trapped beads. Trap 1 has lower stiffness than Trap 2. A DNA molecule is attached to both beads. (c) Measurement of the force-displacement curve for Trap 1. The force is measured in Trap 2 as a function of displacement in Trap 1 (black diamonds) and fit to the derivative of a Gaussian (solid line). The open circles show an independent measurement of the force-displacement curve based on the drag force on an untethered bead, which is obtained by moving the microscope stage and sample at predetermined rates while measuring the displacement from the trap center. Data reproduced from Ref. [148].

movement of RNAP along the DNA. Thus, when the enzyme moves, the extension of the DNA is modified, changing x_{DNA}, but the motion of the bead is actually attenuated by the stiffness of the DNA (k_{DNA}). Thus, this is like a double spring system in series (the DNA itself is like a spring, and the trap of course acts as a Hookean spring)—the stiffness k_{DNA} being much smaller than the stiffness of the trap k_{trap}. Then, we can write, considering that the

DNA is not a linear spring,

$$k_{DNA} = \frac{\partial F(x_{DNA})}{\partial x_{DNA}} |_{F=F_{trap}}, \tag{13.52}$$

and the bead displacement would need to incorporate the compliance due to the DNA and can be written as

$$\Delta x_b = \frac{k_{DNA}}{(k_{trap} + k_{DNA})} \Delta x_{DNA}. \tag{13.53}$$

Thus, when the trap is very stiff, there is virtually no change in displacement of the bead due to the extension of the DNA. It is thus a challenge to maintain an optimum value of k_{trap} in order to get good signal-to-noise in bead displacement due to DNA extension. However, one of the main challenges of such displacement measurements is the Brownian motion of the bead itself, which from Eq. (13.43) is seen to be inversely proportional to trap stiffness. Of course Brownian noise can be reduced by averaging, but too much averaging also seriously limits the bandwidth of the measurement and can actually prevent us from detecting the natural dynamics of proteins (if the averaging time scale is longer than the protein time scales). This presents something of a dilemma — do we sacrifice sensitivity in order to reduce Brownian noise?

The best solution to this problem was devised in Ref. [148], in what was named an *optical force clamp*. The schematics of a passive force clamp is shown in Fig. 13.5(b). In the figure, we see a double trap, basically two traps adjacent to each other. The laser intensity is less in one trap compared to the other, which makes the stiffness of Trap 1 less than that of Trap 2. We have a bead trapped in each trap, with a DNA molecule tethered to both. The task is to measure the dynamics of the DNA with high precision. From Eq. (13.53) it is clear that the highest sensitivity in measuring Δx_{DNA} is to make it equal to Δx_b so that the only motion of the trapped bead is due to that of the DNA. This implies making $k_{trap} = 0$ (see Eq. (13.53)). This is what is achieved in a passive force clamp. Remember that the trapping force is proportional to the gradient of the intensity, so for a Gaussian intensity profile, the nature of the force would be like the derivative of a Gaussian as shown in Fig. 13.53(b). Thus the force varies linearly about the center, but there is a region near the turning point of the derivative, where the force is constant, and where the stiffness (which is like the derivative of the force) is zero. This is where the probe bead is placed, by a constant force applied by the bead in Trap 2. The latter bead is kept in the linear region of the force curve, and a feedback mechanism is applied to keep it in the same location of the force curve. This feedback mechanism is experimentally challenging and essentially consists of a fast measurement of the bead's Brownian motion, and then characterization of the trap stiffness by the power spectrum method. Any change in the trap stiffness represented by the corner frequency obtained by fitting a Lorentzian to the power spectrum of the bead in Trap 2 is immediately compensated by adjusting the intensity of the laser beam or the position of the trap by a scanning mirror or an acousto-optic deflector. In fact, this feedback system is an active force clamp, which maintains a constant trap stiffness or position of the bead with respect to the beam center. The real innovation of the system is the force clamp implemented on the bead in Trap 1, where a constant force with zero stiffness is maintained by adjusting the bead in Trap 2 with the DNA as a tether. To calibrate the force clamp, we measure the force in Trap 2 due to displacement of the bead in Trap 1 as shown in Fig. 13.5(c). The stiffness is constant for small displacements and agrees well with that measured for an untethered bead of the same size using the viscous drag method. What is interesting is the region of the curve where the force is constant for increasing displacements—meaning zero stiffness. This is the region of the force clamp that is around a distance of 240 nm from the trap center. Farther from that the stiffness is negative, meaning that the force approaches zero. The measured force-displacement relation also fits well to the derivative of a Gaussian, as is expected (shown in the solid grey line in Fig. 13.5(c)). Finally, with such a careful and innovative measurement method, Abbondonzieri et al. [154] were able to measure the step size of RNAP to be 3.7 ± 0.6 Å.

13.7.2 Study of interactions

Optical tweezers offer the possibilities of measuring interactions at different levels—namely, between different independently trapped objects, between a trapped object and its surroundings and finally between a trapped object and light itself, which is nothing but the study of light matter interactions at the microscopic scale. We will briefly discuss two types of interactions that are studied: optical binding and hydrodynamic interactions.

13.7.2.1 Optical binding

Optical binding refers to the interaction of optically trapped particles, which causes a reorientation of the stable equilibrium position of the particles due to a change in the intensity distribution of the electric field in their vicinity. The change in the intensity distribution is basically due to scattering by the particles of the light that is incident on them. Thus, if you consider a particle trapped in an intensity maxima, it will also scatter light according to its scattering cross-section. Now, a second particle that comes into the optical trap experiences an intensity distribution due to the trapping laser *and* that due to scattering off the first particle. Thus, the optical potential that it encounters is slightly modified from that due to just the trapping laser. The modified potential also leads to a small force between the particles known as the 'binding' force, that may be both attractive and repulsive. The force is very weak though, and is usually of the order of 1–2 pN. This, however, results in a rearrangement of the spatial distribution of the particles with the interparticle distances even greater than those possible due to electrostatic interactions arising from any surface charges present.

Now, optical binding can either be in the longitudinal direction to the trapping beam or the transverse direction. Transverse optical binding can be achieved by the use of interference patterns projected on the trapping plane of the optical tweezers, so that an optical lattice was formed in that region resulting in the formation of ordered colloidal crystals with a lattice structure mimicking that of the optical lattice. The experiment was redesigned by a single beam shaped like a narrow ribbon so that a similar assembly of particles in the direction transverse to the beam was obtained, and a careful analysis of the experiment revealed the existence of clear potential minima separated by around a single laser wavelength. Longitudinal optical binding is most commonly achieved by the use of counterpropagating beams with the beam foci separated in a way so as to push particles longitudinally to a location between the beam foci due to the resultant scattering force. Note that the distribution was solely due to the scattering force alone since the beams were from different lasers and produced no interference patterns. A comprehensive review of optical binding and its associated physics can be found in Ref. [156].

13.7.2.2 Hydrodynamic interactions

A comparatively recent application of optical tweezers has been in the study of hydrodynamic interactions at the microscopic scale, or in other words, interactions between particles mediated by the properties of the fluid surrounding them. For example, the coupled pendulum is a very well-studied system in classical mechanics. But what would happen if the coupled pendulum were immersed in water, which has a much higher viscosity than air? In fact, it is not just the viscosity that plays the role of a frictional or dissipative force, it is also the fact that liquids flow, and the flow may cause very interesting couplings between two spring-mass systems. Optical tweezers are essentially a spring that is made of light, to which a micro-object is attached. Of course, the natural frequency of the system is completely damped out by the viscous fluid surrounding it, but interesting consequences do arise if we have two such systems adjacent to each other. These consequences are basically expressed in correlations between the movements of these particles. In fact, careful studies of such correlations have even led to the understanding of synchronous motion of flagella that are responsible for the mobility of motile living cells. In fact, such flagella can even be modeled as chains of spherical particles attached to a larger particle representing the cell body [157]. While these are more complex studies, the study of hydrodynamic interactions starts from measurements of the motion of a spherical test bead placed in the velocity field of another bead moving either spontaneously due to Brownian motion or in controlled ways by external forces [55]. These studies can then be extended to more complex shapes of microparticles as well as in diverse fluids (viscous, visco-elastic, etc.). Thus, hydrodynamic interactions have a crucial role to play in the understanding of biophysical phenomena, as well as fluid dynamics at the microscopic level.

13.7.3 Rotational dynamics in particles induced by optical tweezers

Until now we have been dealing only with fundamental Gaussian beams to trap particles. Fundamental Gaussian beams do not possess any intrinsic angular momentum, either spin or orbital. However, it is possible to impart spin angular momentum to a fundamental Gaussian beam by passing it through a circular polarizer or quarter-wave plate (there is an extensive discussion on quarter-wave plates in Chapter 6). The beam thus becomes circularly polarized, which signifies that it has a spin angular momentum $\hbar$ or $-\hbar$ depending on whether the beam is left or right circularly polarized. When a circularly polarized light is incident on a birefringent particle, there is exchange of angular momentum, which causes the particle to spin along its axis. This was first demonstrated in Bethe's classic 1936 experiment where he showed that a quartz plate suspended by a fine string actually underwent a small twist when circular polarized light passed through it, demonstrating the exchange of angular momentum. Thus, trapped birefringent particles may be imparted spin by coupling a circularly polarized fundamental trapping beam. The rate of rotation is directly proportional to the intensity of the beam and the degree of ellipticity of the beam.

From the solutions of the Helmholtz equation, we also know that while the TEM_{00} beam is the fundamental solution and has a Gaussian intensity distribution, higher-order solutions often have a singularity at the center. This is especially true of the Laguerre-Gaussian class of solutions, which are obtained by writing the Helmholtz equation in cylindrical polar coordinates. Even the LG_0^1 has an annular intensity distribution with a central minimum. More importantly, LG beam modes have intrinsic angular momentum, or orbital angular momentum, which is dealt with at length in Chapter 12. Thus, when such a beam is coupled into the trap, it results in the trapping of particles in a ring-like distribution in accordance with the intensity profile, with the particles also revolving around the beam axis due to the intrinsic orbital angular momentum of the beam. A good review of optical tweezers using angular momentum carrying beams is found in Ref. [158].

13.7.4 Nanoparticle trapping and holographic tweezers

We end this chapter with a brief discussion on two of the most recent developments in optical tweezers: trapping nanoparticles and multiparticle trapping using holographic tweezers.

13.7.4.1 Nano-tweezers for nanoparticle trapping

The smaller the particle, the more difficult it is to trap. This happens for two reasons. From Eqs. (13.22) and (13.23), we see that the magnitude of the gradient force scales is a^3, where a is the diameter of the particle, so that the gradient force reduces for small particles. The value of the diffusivity from

Einstein's equation (inversely proportional to a) increases, and the Stokesian drag (proportional to a) reduces, so that the probability of nanoparticles either escaping the trap entirely or jumping out of it due to large amplitude of stochastic Brownian motion kicks is high. In his seminal paper in 1986 [140], Ashkin laid down a rule of thumb that the depth of the optical potential required for stable trapping should be $\sim 10\, k_B T$, where the thermal energy $k_B T$, given by the potential energy $\frac{1}{2}kx^2$, would be higher for particles having large values of x^2. Thus, the well depth should be much higher for nanoparticles compared to micron-sized ones to obtain stable trapping.

Increasing the well depth means increasing the intensity gradient, which can be achieved by either increasing laser power or reducing the spot size. Diffraction prevents reduction of the spot size lower than the wavelength of the laser, while we cannot increase laser powers beyond a certain limit without running into several issues (maintaining the fundamental Gaussian mode, intensity stability, and so on). So how do we increase the gradient force? The answer has been found rather recently, after the development of the field of plasmonics, with which has come the widespread use of evanescent waves. We learned in Chapter 3 that evanescent waves decay within the spatial extent of less than a wavelength in the axial direction, which means that we can actually have spatial confinement of the field within distances far less than those allowed by the diffraction limit. This obviously means much higher intensities, and understandably turns out to be the obvious solution for having stable traps for nanoparticles.

A host of techniques have been developed for plasmonics-based nanotweezers to facilitate nanoparticle trapping. Most of them involve the exciting of surface plasmon polaritons (SPPs) or local plasmon polaritons (LPPs). Since SPPs cannot be coupled to propagating light, they are typically excited by illuminating a gold-dielectric interface in the total internal reflection (TIR) condition for a glass prism having higher refractive index than the dielectric, so that at the surface of the metal, light can be evanescently coupled to SPPs to create intensity distributions much higher than the incident intensity. LPPs can be coupled to propagating light, and they are excited by developing metal structures much smaller than the wavelength of light. Due to their definite dimensions, LPPs have a fixed frequency range of operation for the trapping laser but are really effective in producing very large intensities in their vicinity. A common configuration is to develop gap antennas, or two gold nanorods separated by a few wavelengths, so that nanometric particles are trapped in the middle. Using such an antenna, 10 nm particles have been trapped recently [159]. An elaborate review of plasmonic traps can be found in Ref. [160].

13.7.4.2 Holographic tweezers

As the name suggests, holographic tweezers are constructed by projecting a hologram in the trapping plane. The advantage of such tweezers is the possibility of creating an array of three-dimensional traps with a single laser beam that is incident on a hologram generally created by a computer on a spatial light modulator (SLM). An SLM is essentially a liquid crystal display where the gray levels of all pixels can be addressed by giving adjustable individual voltages. Thus, a lattice of particles can be created and also manipulated individually by addressing each trap separately. Holographic tweezers offer great flexibility and maneuverability to the field of optical trapping, and they are very useful in the study of interactions between the trapped particles, which has deep implications in the understanding of biological processes and hydrodynamics. They also involve the most sophisticated optics technologies in the field of optical tweezers with adaptive optics in the form of SLMs at the heart of their functionality. Ref. [161] is an excellent review of holographic tweezers.

Chapter 14

Pendry lensing and extraordinary transmission of light

This chapter is devoted to a few of the interesting effects that have drawn the attention of researchers in the past fifteen year. There are many counter-intuitive findings that triggered a second look at some of the seemingly well-understood phenomena like imaging and diffraction. We focus on two main effects, namely, the perfect imaging and extraordinary transmission. Note that there have been many other interesting effects like invisibility cloaks, metasurfaces and Fano resonances in nanostructures. The research has been extended to quantum effects in metal-dielectric nanostructures, leading to the birth of nonlinear and quantum plasmonics. We do not cite references on any of these topics since a Google search will reveal the enormous progress that has been made in most of these areas.

14.1 Near-field vs. far-field and resolution

In this section we try to highlight the advantages of near-field measurements over far-field ones in the context of resolution. Let us assume that two point sources are set apart by a small separation d, as shown in Fig. 14.1. We want to resolve this distance by measuring the optical signal at a point P at a distance r from the midpoint between these two sources (the origin). Assuming that both the sources emit spherical waves, the time taken by the optical disturbance $\tau_{\mp}$ to reach P is given by

$$\tau_{\mp} = \frac{k}{\omega}|\mathbf{r} \pm \mathbf{d}/2|, \tag{14.1}$$

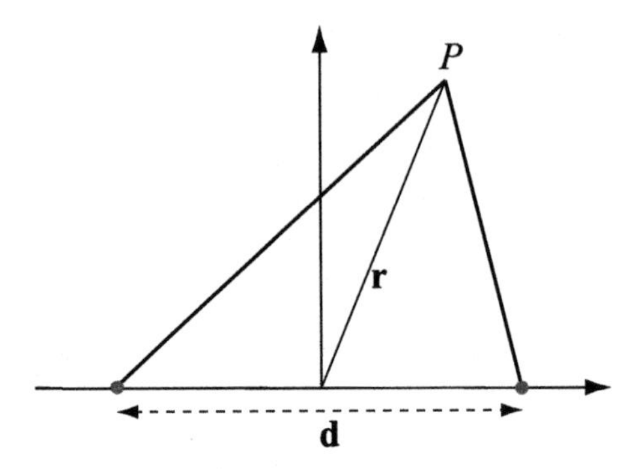

FIGURE 14.1: Schematics of the two point source system.

where the '−' ('+') sign refers to the left (right) source. Thus the amplitude of each wave with amplitude A_0 reaching point P would be

$$E_{\mp}(P) = A_0 \frac{e^{-i\omega(t-\tau_{\mp})}}{|\mathbf{r} \pm \mathbf{d}/2|} = A_0 \frac{e^{-i\omega t + ik|\mathbf{r}\pm\mathbf{d}/2|}}{|\mathbf{r} \pm \mathbf{d}/2|}, \tag{14.2}$$

$$E(P) = E_+ + E_-, \tag{14.3}$$

so that the total disturbance at P is a superposition of both. Let us now distinguish between the far- and near-fields. The inequality $kr \gg kd$ implies a far-field situation (a point of observation is far from the object) while the object (e.g., a set of two sources) may be of sub-wavelength dimensions ($kd \ll 1$) or otherwise. In contrast, the near-field situation holds when $kd \lesssim kr \ll 1$ for a small object. For far-field region for large ($kr \gg kd \gg 1$) or small ($kr \gg 1 \gg kd$) objects Eq. (14.3) with Eq. (14.2) can be simplified to

$$E(P) = \frac{e^{i(kr-\omega t)}}{r}(2\cos(kd(\hat{\mathbf{r}} \cdot \hat{\mathbf{d}})/2)). \tag{14.4}$$

In writing Eq. (14.4) we neglected higher-order terms in d/r. The phase difference between the waves can be made zero by choosing a point $\mathbf{r}_0$ ($r_0 \gg d$) on a large sphere such that $\hat{\mathbf{r}}_0 \cdot \hat{\mathbf{d}} = 0$. The closest local minimum will occur at $\mathbf{r}$ satisfying

$$kd|\hat{\mathbf{r}} \cdot \hat{\mathbf{d}}/2| = \pi/2. \tag{14.5}$$

Measuring two successive local minima in the interference easily leads to the size of the object:

$$d = \frac{\pi}{k|\hat{\mathbf{r}} \cdot \hat{\mathbf{d}}|}. \tag{14.6}$$

However, for small objects ($ka \ll 1$) the phase difference between two points is practically zero and far-field measurements are unable to resolve d. The situation is quite different in the near-field since spatial k-dependent phase does not affect the total field, which can be written as follows:

$$E(P) = A_0 e^{-i\omega t}\left(\frac{1}{|\mathbf{r}+\mathbf{d}/2|} + \frac{1}{|\mathbf{r}-\mathbf{d}/2|}\right). \tag{14.7}$$

Again, for a chosen $\hat{\mathbf{r}}_0$ with $\hat{\mathbf{r}}_0 \cdot \hat{\mathbf{d}} = 0$, we can find d as

$$d = \frac{2\sqrt{4 - r_0^2|A_0|^2}}{|A_0|}. \tag{14.8}$$

Thus the information about the object is contained in the k-dependent phase in the far-field and the k-independent field magnitude in the near-field.

14.2 Angular spectrum decomposition and evanescent waves

The discussion in the previous section did not have any reference to evanescent waves. In this section we highlight the role of evanescent waves in near-field imaging and superresolution. Let a monochromatic wave field at a point $\mathbf{r}$ be given by $\mathbf{E}(\mathbf{r})$. Assume that the wave is propagating predominantly along the z direction. Decomposition of this wave in terms of the spatial harmonics is referred to as the angular spectrum representation [162] and is given by the two-dimensional Fourier transform

$$\mathbf{E}(x, y; z) = \int dk_x \int dk_y\, \boldsymbol{\mathcal{E}}(k_x, k_y; z) e^{i(k_x x + k_y y)}, \tag{14.9}$$

where both the integrations are on the whole real axis. The spatial harmonic amplitude $\boldsymbol{\mathcal{E}}(k_x, k_y; z)$ is given by the transform

$$\boldsymbol{\mathcal{E}}(k_x, k_y; z) = \frac{1}{4\pi^2}\int dx \int dy\, \mathbf{E}(x, y; z) e^{-i(k_x x + k_y y)}. \tag{14.10}$$

Substituting Eq. (14.9) in the Helmholtz equation and starting from the field defined at $z = 0$, we obtain the propagation equation for the spatial harmonic as follows:

$$\boldsymbol{\mathcal{E}}(k_x, k_y; z) = \boldsymbol{\mathcal{E}}(k_x, k_y; 0) e^{\pm i k_z z}, \quad k_z = \pm\sqrt{k^2 - (k_x^2 + k_y^2)}. \tag{14.11}$$

We have to choose the proper sign for the square root so that $Im(k_z) \geq 0$ to ensure causality. Thus the angular spectrum decomposition given by Eq. (14.9) reduces to

$$\mathbf{E}(x, y; z) = \int dk_x \int dk_y\, \boldsymbol{\mathcal{E}}(k_x, k_y; 0) e^{i(k_x x + k_y y)} e^{\pm i k_z z}. \tag{14.12}$$

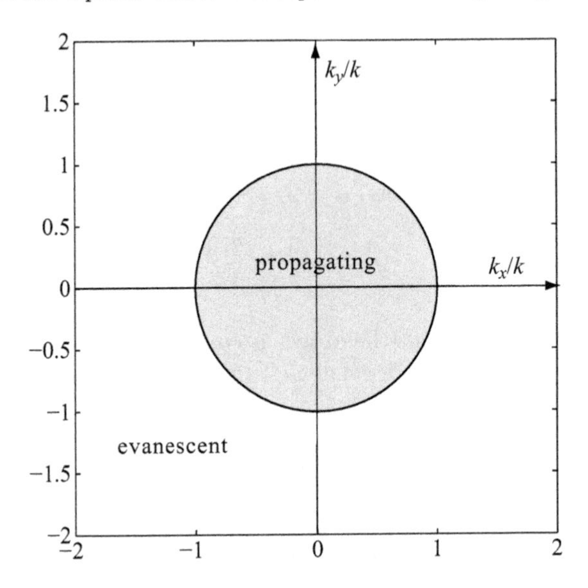

FIGURE 14.2: Propagating and evanescent waves in the $k_x k_y$ plane.

The $\pm$ sign in Eq. (14.12) relates to the forward and backward propagating waves. Analogous results hold for the magnetic field **H**. The amplitudes of the spatial harmonics must obey $\mathbf{k} \cdot \boldsymbol{\mathcal{E}} = \mathbf{k} \cdot \boldsymbol{\mathcal{H}} = 0$ in order to satisfy Maxwell's divergence equations for **E** and **H**. The propagation characteristics of a spatial harmonic depends on the transverse spatial frequencies k_x and k_y, namely, on how they compare with the magnitude of the wave vector k in the medium. We have propagating (evanescent) waves for $k_x^2 + k_y^2 \leqslant k^2$ $(k_x^2 + k_y^2 > k^2)$, i.e., inside (outside) the circle of radius k in the $k_x k_y$ plane (see Fig. 14.2). Thus the propagating harmonics occupy a small parameter space in the transverse k plane while most of the space is dominated by the evanescent waves. Moreover, the angular spectrum decomposition amounts to a superposition of plane waves and evanescent waves. Eq. (14.12) reveals another important aspect of field propagation. In particular, the spatial harmonics at $z = 0$ and at z can be related by a propagator $t(k_x, k_y, z) = e^{\pm i k_z z}$ and Eq. (14.11) can be rewritten as

$$\boldsymbol{\mathcal{E}}(k_x, k_y; z) = t(k_x, k_y, z)\boldsymbol{\mathcal{E}}(k_x, k_y; 0). \tag{14.13}$$

In the standard optics literature, such propagators are also known as the transfer functions. The transfer functions for a stratified medium or for a sequence of optical elements can be written as a product of transfer functions of individual elements as discussed in Sections 9.1 and 9.2. The transfer function is oscillating for propagating spatial harmonics while they imply attenuation for evanescent components. Thus if the image plane is sufficiently separated from the object plane (at $z = 0$), the contribution from the evanescent waves can be ignored and the integration in Eq. (14.12) can be performed within the circle as in Fig. 14.2. As a consequence the information contained in the higher spatial harmonics is lost. Only structures with a lateral dimension d larger than $1/k$ can be resolved. For small objects with $d \leqslant 1/k$, we need to retain as many spatial harmonics as possible going beyond the circle, which necessitates the near-field measurements. The larger the transverse spatial frequency, the stronger is the attenuation and the corresponding fields are more localized to the object plane. In this sense the sub-wavelength resolution necessarily means the inclusion of evanescent waves with larger spatial frequencies.

One of the major discoveries of the last decade was the possibility of amplifying the evanescent waves by a negative index material slab, leading to perfect imaging [18]. Usually a standard lens focuses only the propagating part of the angular spectrum, while the information contained in the evanescent waves is completely lost. A Pendry lens with idealized system parameters, as in the case of a Veselago lens, does focus both the propagating and the evanescent parts of the spectrum irrespective of the value of the transverse wave vector components. In order to have a simplified picture, we look at a two-dimensional version assuming infinite extent in the y direction. Assuming a three-component stratified optical system with transfer functions given by t_1, t_2, t_3, the magnetic field for TM polarised light at the image plane can be written as

$$H_y(x, z = d_1 + d_2 + d_3) = \int dk_x t_1(k_x, d_1) t_2(k_x, d_2) t_3(k_x, d_3) \mathcal{H}_y(k_x, 0) e^{ik_x x}, \tag{14.14}$$

where d_i $(i = 1-3)$ is the spatial extent of the i-th layer along the z direction. Veselago and Pendry lensing at an elementary level has been discussed nicely in Ref. [163]. We briefly recount the arguments in the following section.

14.2.1 Transfer function for a dielectric slab

Consider a dielectric slab with dielectric function and magnetic permeability ε_2 and μ_2 and width d_2 embedded in vacuum (see Fig. 14.3). Let the object and the image planes be at $z = 0$ and $z = z_3 = d_1 + d_2 + d_3$. For a TM-polarized plane monochromatic wave with $\mathbf{k} = (k_x, k_{0z})$ and with $k_x^2 + k_{0z}^2 = \omega^2 \varepsilon_0 \mu_0 = k_0^2$, the propagation from $z = 0$ to $z = z_1 = d_1$ would result in a phase factor of $e^{ik_{0z} d_1}$. Similar arguments hold for propagation in layer 3 from $z = z_2$ to $z = z_3$. Thus the transfer functions t_1 and t_3 of layers 1 and 3 are given, respectively, by

$$t_1 = e^{ik_{0z} d_1}, \qquad t_3 = e^{ik_{0z} d_3}. \tag{14.15}$$

The transfer function for the second layer can be obtained using the results of Section 9.2. Let $m_{ij}, i, j = 1, 2$ denote the elements of the characteristic matrix for the slab of width d_2. The transfer function for this slab is given by

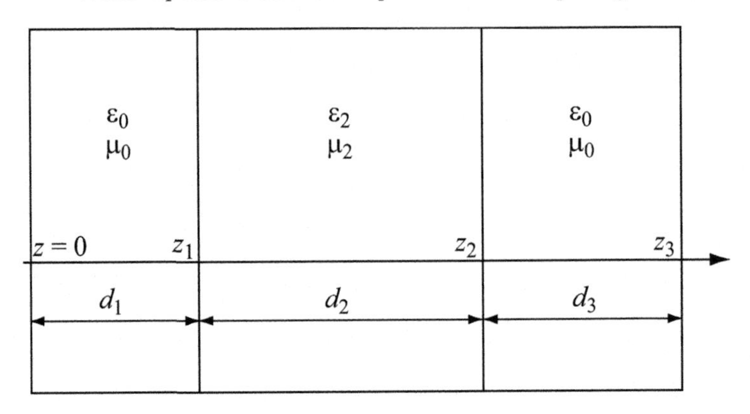

FIGURE 14.3: Schematics of the dielectric slab embedded in vacuum.

$$t_2 = \frac{2p_{0z}}{(m_{11} + m_{12}p_{0z})p_{0z} + (m_{21} + m_{22}p_{0z})}, \tag{14.16}$$

where $p_{0z} = k_{0z}/\varepsilon_0$ and the expressions for m_{ij} were calculated in Section 9.2. In order to reveal the exponential propagation factors, Eq. (14.16) can be recast in the form

$$t_2 = \frac{4\zeta}{(1+\zeta)^2 e^{-ik_{2z}d_2} - (1-\zeta)^2 e^{ik_{2z}d_2}}, \tag{14.17}$$

where $k_{2z} = \sqrt{\omega^2\varepsilon_2\mu_2 - k_x^2}$ and $\zeta = \frac{\varepsilon_0 k_{2z}}{\varepsilon_2 k_{0z}}$ for TM waves. The total transfer function t for propagation from $z = 0$ to $z = z_3$ (to be used in Eq. (14.14)) is the product of the individual transfer functions

$$t = t_1 t_2 t_3. \tag{14.18}$$

For perfect imaging we need to have a flat transfer function irrespective of the value of the argument k_x, which ensures the participation of all the spatial harmonics at equal footing in forming the image. Such is the case for a Veselago lens [17] for propagating waves as well as for a Pendry lens [18] for evanescent waves in the near-field.

14.3 Negative index materials and Pendry lensing

In his seminal paper Veselago discussed far-field imaging with a slab of NIM albeit with the idealized parameters $\varepsilon_2 = \varepsilon_0\varepsilon_r$ and $\mu_2 = \mu_0\mu_r$ with $\varepsilon_r = -1$ and $\mu_r = -1$ [17]. See Fig. 14.4(a), which is a modified version of

Fig. 14.3 with $d_1 = d_3 = d/2$ and $d_2 = d$. For such a system, $k_{2z} = k_{0z}$ and $\varepsilon_2 = -\varepsilon_0$, and ζ reduces to

$$\zeta = \frac{\varepsilon_0 k_{2z}}{\varepsilon_2 k_{0z}} = -1. \tag{14.19}$$

Thus for propagating waves using Eq. (14.19), we have

$$t_1 = e^{ik_{0z}d/2}, \quad t_2 = e^{-ik_{0z}d}, \quad t_3 = e^{ik_{0z}d/2}, \tag{14.20}$$

so that the product equals unity. Indeed for such idealized parameters, all propagating waves can be imaged perfectly by a Veselago lens. It is not difficult to understand Veselago imaging since we know that Snell's laws hold at an interface between a standard medium and a NIM, resulting in negative refraction. An angle of incidence θ would result in a refracted ray at an angle $-\theta$ for the given set of parameters above, and the object placed at $z = 0$ would image in the center of the NIM slab and also at a distance of $d/2$ after the slab (see Fig. 14.4(a)).

Sir Pendry's seminal contribution was his the suggestion that the same system can be used for amplification of evanescent waves, resulting in perfect imaging. Indeed, for an incident evanescent field with **k** outside the circle in

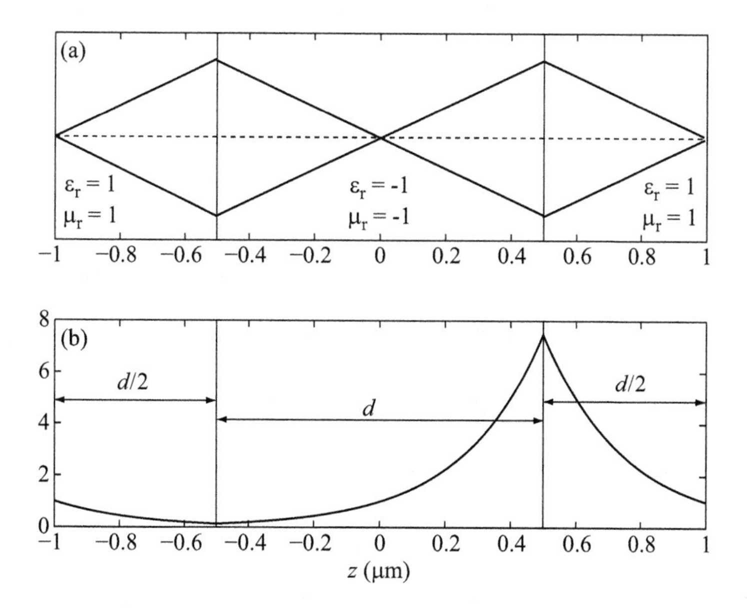

FIGURE 14.4: Focusing of (a) propagating and (b) evanescent waves. The former (latter) is referred to as Veselago (Pendry) lensing.

Fig. 14.2, the propagators in Eq. (14.20) can be rewritten as

$$t_1 = e^{-\kappa_{0z} d/2}, \quad t_2 = e^{\kappa_{0z} d}, \quad t_3 = e^{-\kappa_{0z} d/2}, \tag{14.21}$$

where $\kappa_{0z} = \sqrt{k_x^2 - k_0^2} = -ik_{0z}$. Again the total transfer function $t = t_1 t_2 t_3 = 1$, and it is flat even for evanescent waves. The field emerging from $z = 0$ can be imaged at $z = z_3 = 2d$ (see Fig. 14.4(b)).

14.4 Imaging through an absorption-compensated slab of metamaterial

In reality the situation is much more involved. Losses in the electric and magnetic response play a crucial role in limiting the resolution of a Pendry lens. A detailed analysis of the effects of losses (due to finite imaginary parts in ε and μ) on the total transfer function and how it leads to the cutoff wavelength is presented in Ref. [163]. The existence of the cutoff wavelength is due to deviation from the idealized parameters of a Pendry lens. In light of the cutoff, not all the spatial harmonics are transferred to the image plane and this results in the loss of resolution. The resolution limit has a logarithmic dependence on the NIM losses [164]. There has been a realization of similar lensing with a thin silver slab in the near-field, now referred to as a *poor man's perfect lens* [18, 165]. Recently there have been efforts to compensate for the losses in plasmonic structures by suitable implementation of a gain mechanism [166]. In what follows we show theoretically that even such gain mechanisms may not be adequate to render some of the best experimentally reported metamaterials suitable for perfect imaging if the compensation is not perfect [23].

Most of the experimentally reported metamaterials are highly anisotropic. For theoretical calculations we assume metamaterial to be homogeneous and isotropic. The values for the permittivity (ϵ) and permeability (μ) are taken from experimental work [167]. Further, for the magnetic permeability, μ is fitted to a theoretical model corresponding to a Lorentz-type response. Special attention is paid to the magnetic response since magnetic losses form the dominant mechanism of losses at higher frequencies. For a material with built-in gain, the magnetic response can be written as

$$\mu(\lambda) = \mu_\infty + \frac{\sigma_1}{1 - (\lambda_1/\lambda)^2 - i(\lambda_1/\lambda)\gamma_1} + \frac{f\sigma_2}{1 - (\lambda_2/\lambda)^2 - i(\lambda_2/\lambda)\gamma_2}. \tag{14.22}$$

The last term in Eq. (14.22) gives the contribution of the gain material (with $\sigma_2 < 0$), while the rest is the usual Lorentzian response [168]. The parameter f is a measure of doping concentration such that for $f = 0$ we recover Dolling's material. The parameters $\mu_\infty = 0.6$, $\sigma_1 = 4.425\gamma_1$, $\gamma_1 = 0.028$ and

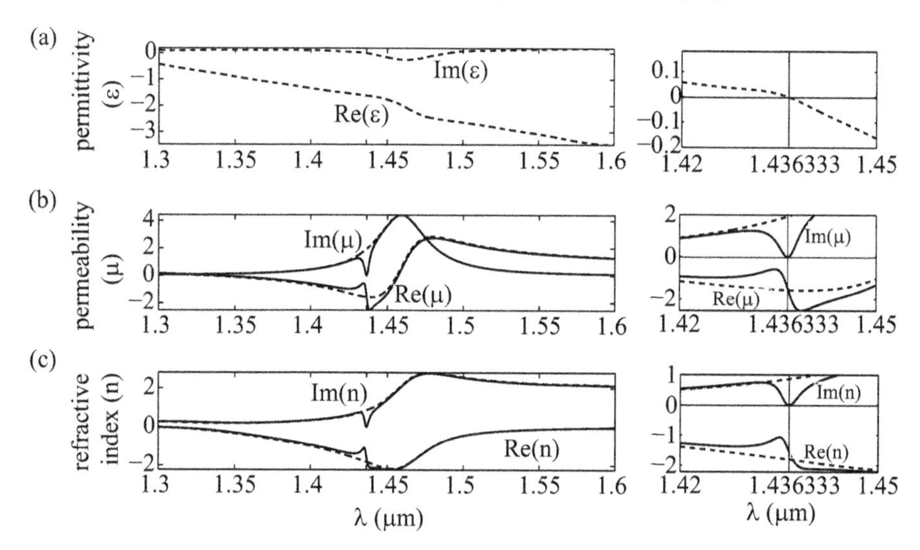

FIGURE 14.5: (a) Permittivity ϵ, (b) permeability μ and (c) refractive index n as functions of λ. The dashed (solid) curves correspond to the NIM without (with) doping. The right panels give expanded views of the left panels around $\lambda_0 = 1.436333\ \mu$m. Adapted from Ref. [23].

$\lambda_1 = 1.459\ \mu$m are chosen to fit the experimental data of Dolling et al. The parameters for the active material are taken to be $\sigma_2 = -0.0054292$, $\gamma_2 = 0.0028$ and $\lambda_2 = 1.43642\ \mu$m such that at $f = 1$, the metamaterial is absorption-compensated at the working wavelength $\lambda_0 \approx 1.436\ \mu$m (see Fig. 14.5). In other words the metamaterial becomes transparent at the free-space working wavelength λ_0.

Let us now probe whether near-perfect imaging can be achieved by such an absorption-compensated NIM slab (with configuration as in Fig. 14.3) with the exception that now we have refractive index-matched ambient medium (ε_0, μ_0 replaced by ε_1, μ_1). In order to probe the superresolution features, let us image two adjacent Gaussians with subwavelength widths. Let the transverse field distribution corresponding to the superposition of the Gaussians of width $\lambda_0/10$ and peaks separated by $\lambda_0/5$ (at the working wavelength $\lambda_0 \approx 1.4363\ \mu$m) define our object. We can calculate the image-plane distribution using the standard Fourier technique and using the transfer functions and the spatial harmonics. In order to reveal the effects of loss compensation by the gain medium, it is useful to compare the cases $f = 1$ (full absorption-compensation), $f = 0.95$ (partial compensation) and $f = 0$ (experimental NIM as is). The image-plane intensity distribution in the transverse plane is shown in Fig. 14.6. Although the partially bleached NIM shows a maximum transmission of about 1.5%, the peaks are indistinguishable (see dash-dotted curve in Fig. 14.6). For the experimental material (without doping) the image

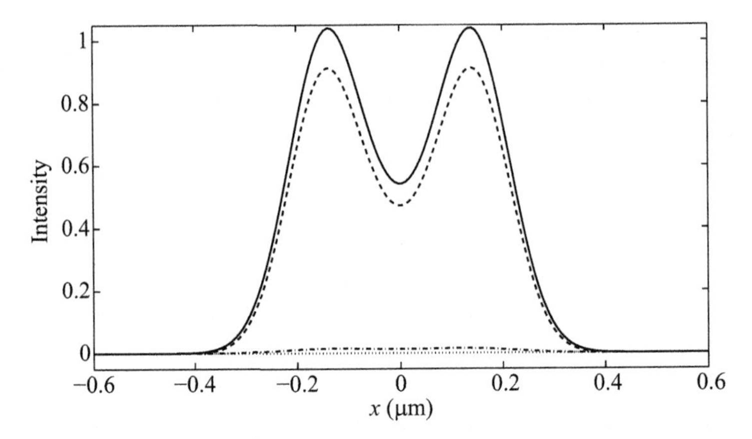

FIGURE 14.6: Effect of loss on imaging. Images obtained at the perfect image plane for $f = 1$ (dashed), 0.95 (dash-dotted) and 0 (dotted) compared with the object (solid). Parameters are for an absorption-compensated NIM slab and refractive index matched ambient medium: $Re(\epsilon_2(\lambda_0)) = -1.61$, $Re(\mu_2(\lambda_0)) = -1.48$, $\mu_1 = 1$, $\epsilon_1 = \epsilon_2\mu_2/\mu_1$. Here $\lambda_0 \approx 1.4363$ μm and $d_2 = 2n_2\lambda_0$, $d_1 = d_3 = n_1/(2n_2)$. Note that for $f = 0$ the curve is flat and near zero. Adapted from Ref. [23].

intensity is near zero. Thus using a loss-compensated NIM should make the current metamaterials usable for much-improved superresolution applications. The future of metamaterials research and their application potentials depend on the development of new technologies to develop such transparent (at least at selected wavelengths) metamaterials.

14.5 Extraordinary transmission

Diffraction of light through small holes or an array of holes in an otherwise opaque screen has been one of the central problems of optics for ages. Any standard textbook deals with diffraction in the framework of Kirchhoff's scalar diffraction theory . However, such a scalar theory [31] or its vector extension [4] has some severe drawbacks in defining the boundary conditions. The shortfall was scrutinized in a seminal paper by Hans Bethe [58]. The following excerpt from Ref. [58] details the major drawbacks of the Kirchhoff theory.

"The available theoretical methods are entirely inadequate for the treatment of our problem. In the usual Kirchhoff method, the diffracted field is expressed in terms of the incident field in the hole. However, the Kirchhoff solution does not satisfy the boundary conditions, viz., it does not give zero tangential component of the electric field on the screen. In most textbooks, the pious hope is expressed that Kirchhoff's method will give at least the first term of a convergent series. This is probably true for the diffraction by an opening, large compared with the wave-length, because then the diffracted field will be relatively small on the screen, thus "almost" fulfilling the boundary conditions. But it is certainly not true for a small hole; in fact, our exact solution of the problem will turn out to be entirely different from Kirchhoff's."

Recall that Kirchhoff's method consists of putting the scalar field ϕ and its normal derivative $\frac{\partial\phi}{\partial z}$ to be zero on the screen (at $z = 0$) everywhere (at an arbitrary x and y) at the opening and replacing ϕ by the field of the incident field at the opening (aperture). For a tiny hole satisfying $a \ll \lambda$, Bethe obtained the first approximation result for the normalized-to-area transmission $\mathcal{T}$ as

$$\mathcal{T} \approx \frac{64}{27\pi^2}\left(\frac{2\pi a}{\lambda}\right)^4 . \tag{14.23}$$

As can be seen from Eq. (14.22), The transmission falls off rapidly with reducing aperture radius a compared to the wavelength λ. Thus there will be negligible transmission from tiny sub-wavelength holes. Both Kirchhoff and Bethe's theories are not adequate to deal with systems where the sub-wavelength aperture is close to the wavelength. A lot of interesting physics takes place when a/λ is not too small and there can be excitation of the localized plasmons and surface plasmons in the metal screen-aperture system. It was a major breakthrough when Ebbesen reported totally counterintuitive extraordinary transmission (EOT) through a regular array of sub-wavelength holes [169]. Let us now focus on the prerequisites for EOT, which makes it so very different from the standard scenario of diffraction. The basic structure where EOT shows up is a two-dimensional periodic array of sub-wavelength holes perforated on an optically thick metal film. The diffraction of light through such a structure is associated with peaks and dips in the transmission spectra. In view of several excellent reviews on EOT both at basic and advanced levels [170, 171, 172], we focus here on the qualitative picture. Note that most of the

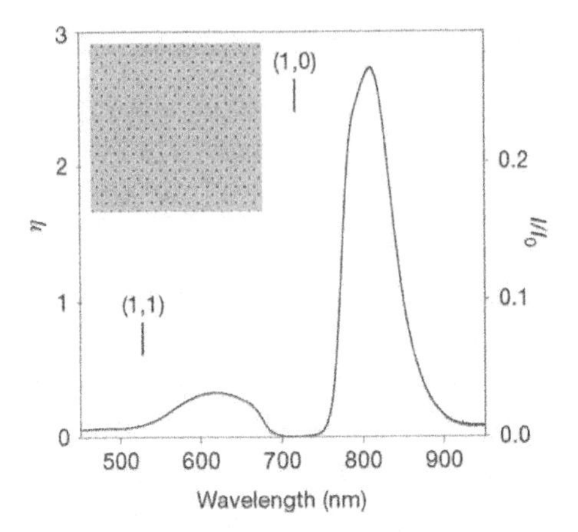

FIGURE 14.7: Transmission spectrum of hole arrays for the triangular hole array (hole diameter 170 nm, period 520 nm) shown in inset. The data are for normal incidence of collimated white light. The ratio I/I_0 on the right vertical axis is the absolute transmission of the array and η on the left axis represents the normalized to the area (occupied by the holes) transmission. Reprinted by permission from Macmillan Publishers Ltd: [Nature] [170], copyright (2007).

quantitative analysis depends heavily on computationally intensive methods like FDTD, which is beyond the scope of this book.

It is now commonly accepted that EOT occurs when normalized-to-area transmission is more than unity. The link between the EOT phenomena and the excitation of the surface plasmons and localized plasmons is now well established. The periodicity in the plane of the metal screen with the holes can provide for the missing momentum mismatch (see Section 10.2) and thus SPs can be excited on both the faces of the metal screen. Moreover, the tiny holes can support the localized plasmon resonances along with guided-along-the-hole modes for specific polarizations. Thus the spectral features arising in the transmission spectra are due to these excitations. Some of these resonances do not show up in the limit of vanishingly small a/λ and infinitely thin, perfectly conducting screen, and thus are missed in Bethe's theory. According to Ebbesen best results in the visible domain for transmission are obtained using noble metals with 150–300 nm holes in films not thicker than 200–300 nm.

Two broad classes of systems have been thoroughly investigated for EOT. The first class has a single hole surrounded by periodic grooves on both sides of the film (for best results), referred to as Ebbesen's bull's eye [173]. The momentum mismatch is compensated by the periodicity of the grooves. The resulting antenna action leads to intense fields near the aperture at the input face due to coupling of the incident light into SPs at a specific wavelength. EOT is mediated when light at the exit face can couple with the modes of the periodic grooves. The modes can then couple with the free space light, which can interfere with the light traveling directly through the hole. Surprisingly narrow spectral features were reported for a hole diameter of 300 nm and a groove period of 650 nm. The beam divergence was shown to be less than a few degrees. Thus for an incident colliemated beam, the output light can be highly collimated, with counterintuitive focusing. Recall that in standard diffraction theory, the narrower the hole, the larger the beam's angular divergence.

The second class refers to a two-dimensional periodic array of holes in the metal screen [170]. EOT here can be understood as a consequence of a three-step process. At step one, the incident light excites the SPs on the front surface. Step two consists of the transmission through the holes to the second surface, while step three consists of re-emission from the exit surface. The peak occurs when the surface supports the standing surface plasmon waves.

In the first approximation, an estimate of the wavelength λ_m where the peak occurs can be made by looking at the momentum-matching condition for a two-dimensional grating as in the Fig. 14.7 inset. For normal incidence the (i, j)-th resonance peak location is given by

$$\lambda_m = \frac{\Lambda}{\sqrt{\frac{4}{3}(i^2 + ij + j^2)}} \sqrt{\frac{\varepsilon_m \varepsilon}{\varepsilon_m + \varepsilon}}, \tag{14.24}$$

where Λ is the grating period and ε_m (ε) is the dielectric function of the metal (dielectric). Note that Eq. 14.24 does not incorporate the effect of holes and the associated losses and hence the results predicted by Eq. (14.24) are blue-shifted from the observed data (see Fig. 14.7).

Appendix A

Elements of complex numbers

A.1 Complex number algebra

Let the imaginary unit number be i. This satisfies

$$i^2 = -1 \,, \quad (i^3 = -i \,, \quad i^4 = 1). \tag{A.1}$$

Let an arbitrary complex number be $z = x + iy$. We define the operations of complex conjugation, addition, multiplication and division as follows:

- **Complex conjugation**: Complex conjugate z^* of z is defined as

$$z^* = x - iy, \tag{A.2}$$

$$x = (z + z^*)/2 \,, \quad y = \frac{z - z^*}{2i}. \tag{A.3}$$

- **Addition**: Let $z_1 = x_1 + iy_1$, $z_2 = x_2 + iy_2$

$$z_1 + z_2 = (x_1 + x_2) + i(y_1 + y_2). \tag{A.4}$$

- **Multiplication**:

$$z_1 z_2 = (x_1 + iy_1)(x_2 + iy_2) = (x_1x_2 - y_1y_2) + i(x_1y_2 + x_2y_1). \tag{A.5}$$

- **Division:**

$$z_1/z_2 = \frac{z_1 z_2^*}{z_2 z_2^*} = (x_1x_2 + y_1y_2)/A_2^2 + i(x_2y_1 - x_1y_2)/A_2^2\ , \quad A_2^2 = x_2^2 + y_2^2. \tag{A.6}$$

A few other important aspects of complex numbers are listed below:

- **Polar representation:**

$$z = Ae^{i\alpha}\ ; \quad A = \sqrt{x^2 + y^2}\ , \quad \tan(\alpha) = \frac{y}{x}\ , x = A\cos(\alpha), y = A\sin(\alpha). \tag{A.7}$$

- **Euler's formula and applications:**

$$\exp(i\alpha) = \cos(\alpha) + i\sin(\alpha), \tag{A.8}$$

$$\sin(\alpha) = \frac{e^{i\alpha} - e^{-i\alpha}}{2i}, \tag{A.9}$$

$$\cos(\alpha) = \frac{e^{i\alpha} + e^{-i\alpha}}{2}. \tag{A.10}$$

- **Products and ratio:** Let $z_1 = A_1 e^{i\alpha_1}$ and $z_2 = A_2 e^{i\alpha_2}$. Then

$$z_1 z_2 = A_1 A_2 e^{i(\alpha_1 + \alpha_2)}, \tag{A.11}$$

$$\frac{z_1}{z_2} = \frac{A_1}{A_2} e^{i(\alpha_1 - \alpha_2)}. \tag{A.12}$$

- **Roots:** If $z^n = Ae^{i\alpha} = Ae^{i(\alpha + q2\pi)}$, then

$$z = A^{(1/n)} e^{i(\alpha/n + q2\pi/n)}, \quad q = 0,\ 1,\ 2, \cdots (n-1). \tag{A.13}$$

A.1.1 Example problems

1. Determine the amplitude and phase angles of (a) $z_1 = 1 + i\sqrt{3}$, (b) $z_2 = \sqrt{3} + i$, (c) $z_1 + z_2$, (d) $z_1 z_2$ and (e) z_1/z_2. In each case indicate the location of these quantities in the complex plane.
2. Show that the amplitude and phase of $\int_0^b \exp(i\beta)d\beta$ are $b\sin(b/2)/(b/2)$ and $b/2$, respectively.

A.2 Harmonic motion and complex representation

The usefulness of complex numbers in the description and analysis of oscillations and waves are linked to Euler's formula. We can express the harmonic

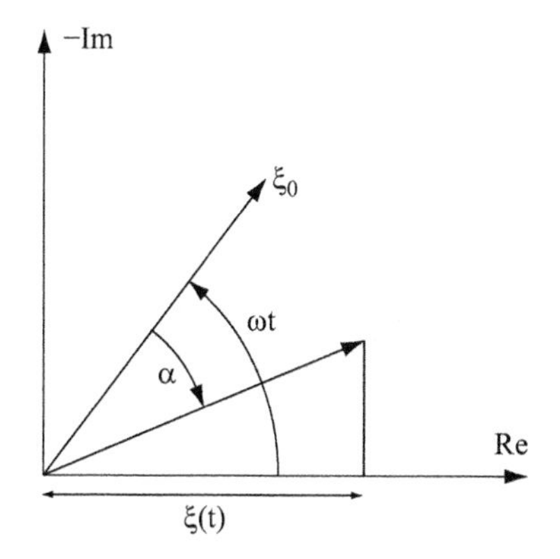

FIGURE A.1: Representation of an oscillation on a complex plane.

function as the real part of a complex exponential function. Consider the displacement of a harmonic oscillator

$$\xi(t) = \xi_0(\omega)\cos(\omega t - \alpha(\omega)), \tag{A.14}$$

where we have assumed that the amplitude as well as the phase angle can, in general, be functions of ω. We can express this function as the real part of the complex number $z = \xi_0 \exp\left[i(\omega t - \alpha)\right]$ or its complex conjugate $z^* = \xi_0 \exp\left[-i(\omega t - \alpha)\right]$. We use the latter option (see Fig. A.1)

$$\xi(t) = \xi_0 \cos(\omega t - \alpha) = Re\{\xi_0 \exp[-i(\omega t - \alpha)]\}, \tag{A.15}$$

$$= Re\{\xi(\omega)\exp[-i\omega t]\}, \tag{A.16}$$

$$\xi(\omega) = \xi_0 \exp(i\alpha). \tag{A.17}$$

Here we have introduced the complex displacement amplitude $\xi(\omega)$. The velocity $u(t)$ and acceleration $a(t)$ are also harmonic functions and can be represented by their corresponding complex amplitudes.

$$u(t) = \frac{d\xi}{dt} = Re\{(-i\omega)\xi(\omega)e^{-i\omega t}\} = Re\{(u(\omega)e^{-i\omega t}\} \tag{A.18}$$

$$a(t) = \frac{d^2\xi}{dt^2} = Re\{(-i\omega)^2\xi(\omega)e^{-i\omega t}\} = Re\{(a(\omega)e^{-i\omega t}\}, \tag{A.19}$$

$$u(\omega) = (-i\omega)\xi(\omega), \tag{A.20}$$

$$a(\omega) = (-i\omega)u(\omega) = (-i\omega)^2\xi(\omega) = -\omega^2\xi(\omega). \tag{A.21}$$

A.3 Demonstration of the usefulness of complex representation

In this section we show how the calculations can be simplified with the complex notation. To this end we consider (a) the superposition of two or more waves and (b) the forced oscillation of a damped harmonic oscillator (relevant physics was discussed in Section 1.4).

A.3.1 Superposition of two or more waves

A.3.1.1 Two waves

Consider two sine wave displacements at the same point in space having the same frequency, but with different amplitudes and phases:

$$E_1(x,t) = A_1\cos(\omega t - kx + \phi_1) = Re\{Z_1 e^{-i(\omega t - kx)}\}, \tag{A.22}$$

$$E_2(x,t) = A_2\cos(\omega t - kx + \phi_2) = Re\{Z_2 e^{-i(\omega t - kx)}\}, \tag{A.23}$$

where

$$Z_1 = A_1 e^{-i\phi_1},\ Z_2 = A_2 e^{-i\phi_2}. \tag{A.24}$$

The resultant field is given by

$$\begin{aligned} E_0(x,t) = E_1(x,t) + E_2(x,t) =& Re\{Z_1 e^{-i\theta} + Z_2 e^{-i\theta}\} \\ =& Re\{(Z_1 + Z_2)e^{-i\theta}\}, \end{aligned} \tag{A.25}$$

where $\theta = \omega t - kx$. The complex number $Z_0 = Z_1 + Z_2$ can be obtained in the complex plane as the vector sum of Z_1 and Z_2. The resultant real field is given by

$$E_0(x,t) = Re\{Z_0 e^{-i\theta}\}, \tag{A.26}$$

$$= A_0\cos(\omega t - kx + \phi_0), \tag{A.27}$$

$$Z_0 = A_0 e^{-i\phi_0}. \tag{A.28}$$

The law of cosines can be applied to obtain

$$|Z_0|^2 = A_0^2 = A_1^2 + A_2^2 + 2A_1A_2\cos(\phi_2 - \phi_1). \tag{A.29}$$

We can also obtain this algebraically:

$$\begin{aligned} |Z_0|^2 &= Z_0 Z_0^* = (Z_1 + Z_2)(Z_1^* + Z_2^*), \\ &= Z_1 Z_1^* + Z_2 Z_2^* + Z_1 Z_2^* + Z_1^* Z_2, \\ &= A_1^2 + A_2^2 + A_1A_2(e^{-i(\phi_1 - \phi_2)} + e^{i(\phi_1 - \phi_2)}). \end{aligned} \tag{A.30}$$

In order to obtain ϕ_0, take the real and imaginary parts of the equation

$$Z_0 = A_0 e^{-i\phi_0} = Z_1 + Z_2 = A_1 e^{-i\phi_1} + A_2 e^{-i\phi_2} \tag{A.31}$$

to get

$$A_0 \cos\phi_0 = A_1 \cos\phi_1 + A_2 \cos\phi_2, \tag{A.32}$$
$$A_0 \sin\phi_0 = A_1 \sin\phi_1 + A_2 \sin\phi_2. \tag{A.33}$$

which can be used to calculate A_0 and ϕ_0.

A.3.1.2 More than two waves

We now generalize this concept to N sine waves. Consider

$$\begin{aligned}
E(x,t) =& \sum_{j=1}^{N} A_j \cos(\omega t - kx + \phi_j), \\
=& \sum_{j=1}^{N} Re\left[A_j e^{-i(\theta+\phi_j)}\right], \\
=& \sum_{j=1}^{N} Re\left(A_j e^{-i\phi_j} e^{-i\theta}\right), \\
=& Re\left[\left(\sum_{j=1}^{N} A_j e^{-i\phi_j}\right) e^{-i\theta}\right], \\
=& Re\left[\left(\sum_{j=1}^{N} Z_j\right) e^{-i\theta}\right] = Re\left[Z_0 e^{-i\theta}\right], \\
=& A_0 \cos(\omega t - kx + \phi_0).
\end{aligned} \tag{A.34}$$

In Eq. (A.34) $Z_0 = A_0 e^{-i\phi_0}$.

$$\begin{aligned}
A_0^2 =& \sum_j \sum_{j'} Z_j Z_{j'}^*, \\
=& \sum_j |Z_j|^2 + \sum_{j<j'} (Z_j Z_{j'}^* + Z_j^* Z_{j'}), \\
=& \sum_j |A_j|^2 + \sum_{j<j'} 2A_j A_{j'} \cos(\phi_j - \phi_{j'}).
\end{aligned} \tag{A.35}$$

Thus Eqs. (A.32) and (A.33) generalize to

$$A_0 \cos\phi_0 = \sum_j A_j \cos\phi_j, \tag{A.36}$$

$$A_0 \sin\phi_0 = \sum_j A_j \sin\phi_j. \tag{A.37}$$

A.3.2 Forced damped oscillator

Consider the equation for the displacement for a forced damped oscillator

$$M\ddot{\xi} + R\dot{\xi} + K\xi = F_0 \cos(\omega t). \tag{A.38}$$

A.3.2.1 Usual method with real variables

The displacement must be harmonic with the same frequency as the driving force, and we assume it to have the form

$$\xi(t) = \xi_0 \cos(\omega t - \alpha). \tag{A.39}$$

Then the equation of motion becomes

$$(K - \omega^2 M)\xi_0 \cos(\omega t - \alpha) - R\omega\xi_0 \sin(\omega t - \alpha) = F_0 \cos(\omega t). \tag{A.40}$$

Next, use

$$\cos(\omega t - \alpha) = \cos(\omega t)\cos(\alpha) + \sin(\omega t)\sin(\alpha), \tag{A.41}$$
$$\sin(\omega t - \alpha) = \sin(\omega t)\cos(\alpha) - \cos(\omega t)\sin(\alpha), \tag{A.42}$$

in Eq. (A.40) and collect terms proportional to $\cos(\omega t)$ and $\sin(\omega t)$ on the left- and right-hand sides to obtain

$$\begin{aligned}[(K - \omega^2 M)\cos(\alpha) + &R\omega \sin(\alpha)]\xi_0 \cos(\omega t) + [(K - \omega^2 M)\sin(\alpha) \\ &- R\omega \cos(\alpha)]\xi_0 \sin(\omega t) = F_0 \cos(\omega t).\end{aligned} \tag{A.43}$$

Since Eq. (A.43) is to be satisfied at all t, we get

$$(K - \omega^2 M)\cos(\alpha) + R\omega \sin(\alpha) = F_0/\xi_0, \tag{A.44}$$
$$(K - \omega^2 M)\sin(\alpha) - R\omega \cos(\alpha) = 0. \tag{A.45}$$

Eq. (A.45) leads to the value of $\tan(\alpha)$, while Eq.(A.44) leads to the solution for the other unknown ξ_0:

$$\tan(\alpha) = \frac{R\omega}{(K - \omega^2 M)}, \tag{A.46}$$
$$\xi_0 = \frac{F_0}{\sqrt{(K - \omega^2 M)^2 + (R\omega)^2}}. \tag{A.47}$$

A.3.2.2 Method with complex amplitudes

In complex notation with $\xi(t) = Re\{\xi(\omega)\exp(-i\omega t)\}$, Eq.(A.38) reads as

$$(-\omega^2 M - i\omega R + K)\xi(\omega) = F(\omega), \tag{A.48}$$

which can be easily solved for $\xi(\omega)$ yielding

$$\xi(\omega) = \xi_0 \exp(i\alpha) = \frac{F(\omega)}{(K - \omega^2 M - i\omega R)}, \tag{A.49}$$

leading to Eqs. (A.46) and (A.47).

Appendix B

Vector spherical harmonics

The vector spherical harmonics are given by

$$\begin{aligned}
\mathbf{M}_{\left(\begin{array}{c} e \\ o \end{array}\right)_{mn}} &= \mp \frac{m}{\sin\theta} z_n(kr) P_n^m(\cos\theta) \left(\begin{array}{c} \sin \\ \cos \end{array}\right) m\phi \boldsymbol{e}_\theta \\
&- z_n(kr) \frac{\partial P_n^m}{\partial\theta} \left(\begin{array}{c} \cos \\ \sin \end{array}\right) m\phi \boldsymbol{e}_\phi, \qquad \text{(B.1)} \\
\mathbf{N}_{\left(\begin{array}{c} e \\ o \end{array}\right)_{mn}} &= \frac{n(n+1)}{kr} z_n(kr) P_n^m(\cos\theta) \left(\begin{array}{c} \cos \\ \sin \end{array}\right) m\phi \boldsymbol{e}_r \\
&+ \frac{1}{kr} [kr z_n(kr)]' \left[\frac{\partial P_n^m}{\partial\theta} \left(\begin{array}{c} \cos \\ \sin \end{array}\right) m\phi \boldsymbol{e}_\theta \right. \\
&\left. \mp \frac{m}{\sin\theta} P_n^m(\cos\theta) \left(\begin{array}{c} \sin \\ \cos \end{array}\right) m\phi \boldsymbol{e}_\phi \right]. \qquad \text{(B.2)}
\end{aligned}$$

Here the primes denote the derivatives with respect to the arguments and z_n represents the spherical Bessel or Hankel function. We append the superscript 1 (3) if the radial dependence is given by the spherical Bessel function $j_n(kr)$ (the spherical Hankel function $h_n^{(1)}(kr)$). Finally, P_n^m represents the associated Legendre function.

Appendix C

MATLAB® case studies

Most of the MATLAB® codes used in Chapter 9 are presented in the CD for this book. We now describe the organization of the CD and the instructions to be followed for running the codes. The CD contains five main folders, namely, `C1, C2, C3, Essential and Misc`. The folders `C1,C2` and `C3` contain the codes described in Sections C.1, C.2 and C.3, respectively, of Appendix C, and the folder `Essential` contains all the essential supporting files needed to compile the main codes. *In order to compile any code, all the contents of* `Essential` *has to be copied to the folder of interest, or else the path to it have to be added by* **addpath**. `Misc` contains all the other miscellaneous codes that can be used to reproduce most of the plots shown in Chapter 9. In `Misc` the name of the folder is associated with the figure number of Chapter 9.

C.1 Stratified media

The code can evaluate the reflection and the transmission coefficients from a stratified medium with isotropic and homogeneous constituent layers. The code is applicable to both continuous and discrete layered media. The code is also applicable to negative index materials, though we have to exercise caution since it has not been tested thoroughly. The first example in the current section is for discrete systems and all the results pertaining to other cases in the text have been obtained using this code. It may be noted that the results for the reflectionless refractive index profiles have been obtained using the same code. This is given in Section C.2.

C.1.1 Main codes

TABLE C.1:

	Filename	Purpose
Main function	reftran.m	Evaluates the reflection and transmission for N isotropic slabs for given λ and θ
calling program 1	angscan.m	as a function of θ
calling program 2	lamscan.m	as a function of λ

C.1.2 Function statement of reftran.m

```
function [r,t]=reftran(lam,thdeg,pol,epsi,mui,epsf,muf,d,eps,mu)
```

Subprograms needed: `charac.m`, `chartotal.m`, `pz.m`

C.1.3 Input/Output variables

Input variable	Description
lam	wavelength λ (in units of μm)
thdeg	angle of incidence $(0 \leq \theta < 90°)$
pol	pol = 0 (pol = 1) corresponds to TE (TM)
epsi	relative permitivity of medium of incidence (ϵ_i/ϵ_0)
mui	relative permeability of medium of incidence (μ_i/μ_0)
epsf	relative permitivity of final medium (ϵ_f/ϵ_0)
muf	relative permeability of final medium (ϵ_f/ϵ_0)
d	array of the widths: d = $[d_1\ d_2\ \cdots\ d_N]$
eps	relative permitivity array: eps = $[\epsilon_1/\epsilon_0\ \epsilon_2/\epsilon_0\ \cdots\ \epsilon_N/\epsilon_0]$
mu	relative permeability array: mu = $[\mu_1/\mu_0\ \mu_2/\mu_0\ \cdots\ \mu_N/\mu_0]$

Output variable	Description
r	$r = A_r/A_{in}$
t	$t = A_t/A_{in}$

C.2 Reflectionless potentials

This code evaluates the refractive index profile based on Kay-Moses prescriptions for arbitrary family of parameters (i.e., $A_i,\ \kappa_i$). For a reflectionless refractive index profile, we discretize the profile and use the characteristic matrix code to evaluate the scattering. To illustrate this we consider the case of one parameter family presented in Section 9.6 (see Eq. (9.93) and the description therein).

C.2.1 Main codes

TABLE C.2:

	Filename	Purpose
Main function	`KMsystem.m`	Evaluates the matrix M_{ij} (see Eq. 9.89)
calling program 1	`angscan.m`	Evaluates scattering as a function of θ
calling proram 2	`lamscan.m`	Evaluates scattering as a function of λ

C.2.2 Function statement of KMsystem.m

```
function [m,c]=kaysystem(n,k,a,z)
```

TABLE C.3:

Input variable	Description
n	size of the parameter family
k	array of κ_i: k=$[\kappa_1, \cdots, \kappa_n,]$
a	array of A_i: a=$[A_1, \cdots, A_2]$
z	spatial coordinate at which M_{ij} is evaluated

TABLE C.4:

Output variable	Description
m	matrix M_{ij} at a specific z
c	RHS of Eq. (9.89) at a specific z

From this continuous version of the profile, we then obtain the discrete version of the refractive index profile, and we make use of `reftan.m` to evaluate the scattering, i.e., reflection and transmission. We can obtain Figs. 9.6 and 9.7 by compiling the codes `angscan.m` and `lamscan.m`, respectively.

C.3 Nonreciprocity

Codes highlight the nonreciprocity in reflection from the layered medium. These codes are all based on the ones developed for stratified media (see Section C.1).

TABLE C.5:

	Filename	**Purpose**
calling program 1	`delay.m`	Evaluates reflection and Wigner delay/advancement in reflection
calling program 2	`pulse.m`	Evaluates the changes undergone by pulse when reflected/transmitted through layered media, verifying the delay/advancement from Wigner time

C.3.1 Subprograms needed

`absorber.m`, which evaluates the dielectric response of the resonant absorbers [see Eq. (9.121)].

We can obtain Figs. 9.15 and 9.16 by compiling the codes `delay.m` and `pulse.m`, respectively.

Bibliography

[1] A.P. French. *Vibrations and Waves.* The M.I.T. Introductory Physics Series. Norton, New York, 1971.

[2] H. Goldstein. *Classical Mechanics.* Addison-Wesley, New York, 3rd edition, 2001.

[3] John D. Jackson. *Classical Electrodynamics.* Wiley, Hoboken, NJ, 3rd edition, 1998.

[4] J. A. Stratton. *Electromagnetic Theory.* McGraw-Hill, New York, 1941.

[5] L. Brillouin. *Wave Propagation and Group Velocity.* Academic Press, New York, 1960.

[6] S. Chu and S. Wong. Linear pulse propagation in an absorbing medium. *Phys. Rev. Lett.*, 48:738–741, 1982.

[7] Robert W. Boyd and Daniel J. Gauthier. Slow and fast light. In E. Wolf, editor, *Progress in Optics*, volume 43, pages 497–530. Elsevier, Amsterdam, 2002.

[8] Eugene Hecht. *Optics.* Addison-Wesley, New York, 2002.

[9] G. S. Agarwal and S. Dutta Gupta. T-matrix approach to the nonlinear susceptibilities of heterogeneous media. *Phys. Rev. A*, 38:5678–5687, Dec 1988.

[10] J. E. Gubernatis. Scattering theory and effective medium approximations to heterogeneous materials. *AIP Conference Proceedings*, 40(1):84–98, 1978.

[11] Wenshan Cai and Vladimir Shalaev. *Optical Metamaterials: Fundamentals and Applications.* Springer, New York, 2010.

[12] Paul N. Butcher and David Cotter. *The Elements of Nonlinear Optics.* Cambridge University Press, Cambridge, 1990.

[13] P. B. Johnson and R. W. Christy. Optical constants of the noble metals. *Physical Review B*, 6(12):4370–4379, 1972.

[14] Edward D. Palik and Gorachand Ghosh. *Handbook of Optical Constants of Solids Five-Volume Set.* Academic Press, New York, 1st edition, 1998.

[15] J. A. Dionne, L. A. Sweatlock, H. A. Atwater, and A. Polman. Planar metal plasmon waveguides: Frequency-dependent dispersion, propagation, localization, and loss beyond the free electron model. *Phys. Rev. B*, 72:075405, Aug 2005.

[16] J. A. Dionne, L. A. Sweatlock, H. A. Atwater, and A. Polman. Plasmon slot waveguides: Towards chip-scale propagation with subwavelength-scale localization. *Phys. Rev. B*, 73:035407, Jan 2006.

[17] V. G. Veselago. The electrodynamics of substances with simultaneously negative values of ε and μ. *Soviet Journal Uspekhi*, 10, 1968.

[18] J. B. Pendry. Negative refraction makes a perfect lens. *Phys. Rev. Lett.*, 85:3966–3969, Oct 2000.

[19] J. B. Pendry, A. J. Holden, W. J. Stewart, and I. Youngs. Extremely low frequency plasmons in metallic mesostructures. *Phys. Rev. Lett.*, 76:4773–4776, Jun 1996.

[20] J. B. Pendry, A. J. Holden, D. J. Robbins, and W. J. Stewart. Magnetism from conductors and enhanced nonlinear phenomena. *IEEE Transactions on Microwave Theory and Techniques*, 47(11):2075–2084, 1999.

[21] D. R. Smith, Willie J. Padilla, D. C. Vier, S. C. Nemat-Nasser, and S. Schultz. Composite medium with simultaneously negative permeability and permittivity. *Phys. Rev. Lett.*, 84:4184–4187, May 2000.

[22] Costas M. Soukoulis, Stefan Linden, and Martin Wegener. Negative refractive index at optical wavelengths. *Science*, 315(5808):47–49, 2007.

[23] Subimal Deb and S. Dutta Gupta. Absorption and dispersion in metamaterials: Feasibility of device applications. *Pramana*, 75(5):837–854, 2010.

[24] D. R. Smith, J. B. Pendry, and M. C. K. Whitshire. Metamaterials and negative refractive index. *Science*, 305:788, 2004.

[25] S. A. Ramakrishna. Physics of negative refractive index materials. *Reports on Progress in Physics*, 68(2):449, 2005.

[26] Yongmin Liu and Xiang Zhang. Metamaterials: A new frontier of science and technology. *Chem. Soc. Rev.*, 40:2494–2507, 2011.

[27] J. C. Bose. On the rotation of plane of polarization of electric waves by a twist structure. *Proc. Royal Soc.*, 63:146–152, 1898.

[28] Edward Collett. *Polarized Light: Fundamentals and Applications.* Marcel Dekker, New York, 1990.

[29] David S. Kliger, James W. Lewis, and C. E. Randall. *Polarized Light in Optics and Spectroscopy.* Academic Press-Harcourt Brace Jovanovich, New York, 2003.

[30] Dennis Goldstein. *Polarized Light.* Marcel Dekker, New York, 2003.

[31] M. Born and E. Wolf. *Principles of Optics: Electromagnetic Theory of Propagation.* Cambridge University Press, Cambridge, 2001.

[32] C. F. Bohren and D. R. Huffman. *Absorption and Scattering of Light by Small Particles.* Wiley, New York, 1983.

[33] R. Clark Jones. A new calculus for the treatment of optical systems. *J. Opt. Soc. Am.*, 37(2):107–112, 1947.

[34] George Gabriel Stokes. On the composition and resolution of streams of polarized light from different sources. *Transactions of the Cambridge Philosophical Society*, 9:339–416, 1852.

[35] Serge Huard. *The Polarization of Light.* Wiley, New York, 1997.

[36] W. S. Bickel and W. M. Bailey. Stokes vectors, Mueller matrices, and polarized scattered light. *American Journal of Physics*, 53(5):468–478, 1985.

[37] Christian Brosseau. *Fundamentals of Polarized Light: A Statistical Optics Approach.* Wiley-Interscience, New York, 1998.

[38] H. Muller. The foundation of optics. *J. Opt. Soc. Am.*, 38:551, 1948.

[39] Russell A. Chipman. Polarimetry. Chap. 22 in Handbook of Optics, 2(2nd edition.):1–37, 1994.

[40] José J. Gil. Characteristic properties of Mueller matrices. *JOSA A*, 17(2):328–334, 2000.

[41] S. R. Cloude. Group theory and polarization algebra. *Optik*, 75(26):26–36, 1986.

[42] Donald G. M. Anderson and Richard Barakat. Necessary and sufficient conditions for a Mueller matrix to be derivable from a Jones matrix. *JOSA A*, 11(8):2305–2319, 1994.

[43] J. J. Gil and E. Bernabeu. A depolarization criterion in Mueller matrices. *Opt. Act.*, 32:259–261, 1985.

[44] Shih-Yau Lu and Russell A. Chipman. Interpretation of Mueller matrices based on polar decomposition. *JOSA A*, 13(5):1106–1113, 1996.

[45] R. M. A. Azzam. Propagation of partially polarized light through anisotropic media with or without depolarization: A differential 4 × 4 matrix calculus. *JOSA*, 68(12):1756–1767, 1978.

[46] Nirmalya Ghosh and I. Alex Vitkin. Tissue polarimetry: Concepts, challenges, applications, and outlook. *Journal of Biomedical Optics*, 16(11):110801, 2011.

[47] B. N. Simon, S. Simon, F. Gori, M. Santarsiero, R. Borghi, N. Mukunda, and R. Simon. Nonquantum entanglement resolves a basic issue in polarization optics. *Physical Review Letters*, 104(2):023901, 2010.

[48] Emil Wolf. Unified theory of coherence and polarization of random electromagnetic beams. *Physics Letters A*, 312(5):263–267, 2003.

[49] Giovanni Milione, H. I. Sztul, D. A. Nolan, and R. R. Alfano. Higher-order Poincaré sphere, Stokes parameters, and the angular momentum of light. *Physical Review Letters*, 107(5):053601, 2011.

[50] Lorenzo Marrucci, Ebrahim Karimi, Sergei Slussarenko, Bruno Piccirillo, Enrico Santamato, Eleonora Nagali, and Fabio Sciarrino. Spin-to-orbital conversion of the angular momentum of light and its classical and quantum applications. *Journal of Optics*, 13(6):064001, 2011.

[51] Edward Collett. Measurement of the four Stokes polarization parameters with a single circular polarizer. *Opt. Commun.*, 52(2):77–80, 1984.

[52] Dennis H Goldstein. Mueller Matrix dual-rotating retarder polarimeter. *Applied Optics*, 31(31):6676–6683, 1992.

[53] Eric Compain and Bernard Drevillon. High-frequency modulation of the four states of polarization of light with a single phase modulator. *Review of Scientific Instruments*, 69(4):1574–1580, 1998.

[54] Matthew H. Smith. Optimization of a dual-rotating-retarder Mueller matrix polarimeter. *Applied optics*, 41(13):2488–2493, 2002.

[55] Blandine Laude-Boulesteix, Antonello De Martino, Bernard Drévillon, and Laurent Schwartz. Mueller polarimetric imaging system with liquid crystals. *Applied Optics*, 43(14):2824–2832, 2004.

[56] Emil Wolf. *Introduction to the Theory of Coherence and Polarization of Light.* Cambridge University Press, 2007.

[57] Bernard C. Kress and Patrick Meyrueis. *Applied Digital Optics: From Micro-Optics to Nanophotonics.* Wiley, Hoboken, NJ, 2009.

[58] H. A. Bethe. Theory of diffraction by small holes. *Phys. Rev.*, 66:163–182, Oct 1944.

[59] A. E. Siegman. *Lasers.* University Science Books, Sausalito, CA, 1986.

[60] Amnon Yariv. *Quantum Electronics.* Wiley, Hoboken, NJ, 3rd edition, 1989.

[61] A. Yariv and P. Yeh. *Optical Waves in Crystals.* Wiley, New York, 1984.

[62] Pochi Yeh. *Optical Waves in Layered Media.* Wiley, New York, 1988.

[63] E. P. Wigner. Lower limit for the energy derivative of the scattering phase shift. *Phys. Rev.*, 98:145–147, 1955.

[64] F. Goos and H. Hänchen. Ein neuer und fundamentaler Versuch zur Totalreflexion. *Ann. Phys.*, 436:333–346, 1947.

[65] S. Dutta Gupta. Nonlinear optics of stratified media. In E. Wolf, editor, *Progress in Optics*, volume 38, chapter 1, pages 1–84. Elsevier, Amsterdam, 1998.

[66] Herbert G. Winful. Nature of "superluminal" barrier tunneling. *Phys. Rev. Lett.*, 90:023901, Jan 2003.

[67] Mahito Kohmoto, Bill Sutherland, and K. Iguchi. Localization of optics: Quasiperiodic media. *Phys. Rev. Lett.*, 58:2436–2438, Jun 1987.

[68] S. Dutta Gupta and Deb Shankar Ray. Localization problem in optics: Nonlinear quasiperiodic media. *Phys. Rev. B*, 41:8047–8053, Apr 1990.

[69] I. Kay and H. E. Moses. Reflectionless transmission through dielectrics and scattering potentials. *J. Appl. Phys.*, 27:1503–1508, 1956.

[70] J. C. Bose. The influence of the thickness of air-space on total reflection of electric radiation. *Bose Institute Transactions*, 42, 1927.

[71] A. Sommerfeld. *Vorlesungen über theoretische Physik., Volume IV OPTIK*, Chapters 34 and 35. Verlag Harri Deutsch Thun, Frankfurt am Main, Germany, 1989.

[72] Raymond Y. Chiao and Aephraim M. Steinberg. Tunneling times and superluminality. In E. Wolf, editor, *Progress in Optics*, volume 37, pages 345– 405. Elsevier, Amsterdam, 1997.

[73] Gunter Nimtz and Winfried Heitmann. Superluminal photonic tunneling and quantum electronics. *Progress in Quantum Electronics*, 21(2):81–108, 1997.

[74] S. K. Sekatskii and V. S. Letokhov. Electron tunneling time measurement by field-emission microscopy. *Phys. Rev. B*, 64:233311, Nov 2001.

[75] Thomas E. Hartman. Tunneling of a wave packet. *Journal of Applied Physics*, 33(12):3427–3433, Dec 1962.

[76] A. Enders and G. Nimtz. Evanescent-mode propagation and quantum tunneling. *Phys. Rev. E*, 48:632–634, Jul 1993.

[77] C. K. Carniglia and L. Mandel. Differential phase shifts of T.E. and T.M. evanescent waves. *J. Opt. Soc. Am.*, 61(10):1423–1424, 1971.

[78] V. S. C. Manga Rao. *Planar and Spherical Microstructures for the Study of Strong Coupling and Propagation Aspects.* PhD thesis, School of Physics, Hyderabad, University of Hyderabad, 2005.

[79] Stefano Longhi. Superluminal pulse reflection in asymmetric one-dimensional photonic band gaps. *Phys. Rev. E*, 64:037601, Aug 2001.

[80] S. Esposito. Multibarrier tunneling. *Phys. Rev. E*, 67:016609, Jan 2003.

[81] Siegfried Flügge. *Practical Quantum Mechanics.* Springer Verlag, Berlin, 1994.

[82] J. Lekner. Nonreflecting stratifications. *Am. J. Phys.*, 75:1151–1157, 2007.

[83] S. Dutta Gupta and G. S. Agarwal. A new approach for broad-band omnidirectional antireflection coatings. *Opt. Express*, 15:9614–9624, 2007.

[84] L. V. Thekkekara, Venu Gopal Achanta, and S. Dutta Gupta. Optical reflectionless potentials for broadband, omnidirectional antireflection. *Opt. Express*, 22:17382–17386, 2014.

[85] Wenjie Wan, Yidong Chong, Li Ge, Heeso Noh, A. Douglas Stone, and Hui Cao. Time-reversed lasing and interferometric control of absorption. *Science*, 331(6019):889–892, 2011.

[86] Frederick C. Evering, Jr. Artificial diffraction anomalies for gratings of rectangular profile. *Appl. Opt.*, 5:1313, 1966.

[87] A. Yariv. Universal relations for coupling of optical power between microresonators and dielectric waveguides. *Electronics Letters*, 36(4):321–322, Feb 2000.

[88] Ming Cai, Oskar Painter, and Kerry J. Vahala. Observation of critical coupling in a fiber taper to a silica-microsphere whispering-gallery mode system. *Phys. Rev. Lett.*, 85:74–77, Jul 2000.

[89] Shourya Dutta-Gupta, O. J. F. Martin, S. Dutta Gupta, and G. S. Agarwal. Controllable coherent perfect absorption in a composite film. *Opt. Express*, 20(2):1330–1336, Jan 2012.

[90] R. J. Potton. Reciprocity in optics. *Reports on Progress in Physics*, 67(5):717, 2004.

[91] G. S. Agarwal and S. Dutta Gupta. Reciprocity relations for reflected amplitudes. *Opt. Lett.*, 27(14):1205–1207, Jul 2002.

[92] A. Armitage, M. S. Skolnick, A. V. Kavokin, D. M. Whittaker, V. N. Astratov, G. A. Gehring, and J. S. Roberts. Polariton-induced optical asymmetry in semiconductor microcavities. *Phys. Rev. B*, 58:15367–15370, Dec 1998.

[93] Zafar Ahmed. Schrödinger transmission through one-dimensional complex potentials. *Phys. Rev. A*, 64:042716, Sep 2001.

[94] J. J. Sanchez-Mondragon, N. B. Narozhny, and J. H. Eberly. Theory of spontaneous-emission line shape in an ideal cavity. *Phys. Rev. Lett.*, 51:550–553, Aug 1983.

[95] G. S. Agarwal. Vacuum-field rabi splittings in microwave absorption by Rydberg atoms in a cavity. *Phys. Rev. Lett.*, 53:1732–1734, Oct 1984.

[96] V. S. C. Manga Rao, S. Dutta Gupta, and G. S. Agarwal. Atomic absorbers for controlling pulse propagation in resonators. *Opt. Lett.*, 29:307, 2004.

[97] H. G. Winful. Mechanism for superluminal tunneling. *Nature*, 424:638, 2003.

[98] H. Raether. *Surface Plasmons on Smooth and Rough Surfaces and on Gratings.* Springer, New York, 1988.

[99] Stefan Alexander Maier. *Plasmonics: Fundamentals and Applications.* Springer, New York, 1st edition, 2007.

[100] S. Dutta Gupta. Slow light with symmetric gap plasmon guides with finite width metal claddings. *Pramana - J. Phys.*, 72:303–314, 2009.

[101] Michael K. Barnoski (ed.). *Introduction to Integrated Optics.* Springer, New York, 1974.

[102] K. Nireekshan Reddy and S. Dutta Gupta. Light-controlled perfect absorption of light. *Opt. Lett.*, 38(24):5252–5255, 2013.

[103] Pavel Ginzburg, David Arbel, and Meir Orenstein. Gap plasmon polariton structure for very efficient microscale-to-nanoscale interfacing. *Opt. Lett.*, 31(22):3288–3290, Nov 2006.

[104] Dror Sarid. Long-range surface-plasma waves on very thin metal films. *Phys. Rev. Lett.*, 47:1927–1930, Dec 1981.

[105] T. Inagaki, M. Motosuga, E. T. Arakawa, and J. P. Goudonnet. Coupled surface plasmons in periodically corrugated thin silver films. *Phys. Rev. B*, 32:6238–6245, Nov 1985.

[106] S. Dutta Gupta, G. V. Varada, and G. S. Agarwal. Surface plasmons in two-sided corrugated thin films. *Phys. Rev. B*, 36:6331–6335, Oct 1987.

[107] Dror Sarid, R. T. Deck, and J. J. Fasanot. Enhanced nonlinearity of the propagation constant of a long-range surface-plasma wave. *J. Opt. Soc. Am.*, 72(10):1345–1347, Oct 1982.

[108] J. C. Quail, J. G. Rako, H. J. Simon, and R. T. Deck. Optical second-harmonic generation with long-range surface plasmons. *Phys. Rev. Lett.*, 50:1987–1989, Jun 1983.

[109] G. S. Agarwal and S. Dutta Gupta. Exact results on optical bistability with surface plasmons in layered media. *Phys. Rev. B*, 34:5239–5243, Oct 1986.

[110] K. M. Leung. p-polarized nonlinear surface polaritons in materials with intensity-dependent dielectric functions. *Phys. Rev. B*, 32:5093–5101, Oct 1985.

[111] Giovanna Panzarini, Lucio Claudio Andreani, A. Armitage, D. Baxter, M. S. Skolnick, V. N. Astratov, J. S. Roberts, Alexey V. Kavokin, Maria R. Vladimirova, and M. A. Kaliteevski. Exciton-light coupling in single and coupled semiconductor microcavities: Polariton dispersion and polarization splitting. *Phys. Rev. B*, 59:5082–5089, Feb 1999.

[112] M. L. Gorodetsky, A. A. Savchenkov, and V. S. Ilchenko. Ultimate q of optical microsphere resonators. *Opt. Lett.*, 21:453, 1996.

[113] M. H. Fields, J. Popp, and R. K. Chang. Nonlinear optics in microspheres. In E. Wolf, editor, *Progress in Optics*, volume 41, pages 3–89. North-Holland, Amsterdam, 2000.

[114] Larry D. Travis Michael I. Mishchenko, Joachim W. Hovenier. *Light Scattering by Nonspherical Particles.* Academic Press, New York, 1st edition, 1999.

[115] Lina He, Sahin Kaya Ozdemir, Jiangang Zhu, Woosung Kim, and Lan Yang. Detecting single viruses and nanoparticles using whispering gallery microlasers. *Nature Nanotechnology*, 6(7):428–432, Jul 2011.

[116] H. M. Lai, P. T. Leung, K. Young, P. W. Barber, and S. C. Hill. Time-dependent perturbation for leaking electromagnetic modes in open systems with application to resonance in microdroplets. *Phys. Rev. A*, 41:5187, 1990.

[117] J. C. Knight, N. Dubreuil, V. Sandoghdar, J. Hare, V. Lefevre-Seguin, J. M. Raimond, and S. Haroche. Characterizing whispering-gallery modes in microspheres by direct observation of the optical standing-wave pattern in the near field. *Opt. Lett.*, 21:698, 1996.

[118] Richard A. Beth. Mechanical detection and measurement of the angular momentum of light. *Physical Review*, 50(2):115–125, 1936.

[119] John David Jackson and Ronald F. Fox. Classical electrodynamics. *American Journal of Physics*, 67(9):841–842, 1999.

[120] J. H. Poynting. The wave motion of a revolving shaft, and a suggestion as to the angular momentum in a beam of circularly polarised light. *Proceedings of the Royal Society of London*, 82(557):560–567, 1909.

[121] Les Allen, Marco W Beijersbergen, R.J.C. Spreeuw, and J.P. Woerdman. Orbital angular momentum of light and the transformation of laguerre-gaussian laser modes. *Physical Review A*, 45(11):8185–8189, 1992.

[122] Leslie Allen, Stephen M Barnett, and Miles J. Padgett. *Optical Angular Momentum.* Taylor & Francis, London, 2003.

[123] A.T. O'Neil, I. MacVicar, L. Allen, and M.J. Padgett. Intrinsic and extrinsic nature of the orbital angular momentum of a light beam. *Physical Review Letters*, 88(5):053601, 2002.

[124] Alison M. Yao and Miles J. Padgett. Orbital angular momentum: Origins, behavior and applications. *Adv. Opt. Photon.*, 3(2):161–204, 2011.

[125] H. He, M. E. J. Friese, N. R. Heckenberg, and H. Rubinsztein-Dunlop. Direct observation of transfer of angular momentum to absorptive particles from a laser beam with a phase singularity. *Physical Review Letters*, 75(5):826–829, 1995.

[126] Konstantin Y. Bliokh. Geometrodynamics of polarized light: Berry phase and spin Hall effect in a gradient-index medium. *Journal of Optics A: Pure and Applied Optics*, 11(9):094009, 2009.

[127] Lorenzo Marrucci, C. Manzo, and D. Paparo. Optical spin-to-orbital angular momentum conversion in inhomogeneous anisotropic media. *Physical Review Letters*, 96(16):163905, 2006.

[128] David L. Andrews and Mohamed Babiker. *The Angular Momentum of Light.* Cambridge University Press, Cambridge, 2012.

[129] Michael V. Berry. Quantal phase factors accompanying adiabatic changes. *Proceedings of the Royal Society of London A*, 392:45–57, 1984.

[130] M. V. Berry. Interpreting the anholonomy of coiled light. *Nature*, 326:277–278, 1987.

[131] Shivaramakrishnan Pancharatnam. Generalized theory of interference, and its applications. Part I. Coherent pencils. In *Proceedings of the Indian Academy of Sciences, Section A*, volume 44, pages 247–262. Indian Academy of Sciences, Bengaluru, 1956.

[132] Rajendra Bhandari and Joseph Samuel. Observation of topological phase by use of a laser interferometer. *Physical Review Letters*, 60(13):1211–1213, 1988.

[133] R. Simon, H. J. Kimble, and E. C. G. Sudarshan. Evolving geometric phase and its dynamical manifestation as a frequency shift: An optical experiment. *Physical Review Letters*, 61(1):19–22, 1988.

[134] M. J. Padgett and J. Courtial. Poincaré-sphere equivalent for light beams containing orbital angular momentum. *Optics Letters*, 24(7):430–432, 1999.

[135] J. Soni, S. Ghosh, S. Mansha, A. Kumar, S. Dutta Gupta, A. Banerjee, and N. Ghosh. Enhancing spin-orbit interaction of light by plasmonic nanostructures. *Optics Letters*, 38(10):1748–1750, 2013.

[136] B. Richards and E. Wolf. Electromagnetic diffraction in optical systems. ii. structure of the image field in an aplanatic system. *Proceedings of the Royal Society of London. Series A*, 253:358–379, 1959.

[137] Konstantin Y. Bliokh, Avi Niv, Vladimir Kleiner, and Erez Hasman. Geometrodynamics of spinning light. *Nature Photonics*, 2(12):748–753, 2008.

[138] Onur Hosten and Paul Kwiat. Observation of the Spin Hall effect of light via weak measurements. *Science*, 319:787–790, 2008.

[139] A. Ashkin. Optical trapping and manipulation of neutral particles using lasers. *Proceedings of the National Academy of Sciences, USA*, 94:4853–4860, 1997.

[140] A. Ashkin. Observation of a single-beam gradient force optical trap for dielectric particles. *Optics Letters*, 11:288–290, 1986.

[141] A. Ashkin. Forces of a single-beam gradient laser trap on a dielectric sphere in the ray optics regime. *Biophysical Journal*, 61:569–582, 1992.

[142] F. Czerwinski, A. C. Richardson, and L. B. Oddershede. Quantifying noise in optical tweezers by allan variance. *Optics Express*, 17:13255–13269, 2009.

[143] K. C. Neumann and A. Nagy. Single-molecule force spectroscopy: optical tweezers, magnetic tweezers and atomic force microscopy. *Nature Methods*, 5:491–505, 1999.

[144] L. D. Landau and E. M. Lifshitz. *Fluid Mechanics*. Butterworth Heinemann, Oxford, 2nd edition, 1987.

[145] D. Bedeaux and P. Mazur. Brownian motion and fluctuating hydrodynamics. *Physica*, 76:569–582, 1974.

[146] K. Berg-Sorensen and H. Flyvbjerg. Power spectrum analysis for optical tweezers. *Review of Scientific Instruments*, 75:594–612, 2004.

[147] B. Roy, S. Bera, and A. Banerjee. Simultaneous detection of rotational and translational motion in optical tweezers by measurement of backscattered intensity. *Optics Letters*, 39:3316–3319, 2014.

[148] W. J. Greenleaf, M. T. Woodside, E. A. Abbondanzieri, and S. M. Block. Passive all-optical force clamp for high resolution laser trapping. *Physical Review Letters*, 95:208102, 2005.

[149] T. T. Perkins, H.W. Li, R.V. Dalal, J. Gelles, and S. M. Block. Forward and reverse motion of single RECBCD molecules on DNA. *Biophysical Journal*, 86:1640–1648, 2004.

[150] A. M. van Oijen, P. C. Blainey, D. J. Crampton, C. C. Richardson, T. Ellenberger, and X. S. Xie. Single-molecule kinetics of lambda exonuclease reveal base dependence and dynamic disorder. *Science*, 301:1235–1238, 2003.

[151] T. T. Perkins. Optical traps for single molecule biophysics: A primer. *Lasers and Photonics Review*, pages 1–18, 2008.

[152] K. Svoboda, C. F. Schmidt, B. J. Schnapp, and S. M. Block. Direct observation of kinesin stepping by optical trapping interferometry. *Nature*, 365:721727, 1993.

[153] J. E. Molloy, J. E. Burns, J. Kendrick-Jones, R. T. Tregear, and D. C. White. Movement and force produced by a single myosin head. *Nature*, 378:209–212, 1995.

[154] E. A. Abbondanzieri, W. J. Greenleaf, J. W. Shaevitz, R. Landick, and S. M. Block. Direct observation of base-pair stepping by RNA polymerase. *Nature*, 438:460–465, 2005.

[155] D. A. Schafer, J. Gelles, M. P. Sheetz, and R. Landick. Transcription by single molecules of RNA polymerase observed by light microscopy. *Nature*, 352:444–448, 1991.

[156] K. Dholakia and P. Zemanek. Gripped by light: Optical binding. *Reviews of Modern Physics*, 82:1767–1791, 2010.

[157] A. D. Mehta, M. Rief, J. A. Spudich, D. A. Smith, and R. M. Simmons. Single-molecule biomechanics with optical methods. *Science*, 283:1689–1695, 1999.

[158] M. Padgett and R. Bowman. Tweezers with a twist. *Nature Photonics*, 5:343–348, 2011.

[159] Weihua Zhang, Lina Huang, Christian Santschi, and Olivier J. F. Martin. Trapping and sensing 10 nm metal nanoparticles using plasmonic dipole antennas. *Nano Letters*, 10(3):1006–1011, 2010. PMID: 20151698.

[160] M. L. Juan, M. Rrighini, and R. Quidant. Plasmon nano-optical tweezers. *Nature Photonics*, 5:349–356, 2011.

[161] D. Grier and Y. Roichman. Holographic optical trapping. *Applied Optics*, 5:880–887, 2006.

[162] Lukas Novotny and Bert Hecht. *Principles of Nano-optics*. Cambridge University Press, Cambridge, 2nd edition, 2006.

[163] Laszlo Solymar and Ekaterina Shamonina. *Waves in Metamaterials*. Oxford University Press, Oxford, UK, 2009.

[164] David R. Smith, David Schurig, Marshall Rosenbluth, Sheldon Schultz, S. Anantha Ramakrishna, and John B. Pendry. Limitations on subdiffraction imaging with a negative refractive index slab. *Applied Physics Letters*, 82(10):1506–1508, 2003.

[165] Nicholas Fang and Xiang Zhang. Imaging properties of a metamaterial superlens. *Applied Physics Letters*, 82(2):161–163, 2003.

[166] Israel De Leon and Pierre Berini. Amplification of long-range surface plasmons by a dipolar gain medium. *Nat. Photon*, 4(6):382–387, 2010.

[167] Gunnar Dolling, Christian Enkrich, Martin Wegener, Costas M. Soukoulis, and Stefan Linden. Low-loss negative-index metamaterial at telecommunication wavelengths. *Opt. Lett.*, 31(12):1800–1802, Jun 2006.

[168] Shuang Zhang, Wenjun Fan, K. J. Malloy, S.R. Brueck, N. C. Panoiu, and R. M. Osgood. Near-infrared double negative metamaterials. *Opt. Express*, 13(13):4922–4930, Jun 2005.

[169] T. W. Ebbesen, H. J. Lezec, H. F. Ghaemi, T. Thio, and P. A. Wolff. Extraordinary optical transmission through sub-wavelength hole arrays. *Nature*, 391(6668):667–669, 1998.

[170] C. Genet and T. W. Ebbesen. Light in tiny holes. *Nature*, 445(7123):39–46, 2007.

[171] A. S. Vengurlekar. Extraordinary optical transmission through metal films with subwavelength holes and slits. *Current Science*, 98:1020–1032, 2010.

[172] F. J. Garcia de Abajo. Colloquium: Light scattering by particle and hole arrays. *Rev. Mod. Phys.*, 79:1267, 2007.

[173] H. J. Lezec, A. Degiron, E. Devaux, R. A. Linke, L. Martin-Moreno, F. J. Garcia-Vidal, and T. W. Ebbesen. Beaming light from a subwavelength aperture. *Science*, 297(5582):820–822, 2002.

Subject index